FLUID DYNAMICS

Fluid Dynamics

Part 4: Hydrodynamic Stability Theory

Anatoly I. Ruban

Imperial College London

Jitesh S. B. Gajjar

University of Manchester

Andrew G. Walton

Imperial College London

OXFORD

UNIVERSITY PRESS

OXFORD
UNIVERSITY PRESS

Great Clarendon Street, Oxford, OX2 6DP,
United Kingdom

Oxford University Press is a department of the University of Oxford.
It furthers the University's objective of excellence in research, scholarship,
and education by publishing worldwide. Oxford is a registered trade mark of
Oxford University Press in the UK and in certain other countries

Published in the United States of America by Oxford University Press
198 Madison Avenue, New York, NY 10016, United States of America

British Library Cataloguing in Publication Data
Data available

Library of Congress Control Number: 2023930926

ISBN 978-0-19-886994-8

DOI: 10.1093/oso/9780198869948.001.0001

Printed and bound by
CPI Group (UK) Ltd, Croydon, CR0 4YY

Dedicated to Prof. P. Hall, Prof. F.T. Smith FRS, Prof. J.T. Stuart FRS,
and also to Joseph and Isaac,
and Andrei.

Preface

This is Part 4 of a book series on fluid dynamics that is comprised of the following four parts:

Part 1. Classical Fluid Dynamics
Part 2. Asymptotic Problems of Fluid Dynamics
Part 3. Boundary Layers
Part 4. Hydrodynamic Stability Theory

The series is designed to give a comprehensive and coherent description of fluid dynamics, starting with chapters on classical theory suitable for an introductory undergraduate lecture course, and then progressing through more advanced material up to the level of modern research in the field. Our main attention is on high-Reynolds-number flows, both incompressible and compressible. Correspondingly, the target reader groups are undergraduate and masters students reading mathematics, aeronautical engineering, or physics, as well as doctoral students and established researchers working in the field.

In Part 1, we started with a discussion of the fundamental concepts of fluid dynamics, based on the *continuum hypothesis*. We then analysed the forces acting inside a fluid, and deduced the Navier–Stokes equations for incompressible and compressible fluids in Cartesian and curvilinear coordinates. These were deployed to study the properties of a number of flows that are represented by the so-called *exact solutions* of the Navier–Stokes equations. This was followed by a detailed discussion of the theory of inviscid flows for incompressible and compressible fluids. When dealing with incompressible inviscid flows, particular attention was paid to two-dimensional potential flows. These can be described in terms of the *complex potential*, allowing for the full power of the theory of functions of a complex variable to be employed. We demonstrated how the method of conformal mapping can be used to study various flows of interest, such as flows past *Joukovskii aerofoils* and separated flows. For the latter the *Kirchhoff model* was adopted. The final chapter of Part 1 was devoted to compressible flows of a perfect gas, including supersonic flows. Particular attention was given to the theory of characteristics, which was used, for example, to analyse the *Prandtl–Meyer flow* over a body surface with a bend or a corner. The properties of shock waves were also discussed in detail for steady and unsteady flows.

In Part 2 we introduced the reader to *asymptotic methods*. Also termed *perturbation methods*, they are now an inherent part of fluid dynamics. We started with a discussion of the mathematical aspects of asymptotic theory. This was followed by an exposition of the results of application of the theory to various fluid-dynamic problems. The first of these was the *thin aerofoil theory* for incompressible and subsonic flows, both steady and unsteady. In particular, it was shown that this theory allowed us to reduce the task of calculating the lift force to the evaluation of a simple integral. We then turned our attention to supersonic flows. We first analysed the linear approximation to the governing Euler equations, which led to a remarkably simple relationship between the slope of the aerofoil surface and the pressure, known as the *Ackeret formula*. We then considered the second-order *Buzemann approximation*, and performed the analysis of a rather slow process of attenuation of the perturbations in the far-field. Part 2 also contained a detailed discussion of the properties of inviscid *transonic* and *hypersonic flows*. We concluded Part 2 with analysis of viscous low-Reynolds-number flows. Two classical problems of the low-Reynolds-number flow theory were considered: the flow past a sphere and the flow past a circular cylinder. In both cases the flow analysis led to a difficulty, known as *Stokes paradox*. We showed how this paradox could be resolved using the formalism of matched asymptotic expansions.

Part 3 was devoted to high-Reynolds-number flows. We began with the analysis of the flows that could be described in the framework of the *classical boundary-layer theory* put forward by Prandtl in 1904. To this category belong the Blasius boundary layer on a flat plate and the Falkner–Skan solutions for the boundary layer on a wedge surface. We also presented Schlichting's solution for the laminar jet and Tollmien's solution for the viscous wake. These were followed by analysis of Chapman's shear layer that was performed with the help of Prandtl's transposition theorem. We also considered the boundary layer on the surface of a rapidly rotating cylinder with the purpose of linking the circulation around the cylinder with the speed of its rotation. We concluded the discussion of classical boundary-layer theory with analysis of compressible boundary layers, including the interactive boundary layers in hypersonic flows. We then turned our attention to separated flows. These could not be described in the framework of classical boundary-layer theory. Instead the concept of *viscous-inviscid interaction* should be used. We started with the so-called *self-induced separation* in supersonic flow. The theory of self-induced separation was developed by Stewartson and Williams (1969) and Neiland (1969), and led to the formulation of the *triple-deck model*. We then presented Sychev's (1972) theory of the boundary-layer separation in an incompressible fluid flow past a circular cylinder. This was followed by a discussion of the triple-deck flow near the trailing edge of a flat plate first investigated by Stewartson (1969) and Messiter (1970). Then the incipience of the separation at corner points of the body surface was analysed based on triple-deck theory. Part 3 concludes with analysis of the formation and bursting of short separation bubbles at the leading edge of a thin aerofoil, for which purpose a special version of triple-deck theory, referred to as *marginal separation theory*, was developed by Ruban (1981, 1982) and Stewartson *et al.* (1982).

Part 4 is devoted to *hydrodynamic stability theory* which serves to predict the onset of *laminar-turbulent transition* in fluid flows. We start with the classical results of the theory. In Chapter 1 we introduce the concept of linear instability of fluid flows, and formulate the *Orr–Sommerfeld equation* describing the stability properties of *parallel* and *quasi-parallel* flows. In the latter category are two-dimensional boundary layers where the Orr–Sommerfeld equation describes the instability in the form of *Tollmien–Schlichting waves*. We then consider the stability of 'inviscid flows' governed by the *Rayleigh equation*. In addition to describing the general properties of the Rayleigh equation, we present a numerical solution of this equation for a laminar jet. This is followed by a discussion of the Kelvin–Helmholtz instability of the shear layers that form, for example, when the boundary layer separates from a rigid body surface. We conclude Chapter 1 with a discussion of two other modes of instability, first, the *Cross-Flow instability* that is known to dominate laminar-turbulent transition on swept wings and, second, the centrifugal instability on a concave surface, taking the form of *Taylor–Görtler vortices*.

In Chapter 2 we concentrate our attention on parallel shear flows. The linear stability of such flows is governed by the Orr–Sommerfeld equation. Typically, computational solutions of this equation, when plotted in the wave number–Reynolds number plane, demonstrate the existence of two distinct branches along which disturbances neither grow nor decay but remain in a so-called neutral state, with these branches bounding a region of instability. In Chapter 2 we employ asymptotic analysis, and the method of matched asymptotic expansions, to uncover the nature of these modes in the vicinity of the two branches in the limit of large Reynolds number. The analysis, inspired by the work of Lin (1946), is carried out specifically for boundary-layer flows under the parallel flow assumption, and also for plane Poiseuille flow. For the lower branch we recover the *triple-deck* flow model discussed in detail in Part 3 of this book series. The mode structure in the vicinity of the upper branch is particularly complicated, with viscous effects not simply confined to near-wall regions, but also playing a vital role within an internal layer centred around the location where the disturbance phase speed is equal to the basic flow under consideration. This layer is known as a *critical layer* and its properties are studied in some detail, including the changes in the internal dynamics that arise as the disturbance size is increased, using ideas developed by Benney and Bergeron (1969), Haberman (1972), and Smith and Bodonyi (1982*b*).

We then turn our attention to more recent developments in the field. In Chapter 3 we introduce the reader to the *receptivity theory* that has now become an integral part of the theoretical predictions of laminar-turbulent transition in aerodynamic flows. The theory studies the process of excitation of instability modes in the boundary layers by various 'external perturbations', such as free-stream turbulence, acoustic noise, and body surface roughness. In this presentation, we use the triple-deck theory to describe the receptivity phenomena. We start with Terent'ev's (1981, 1987) theoretical model of earlier experiments of Schubauer and Skramstad (1948), where Tollmien–Schlichting waves were generated by a vibrating ribbon. We first study harmonic oscillations of the ribbon, and calculate the amplitude of the generated Tollmien–Schlichting wave. Then, an initial-value problem is considered where the ribbon starts to oscillate at a certain time, creating a wave packet in the boundary layer. Chapter 3 concludes with an analysis of the receptivity of the boundary layer to acoustic perturbations. When describing this form of receptivity we follow the papers by Ruban (1984) and Goldstein (1985). These authors demonstrated that when an acoustic wave interacts with an isolated roughness on the body surface, a Tollmien–Schlichting wave forms in the boundary layer behind the roughness. The theory allows us to predict the initial amplitude of the Tollmien–Schlichting wave.

In Chapter 4 we discuss the *weakly nonlinear stability theory* that is aimed at predicting how the growth of the amplitude of the perturbations affects the critical Reynolds number. The governing equation of weakly nonlinear theory was first formulated by Landau (1944) based on physical arguments. A formal derivation of this equation was given by Stuart (1960) and Watson (1960) with the amplitude of the perturbations assumed to be a small but non-zero parameter. Since then this equation is referred to as the *Landau–Stuart equation*. It appears that fluid flows can be subdivided into two classes, *subcritically unstable* and *supercritically unstable*. To the first category belongs, for example, plane Poiseuille flow. For this flow, an increase in the amplitude of the perturbation leads to a decrease of the critical Reynolds number. Contrary to that, the Blasius boundary layer shows a supercritical behaviour where the critical Reynolds number increases with the amplitude of the perturbations. We conclude Chapter 4 with numerical analysis of finite amplitude perturbations that leads to the concept of a *neutral surface*.

Finally, in Chapter 5 we introduce the reader to the concept of a self-sustaining process within a viscous fluid. This is a series of interactions by which certain distinguishable flow structures, which would usually decay due to the action of viscosity, are maintained by a transfer of energy via long-scale/short-scale interplay from other disturbances present in the flow. In turn, these disturbances are themselves supported by the very structure they help to preserve. This type of process, which is inherently nonlinear and typically three-dimensional in nature, is found to be responsible for the generation of so-called exact coherent structures in shear flows. For moderate to large Reynolds numbers, these equilibrium solutions correspond to certain coherent states visited by turbulent flows and have been observed experimentally (e.g. see Hof *et al.*, 2004). We will investigate the mathematical nature of these equilibrium states at both finite Reynolds number, where certain approximations need to be made, and at asymptotically large Reynolds number, where the theory can be given a rigorous mathematical foundation. We concentrate in the main on applications to channel flow, while indicating how the analysis can be applied more broadly.

The material presented in this book is based on lecture courses given by the authors at Imperial College London, the University of Manchester, and Moscow Institute of Physics and Technology.

Contents

Introduction

Hydrodynamic stability theory is concerned with the important question of how (and why) a laminar flow undergoes a transition to a turbulent state. Reynolds (1883) was the first to perform a careful experimental investigation of the laminar-turbulent transition process in the Hagen–Poiseuille flow through a circular tube. His apparatus is shown in Figure I.1. The tube was placed horizontally inside a large glass tank filled with water. One end of the tube was connected through a tap to a sink; the tap was used to regulate the amount of water passing through the tube. The other end of the tube was open to the surrounding water, allowing the water to enter the tube once the tap was opened. In order to reduce the disturbances in the flow through the tube, the open end of the tube was fitted with a trumpet mouthpiece. The flow visualization was performed with highly coloured water added to the flow in front of the trumpet. It was supplied through a thin tube connected to a reservoir on the top of the water tank (see Figure I.1).

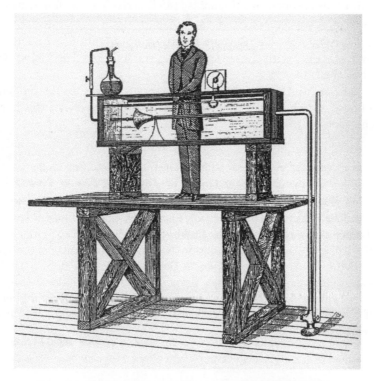

Fig. I.1: Reynolds' apparatus (see Reynolds, 1883).

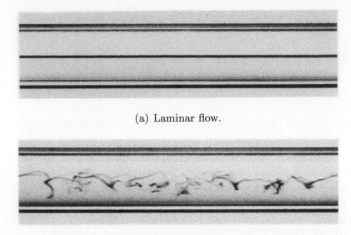

(a) Laminar flow.

(b) Turbulent flow.

Fig. I.2: Laminar-turbulent transition in the Hagen–Poiseuille flow through a circular tube. These photographs were taken by N. H. Johanneses and C. Lowe using the original Reynolds' pipe.

As a result of his observations Reynolds (1883) arrived at the following conclusions:

1. When the fluid velocity was sufficiently small, the coloured dye streak extended in a perfectly straight line through the tube; see Figure I.2(a). This means that the flow was steady with the trajectories of the fluid particles being straight lines parallel to the tube axis.

2. As the velocity was increased in small steps, the flow would suddenly develop unsteadiness; see Figure I.2(b). Reynolds (1883) reported that this happens when the dimensionless parameter

$$Re = \frac{\hat{u}_{\max} a}{\nu},$$

now called the Reynolds number, reaches the critical value $Re_c \simeq 13\,000$. Here $\hat{u}_{\max}$ is the fluid velocity along the tube axis, a denotes the tube radius and ν the kinematic viscosity of water.

As the transition from the *laminar* flow regime (Figure I.2a) to the *turbulent* one (Figure I.2b) takes place, the fluid motion becomes significantly more complicated. In addition to the primary flow parallel to the tube axis, the fluid particles are now involved in secondary motions in planes perpendicular to the axis; the secondary flow is unsteady and rather irregular. This leads to a mixing of the fluid and enhances the exchange of momentum between the fluid particles. As a consequence, instead of the parabolic velocity profile[1]

$$\frac{\hat{u}}{\hat{u}_{\max}} = 2\left(1 - \frac{\hat{r}^2}{a^2}\right),$$

[1]See Section 2.1.3 in Part 1 of this book series.

(a) Laminar flow.   (b) Turbulent flow.

Fig. I.3: Velocity distribution across the tube for laminar and turbulent flow regimes; in (b) the dashed line reproduces the laminar velocity profile.

characteristic of the laminar flow (see Figure I.3a), a more uniform distribution of the velocity is observed in the core of flow near the tube axis (see Figure I.3b). Assuming that the fluid flux through the tube remains unchanged, the velocity then exhibits a steeper rise from zero near the tube wall. This explains why the resistance of the tube to the flow appears to increase in the turbulent flow regime. It is known that the shear stress produced by the flow on the tube wall may be calculated as[2]

$$\tau_w = \mu \frac{\partial \hat{u}}{\partial \hat{r}}\bigg|_{\hat{r}=a}.$$

Here μ is the dynamic viscosity coefficient, $\hat{u}$ is the axial velocity component, and $\hat{r}$ is the radial coordinate measured from the tube axis; in this presentation we use 'hat' to denote dimensional variables. Comparing the velocity distributions in the laminar and turbulent flows (see Figure I.3) it is easy to see that τ_w is, indeed, larger in the turbulent flow.

Of course, Hagen–Poiseuille flow is not the only form of fluid motion that is subject to laminar-turbulent transition. Following Reynolds' (1883) discovery, various other flows were investigated. In particular, the transition in the boundary-layer flow on a flat plate was first observed by Burgers (1924) and later studied in more detail by Dryden (1947) and Klebanoff and Tidstrom (1959). They found that near the leading edge the flow is always laminar, and may be described by the Blasius solution.[3] However, at a

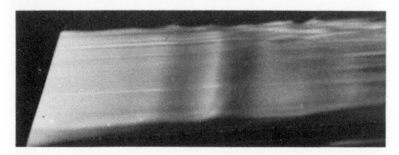

Fig. I.4: Tollmien–Schlichting waves in the boundary layer on a flat plate. Flow visualization by Werlé (1980).

[2]See the fifth equation in (1.8.65) on page 92 in Part 1 of this book series.
[3]See Section 1.1 in Part 3 of this book series.

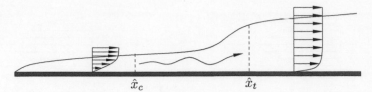

Fig. I.5: Laminar-turbulent transition in the boundary layer on a flat plate.

certain distance $\hat{x}_c$ from the leading edge the unsteadiness starts to develop in the flow in the form of so-called *Tollmien–Schlichting waves* that are superimposed on the steady Blasius flow (Figure I.4). Typically the initial amplitude of these waves is too small to cause noticeable changes in the velocity field, but they grow downstream, and there exists a second critical point $\hat{x}_t$ near which laminar-turbulent transition takes place. As a result of the transition the thickness of the boundary layer increases significantly, and the velocity profiles becomes 'fuller' which leads to an increase in the shear stress on the plate surface; see Figure I.5.

If we now define the Reynolds number as

$$Re = \frac{V_\infty \hat{x}_t}{\nu},$$

where V_∞ is the velocity at the outer edge of the boundary layer, then transition is typically observed when Re reaches a value of $\simeq 5 \cdot 10^5$.

1

Classical Hydrodynamic Stability Theory

1.1   Linear Stability Theory

The theory of Hydrodynamic Stability is aimed at predicting if, and when, the transition from a laminar to a turbulent state should be expected for a particular flow, and how the flow changes through the transition region. To outline the mathematical approach used in the stability theory, we shall start with a fairly general flow past a rigid body as depicted in Figure 1.1. We use a Cartesian coordinate system $(\hat{x}, \hat{y}, \hat{z})$ and denote the corresponding components of the velocity vector as $(\hat{u}, \hat{v}, \hat{w})$. We further denote the time by $\hat{t}$, the pressure by $\hat{p}$, the fluid density by ρ and the kinematic viscosity by ν. Assuming the flow incompressible and disregarding the body force, we can write the Navier–Stokes equations that describe the fluid motion as[1]

$$\frac{\partial \hat{u}}{\partial \hat{t}} + \hat{u}\frac{\partial \hat{u}}{\partial \hat{x}} + \hat{v}\frac{\partial \hat{u}}{\partial \hat{y}} + \hat{w}\frac{\partial \hat{u}}{\partial \hat{z}} = -\frac{1}{\rho}\frac{\partial \hat{p}}{\partial \hat{x}} + \nu\left(\frac{\partial^2 \hat{u}}{\partial \hat{x}^2} + \frac{\partial^2 \hat{u}}{\partial \hat{y}^2} + \frac{\partial^2 \hat{u}}{\partial \hat{z}^2}\right), \qquad (1.1.1a)$$

$$\frac{\partial \hat{v}}{\partial \hat{t}} + \hat{u}\frac{\partial \hat{v}}{\partial \hat{x}} + \hat{v}\frac{\partial \hat{v}}{\partial \hat{y}} + \hat{w}\frac{\partial \hat{v}}{\partial \hat{z}} = -\frac{1}{\rho}\frac{\partial \hat{p}}{\partial \hat{y}} + \nu\left(\frac{\partial^2 \hat{v}}{\partial \hat{x}^2} + \frac{\partial^2 \hat{v}}{\partial \hat{y}^2} + \frac{\partial^2 \hat{v}}{\partial \hat{z}^2}\right), \qquad (1.1.1b)$$

$$\frac{\partial \hat{w}}{\partial \hat{t}} + \hat{u}\frac{\partial \hat{w}}{\partial \hat{x}} + \hat{v}\frac{\partial \hat{w}}{\partial \hat{y}} + \hat{w}\frac{\partial \hat{w}}{\partial \hat{z}} = -\frac{1}{\rho}\frac{\partial \hat{p}}{\partial \hat{z}} + \nu\left(\frac{\partial^2 \hat{w}}{\partial \hat{x}^2} + \frac{\partial^2 \hat{w}}{\partial \hat{y}^2} + \frac{\partial^2 \hat{w}}{\partial \hat{z}^2}\right), \qquad (1.1.1c)$$

$$\frac{\partial \hat{u}}{\partial \hat{x}} + \frac{\partial \hat{v}}{\partial \hat{y}} + \frac{\partial \hat{w}}{\partial \hat{z}} = 0. \qquad (1.1.1d)$$

These have to be solved with the no-slip condition on the body surface S

$$\hat{u} = \hat{v} = \hat{w} = 0 \quad \text{on} \quad S, \qquad (1.1.2a)$$

and the following conditions in the free-stream flow far from the body:

$$\left.\begin{array}{r} u \to u_\infty, \\ \hat{v} \to v_\infty, \\ \hat{w} \to w_\infty, \\ \hat{p} \to p_\infty \end{array}\right\} \quad \text{as} \quad \hat{x}^2 + \hat{y}^2 + \hat{z}^2 \to \infty. \qquad (1.1.2b)$$

---

[1]See equations (1.7.6) on page 62 in Part 1 of this book series. Here we shall denote the dimensional variables by 'hat'.

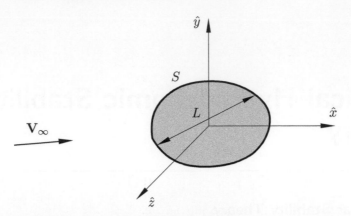

Fig. 1.1: Flow layout.

Here $(u_\infty, v_\infty, w_\infty)$ are the components of the free-stream velocity vector $\mathbf{V}_\infty$, and p_∞ denotes the free-stream pressure.

Let us now express the governing equations (1.1.1) and boundary conditions (1.1.2) in dimensionless form. For this purpose we introduce the characteristic scale L of the body, and denote the modulus of the free-stream velocity vector $\mathbf{V}_\infty$ by V_∞. Using these quantities, the non-dimensional independent and dependent variables are introduced through the equations

$$\left.\begin{aligned} \hat{x} = Lx, \qquad \hat{y} = Ly, \qquad \hat{z} = Lz, \qquad \hat{t} = \frac{L}{V_\infty}t, \\ \hat{u} = V_\infty u, \qquad \hat{v} = V_\infty v, \qquad \hat{w} = V_\infty w, \qquad \hat{p} = p_\infty + \rho V_\infty^2 p. \end{aligned}\right\} \tag{1.1.3}$$

Substitution of (1.1.3) into (1.1.1) results in

$$\frac{\partial u}{\partial t} + u\frac{\partial u}{\partial x} + v\frac{\partial u}{\partial y} + w\frac{\partial u}{\partial z} = -\frac{\partial p}{\partial x} + \frac{1}{Re}\left(\frac{\partial^2 u}{\partial x^2} + \frac{\partial^2 u}{\partial y^2} + \frac{\partial^2 u}{\partial z^2}\right), \tag{1.1.4a}$$

$$\frac{\partial v}{\partial t} + u\frac{\partial v}{\partial x} + v\frac{\partial v}{\partial y} + w\frac{\partial v}{\partial z} = -\frac{\partial p}{\partial y} + \frac{1}{Re}\left(\frac{\partial^2 v}{\partial x^2} + \frac{\partial^2 v}{\partial y^2} + \frac{\partial^2 v}{\partial z^2}\right), \tag{1.1.4b}$$

$$\frac{\partial w}{\partial t} + u\frac{\partial w}{\partial x} + v\frac{\partial w}{\partial y} + w\frac{\partial w}{\partial z} = -\frac{\partial p}{\partial z} + \frac{1}{Re}\left(\frac{\partial^2 w}{\partial x^2} + \frac{\partial^2 w}{\partial y^2} + \frac{\partial^2 w}{\partial z^2}\right), \tag{1.1.4c}$$

$$\frac{\partial u}{\partial x} + \frac{\partial v}{\partial y} + \frac{\partial w}{\partial z} = 0, \tag{1.1.4d}$$

where Re is the Reynolds number:

$$Re = \frac{V_\infty L}{\nu}.$$

The boundary conditions (1.1.2) become

$$u = v = w = 0 \quad \text{on} \quad S, \tag{1.1.5a}$$

and

$$
\left.
\begin{aligned}
u &\to u_\infty/V_\infty, \\
v &\to v_\infty/V_\infty, \\
w &\to w_\infty/V_\infty, \\
p &\to 0
\end{aligned}
\right\}
\quad \text{as} \quad x^2 + y^2 + z^2 \to \infty.
\qquad (1.1.5b)
$$

Let us now assume that the body is motionless, and the boundary-value problem (1.1.4), (1.1.5) admits a steady solution

$$
u = U(x,y,z), \quad v = V(x,y,z), \quad w = W(x,y,z), \quad p = P(x,y,z) \qquad (1.1.6)
$$

for a range of values of the Reynolds number *Re*. We shall call (1.1.6) the *basic solution*. The existence of this solution does not, however, guarantee that the corresponding flow can be actually observed in nature. For this to happen the flow has to be stable, that is, if we superimpose on the basic state (1.1.6) a perturbation of a small amplitude ε, so that

$$
\left.
\begin{aligned}
u &= U(x,y,z) + \varepsilon u'(t,x,y,z), & v &= V(x,y,z) + \varepsilon v'(t,x,y,z), \\
w &= W(x,y,z) + \varepsilon w'(t,x,y,z), & p &= P(x,y,z) + \varepsilon p'(t,x,y,z),
\end{aligned}
\right\}
\qquad (1.1.7)
$$

then the perturbation has to decay in time returning the solution to its basic state (1.1.6).

Substituting (1.1.7) into the Navier–Stokes equations (1.1.4) and working with the $O(\varepsilon)$ terms, yields the following set of linear equations for the perturbation functions:

$$
\frac{\partial u'}{\partial t} + U\frac{\partial u'}{\partial x} + \frac{\partial U}{\partial x}u' + V\frac{\partial u'}{\partial y} + \frac{\partial U}{\partial y}v' + W\frac{\partial u'}{\partial z} + \frac{\partial U}{\partial z}w'
$$
$$
= -\frac{\partial p'}{\partial x} + \frac{1}{Re}\left(\frac{\partial^2 u'}{\partial x^2} + \frac{\partial^2 u'}{\partial y^2} + \frac{\partial^2 u'}{\partial z^2}\right), \qquad (1.1.8a)
$$

$$
\frac{\partial v'}{\partial t} + U\frac{\partial v'}{\partial x} + \frac{\partial V}{\partial x}u' + V\frac{\partial v'}{\partial y} + \frac{\partial V}{\partial y}v' + W\frac{\partial v'}{\partial z} + \frac{\partial V}{\partial z}w'
$$
$$
= -\frac{\partial p'}{\partial y} + \frac{1}{Re}\left(\frac{\partial^2 v'}{\partial x^2} + \frac{\partial^2 v'}{\partial y^2} + \frac{\partial^2 v'}{\partial z^2}\right), \qquad (1.1.8b)
$$

$$
\frac{\partial w'}{\partial t} + U\frac{\partial w'}{\partial x} + \frac{\partial W}{\partial x}u' + V\frac{\partial w'}{\partial y} + \frac{\partial W}{\partial y}v' + W\frac{\partial w'}{\partial z} + \frac{\partial W}{\partial z}w'
$$
$$
= -\frac{\partial p'}{\partial z} + \frac{1}{Re}\left(\frac{\partial^2 w'}{\partial x^2} + \frac{\partial^2 w'}{\partial y^2} + \frac{\partial^2 w'}{\partial z^2}\right), \qquad (1.1.8c)
$$

$$
\frac{\partial u'}{\partial x} + \frac{\partial v'}{\partial y} + \frac{\partial w'}{\partial z} = 0. \qquad (1.1.8d)
$$

The corresponding boundary conditions are obtained by substituting (1.1.7) into (1.1.5a) and (1.1.5b). We have

$$u' = v' = w' = 0 \qquad \text{on} \quad S, \tag{1.1.9a}$$

$$u' = v' = w' = p' = 0 \quad \text{at} \quad x^2 + y^2 + z^2 = \infty. \tag{1.1.9b}$$

What one would like to do now is to construct the solution to the initial-value problem where, given an initial distribution of the perturbation in space, the subsequent behaviour of the functions u', v', w', and p' is predicted. Notice that all the coefficients in equations (1.1.8) are expressed through the solution (1.1.6) of the steady problem, and therefore, do not depend on time t. This means that the initial-value problem can be solved using the method of Laplace transforms. Recall that if we consider, for example, the function $u'(t, x, y, z)$, then its Laplace transform $\tilde{u}(\sigma, x, y, z)$ is calculated using the integral

$$\tilde{u}(\sigma, x, y, z) = \int_0^\infty u'(t, x, y, z) e^{-\sigma t} dt, \tag{1.1.10}$$

with σ being a complex parameter, $\sigma = \sigma_r + i\sigma_i$. It is known that if there exists a real constant $M > 0$ such that $|u'| < M e^{\lambda t}$ for all $t > 0$, where λ is another real constant, then the Laplace transform (1.1.10) exists for all σ situated in the complex plane on the right-hand side of the vertical line crossing the real axis at the point $\sigma_r = \lambda$; see Figure 1.2.

For known $\tilde{u}(\sigma, x, y, z)$ the original function $u'(t, x, y, z)$ may be recovered using the Bromwich integral

$$u'(t, x, y, z) = \frac{1}{2\pi i} \int_{a-i\infty}^{a+i\infty} \tilde{u}(\sigma, x, y, z) e^{\sigma t} d\sigma, \tag{1.1.11}$$

where the integration is performed in the complex σ-plane along any vertical line $\sigma_r = a$ situated to the right of the line $\sigma_r = \lambda$.

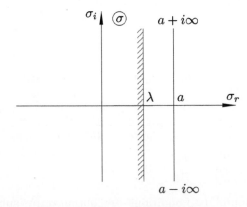

Fig. 1.2: The complex σ-plane.

1.1.1 Global stability analysis

The form of the inversion integral (1.1.11) suggests that the solution of the linearized equations (1.1.8) can represented as a superposition of so-called *normal modes*, that are written as

$$
\left. \begin{array}{ll}
u' = \tilde{u}(x,y,z)\,e^{\sigma t}, & v' = \tilde{v}(x,y,z)\,e^{\sigma t}, \\[2mm]
w' = \tilde{w}(x,y,z)\,e^{\sigma t}, & p' = \tilde{p}(x,y,z)\,e^{\sigma t}.
\end{array} \right\}
\tag{1.1.12}
$$

Substitution of (1.1.12) into equations (1.1.8) yields

$$
\sigma\tilde{u} + U\frac{\partial \tilde{u}}{\partial x} + \frac{\partial U}{\partial x}\tilde{u} + V\frac{\partial \tilde{u}}{\partial y} + \frac{\partial U}{\partial y}\tilde{v} + W\frac{\partial \tilde{u}}{\partial z} + \frac{\partial U}{\partial z}\tilde{w}
$$
$$
= -\frac{\partial \tilde{p}}{\partial x} + \frac{1}{Re}\left(\frac{\partial^2 \tilde{u}}{\partial x^2} + \frac{\partial^2 \tilde{u}}{\partial y^2} + \frac{\partial^2 \tilde{u}}{\partial z^2} \right),
\tag{1.1.13a}
$$

$$
\sigma\tilde{v} + U\frac{\partial \tilde{v}}{\partial x} + \frac{\partial V}{\partial x}\tilde{u} + V\frac{\partial \tilde{v}}{\partial y} + \frac{\partial V}{\partial y}\tilde{v} + W\frac{\partial \tilde{v}}{\partial z} + \frac{\partial V}{\partial z}\tilde{w}
$$
$$
= -\frac{\partial \tilde{p}}{\partial y} + \frac{1}{Re}\left(\frac{\partial^2 \tilde{v}}{\partial x^2} + \frac{\partial^2 \tilde{v}}{\partial y^2} + \frac{\partial^2 \tilde{v}}{\partial z^2} \right),
\tag{1.1.13b}
$$

$$
\sigma\tilde{w} + U\frac{\partial \tilde{w}}{\partial x} + \frac{\partial W}{\partial x}\tilde{u} + V\frac{\partial \tilde{w}}{\partial y} + \frac{\partial W}{\partial y}\tilde{v} + W\frac{\partial \tilde{w}}{\partial z} + \frac{\partial W}{\partial z}\tilde{w}
$$
$$
= -\frac{\partial \tilde{p}}{\partial z} + \frac{1}{Re}\left(\frac{\partial^2 \tilde{w}}{\partial x^2} + \frac{\partial^2 \tilde{w}}{\partial y^2} + \frac{\partial^2 \tilde{w}}{\partial z^2} \right),
\tag{1.1.13c}
$$

$$
\frac{\partial \tilde{u}}{\partial x} + \frac{\partial \tilde{v}}{\partial y} + \frac{\partial \tilde{w}}{\partial z} = 0.
\tag{1.1.13d}
$$

One also needs to substitute (1.1.12) into the boundary conditions (1.1.9), which leads to

$$
\tilde{u} = \tilde{v} = \tilde{w} = 0 \qquad \text{on} \quad S,
\tag{1.1.14a}
$$

$$
\tilde{u} = \tilde{v} = \tilde{w} = \tilde{p} = 0 \quad \text{at} \quad x^2 + y^2 + z^2 = \infty.
\tag{1.1.14b}
$$

Notice that the boundary-value problem (1.1.13), (1.1.14) admits a trivial solution

$$
\tilde{u} \equiv 0, \quad \tilde{v} \equiv 0, \quad \tilde{w} \equiv 0, \quad \tilde{p} \equiv 0.
$$

In order for a non-trivial solution to exist, the parameter σ should be chosen appropriately. Hence, the boundary-value problem (1.1.13), (1.1.14) is an eigen-value problem. The solution of this problem is a difficult numerical task referred to as the *global stability analysis*. In this presentation we will illustrate this approach using, as an example, the flow past a circular cylinder.

Before turning to the cylinder problem we shall make the following comment. Suppose that as a result of the calculations, a spectrum of eigen-values of σ is found. Then one can judge if the flow considered is stable or unstable by the position of the eigen-values in the complex σ-plane. Indeed, the time dependence of each normal mode (1.1.12) is given by the exponential

$$e^{\sigma t} = e^{(\sigma_r + i\sigma_i)t} = e^{\sigma_r t}\big[\cos(\sigma_i t) + i\sin(\sigma_i t)\big],$$

which shows that the flow perturbations oscillate with the frequency σ_i and grow or decay with the amplitude $e^{\sigma_r t}$. If all the eigen-values are such that σ_r is negative, then the flow is stable with respect to small perturbations. If, on the other hand, there exists at least one eigen-value with positive real part of σ, then the flow is unstable, namely, the perturbations u', v', w', and p' in (1.1.7) grow with time, and the flow will evolve away from the initial laminar state (U, V, W, P).

Flow past a circular cylinder

Let a circular cylinder of radius a be placed in a uniform flow of an incompressible viscous fluid; see Figure 1.3. When dealing with this flow it is convenient to use a Cartesian coordinate system with its origin at the centre of the cylinder, the x-axis parallel to the free-stream velocity vector, the z-axis along the axis of the cylinder, and the y-axis in the perpendicular direction. Non-dimensional variables can be introduced by once again using the scalings (1.1.3) where L is replaced by the cylinder radius a. Accordingly, the Reynolds number is now defined as $Re = V_\infty a/\nu$.

We shall assume that the free-stream velocity vector is perpendicular to the generatrix of the cylinder. In this case, one can expect the flow past the cylinder to be two-dimensional; that is, the 'spanwise' velocity component w is zero and none of the fluid-dynamic functions depend on z, which renders the Navier–Stokes equations (1.1.4) in the form

$$\frac{\partial u}{\partial t} + u\frac{\partial u}{\partial x} + v\frac{\partial u}{\partial y} = -\frac{\partial p}{\partial x} + \frac{1}{Re}\left(\frac{\partial^2 u}{\partial x^2} + \frac{\partial^2 u}{\partial y^2}\right), \qquad (1.1.15a)$$

$$\frac{\partial v}{\partial t} + u\frac{\partial v}{\partial x} + v\frac{\partial v}{\partial y} = -\frac{\partial p}{\partial y} + \frac{1}{Re}\left(\frac{\partial^2 v}{\partial x^2} + \frac{\partial^2 v}{\partial y^2}\right), \qquad (1.1.15b)$$

$$\frac{\partial u}{\partial x} + \frac{\partial v}{\partial y} = 0. \qquad (1.1.15c)$$

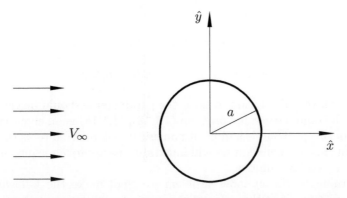

Fig. 1.3: Flow past a circular cylinder.

These equations have to be solved subject to the free-stream conditions,

$$u = 1, \quad v = 0, \quad p = 0 \quad \text{at} \quad x^2 + y^2 = \infty, \qquad (1.1.16)$$

and the no-slip conditions on the cylinder surface,

$$u = v = 0 \quad \text{if} \quad x^2 + y^2 = 1. \qquad (1.1.17)$$

The results of calculations of the steady basic flow

$$u = U(x, y), \quad v = V(x, y), \quad p = P(x, y)$$

are displayed in Figure 1.4 in the form of streamline plots. For small values of the Reynolds number $Re = V_\infty a/\nu$ the flow remains attached to the cylinder surface as Figure 1.4(a) plotted for $Re = 2.5$ shows. However, an increase of the Reynolds number leads to flow separation from the cylinder surface. As a result, two eddies form behind the cylinder. These are clearly seen in Figure 1.4(b) plotted for $Re = 5$. When the eddies first appear, they occupy a small vicinity of the rear stagnation point.

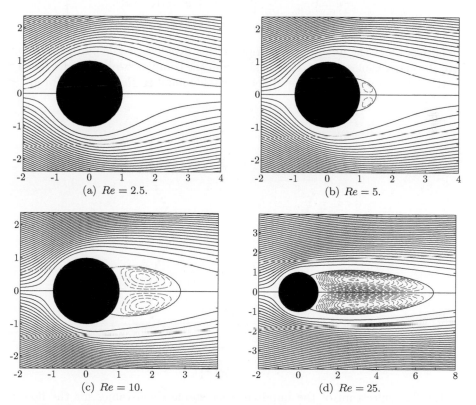

(a) $Re = 2.5$. (b) $Re = 5$. (c) $Re = 10$. (d) $Re = 25$.

Fig. 1.4: Steady flow past a circular cylinder for Reynolds numbers $Re = 2.5$, 5, 10, and 25. The dashed lines show the streamlines in the recirculation eddies.

Then, as the Reynolds number increases, the eddies become progressively larger; see Figures 1.4(c) and 1.4(d).

To study the stability of this flow we shall assume that the perturbations are also two-dimensional, and then, combining (1.1.12) and (1.1.7), we will have

$$
\left.
\begin{aligned}
u = U(x,y) + \varepsilon e^{\sigma t}\tilde{u}(x,y), \qquad v = V(x,y) + \varepsilon e^{\sigma t}\tilde{v}(x,y), \\
p = P(x,y) + \varepsilon e^{\sigma t}\tilde{p}(x,y).
\end{aligned}
\right\}
\tag{1.1.18}
$$

Substitution of (1.1.18) into (1.1.15)–(1.1.17) renders the eigen-value problem (1.1.13), (1.1.14) in the form

$$
\left.
\begin{aligned}
\sigma\tilde{u} + U\frac{\partial\tilde{u}}{\partial x} + \frac{\partial U}{\partial x}\tilde{u} + V\frac{\partial\tilde{u}}{\partial y} + \frac{\partial U}{\partial y}\tilde{v} &= -\frac{\partial\tilde{p}}{\partial x} + \frac{1}{Re}\left(\frac{\partial^2\tilde{u}}{\partial x^2} + \frac{\partial^2\tilde{u}}{\partial y^2}\right), \\
\sigma\tilde{v} + U\frac{\partial\tilde{v}}{\partial x} + \frac{\partial V}{\partial x}\tilde{u} + V\frac{\partial\tilde{v}}{\partial y} + \frac{\partial V}{\partial y}\tilde{v} &= -\frac{\partial\tilde{p}}{\partial y} + \frac{1}{Re}\left(\frac{\partial^2\tilde{v}}{\partial x^2} + \frac{\partial^2\tilde{v}}{\partial y^2}\right), \\
\frac{\partial\tilde{u}}{\partial x} + \frac{\partial\tilde{v}}{\partial y} &= 0, \\
\tilde{u} = \tilde{v} = 0 \qquad \text{at} \qquad x^2 + y^2 &= 1, \\
\tilde{u} = \tilde{v} = \tilde{p} = 0 \qquad \text{at} \qquad x^2 + y^2 &= \infty.
\end{aligned}
\right\}
\tag{1.1.19}
$$

The results of the numerical solution of this problem are displayed in Figure 1.5 where the eigen-values of (1.1.19) are shown in the complex σ-plane. It has been found that there exists a critical value of the Reynolds number $Re_c \approx 24.4$. For $Re < Re_c$

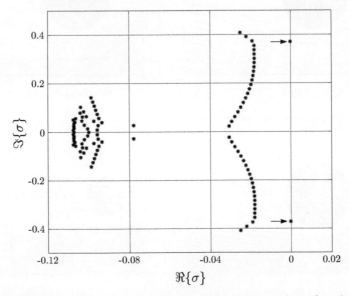

Fig. 1.5: The spectrum of the hundred least stable eigen-values for the flow past a circular cylinder at $Re = Re_c$. Arrows highlight the eigen-values which cross the imaginary axis.

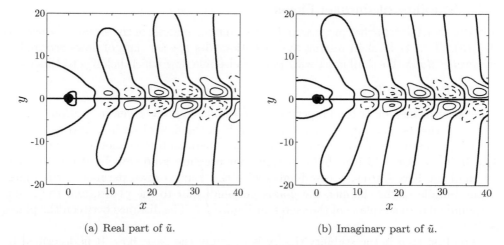

(a) Real part of $\tilde{u}$. (b) Imaginary part of $\tilde{u}$.

Fig. 1.6: Contour plots of the real and imaginary parts of the streamwise velocity perturbation eigenfunction $\tilde{u}$, at the onset of instability $Re = Re_c$. Contour levels are in intervals of 0.1, and the positive and negative levels are shown by solid and dashed lines, respectively. Also we use thicker solid lines for zero levels of $\Re\{\tilde{u}\}$ and $\Im\{\tilde{u}\}$.

(not shown) the flow is stable with all eigen-solutions lying to the left of the imaginary axis, where $\sigma_r = \Re\{\sigma\} < 0$. Figure 1.5 'captures the moment' when the Reynolds number reaches its critical value Re_c. It displays the spectrum of the hundred least stable eigen-values at the critical Reynolds number. Notice that they occur in complex conjugate pairs. When the Reynolds number approaches Re_c, one pair approach the imaginary axis (these solutions are shown by arrows in Figure 1.5), and the flow becomes unstable. In fact, at $Re = Re_c$ the flow performs harmonic oscillations with the frequency $\sigma_i = \Im\{\sigma\} = 0.3709$. The corresponding eigen-functions are also complex conjugate to one another. In Figure 1.6 we show the real and imaginary parts of the perturbed longitudinal velocity component $\tilde{u}(x, y)$. The perturbation field develops a rather long wake behind the cylinder. For the attenuation condition

$$\tilde{u} = \tilde{v} = \tilde{p} = 0 \quad \text{at} \quad x^2 + y^2 = \infty$$

to hold, the computation region had to be extended to $x \approx 300$. In Figure 1.6 only a part of the wake is shown. It is easily seen that $\tilde{u}(x, y)$ is anti-symmetric with respect to the x-axis. These results are in agreement with laboratory observations. It is known from experiment that when the Reynolds number reaches a critical value, the flow becomes oscillatory, with vortices periodically shedding from the upper and lower sides of the cylinder. This vortex system is known as the *Kármán vortex street*.

The global stability analysis may be used to investigate the stability properties of various other flows. Unfortunately, it is not clear how the method can be applied to boundary layers. As we can see from Figures (I.4) and (I.5), in the boundary layer the instabilities grow with the downstream coordinate x, not with time t, which makes the attenuation condition (1.1.14b) inapplicable. An alternative approach is needed (see Section 1.3).

1.2 Stability of Parallel Flows

In order to achieve further progress in the theoretical analysis of the stability equations (1.1.13), (1.1.14) we shall restrict our attention to a special class of flows referred to as *parallel flows*. We shall start with plane Poiseuille flow which belongs to this class.

1.2.1 Poiseuille flow

A detailed discussion of the properties of Poiseuille flow was given in Section 2.1.2 in Part 1 of this book series. Remember that this is the flow of an incompressible fluid through a channel formed by two infinite planes parallel to one another (Figure 1.7). The fluid motion is driven by a longitudinal pressure gradient. To study the flow we shall use here Cartesian coordinates with the $\hat{x}$-axis drawn parallel to the plates in the middle of the channel, the $\hat{y}$-axis perpendicular to the plates, and the $\hat{z}$-axis perpendicular to the plane of the sketch in Figure 1.7. The distance between the plates is $2h$.[2]

The first step in the stability theory is to study the basic flow. It is described by the steady version of the Navier–Stokes equations (1.1.1):

$$\hat{u}\frac{\partial \hat{u}}{\partial \hat{x}} + \hat{v}\frac{\partial \hat{u}}{\partial \hat{y}} + \hat{w}\frac{\partial \hat{u}}{\partial \hat{z}} = -\frac{1}{\rho}\frac{\partial \hat{p}}{\partial \hat{x}} + \nu\left(\frac{\partial^2 \hat{u}}{\partial \hat{x}^2} + \frac{\partial^2 \hat{u}}{\partial \hat{y}^2} + \frac{\partial^2 \hat{u}}{\partial \hat{z}^2}\right), \qquad (1.2.1a)$$

$$\hat{u}\frac{\partial \hat{v}}{\partial \hat{x}} + \hat{v}\frac{\partial \hat{v}}{\partial \hat{y}} + \hat{w}\frac{\partial \hat{v}}{\partial \hat{z}} = -\frac{1}{\rho}\frac{\partial \hat{p}}{\partial \hat{y}} + \nu\left(\frac{\partial^2 \hat{v}}{\partial \hat{x}^2} + \frac{\partial^2 \hat{v}}{\partial \hat{y}^2} + \frac{\partial^2 \hat{v}}{\partial \hat{z}^2}\right), \qquad (1.2.1b)$$

$$\hat{u}\frac{\partial \hat{w}}{\partial \hat{x}} + \hat{v}\frac{\partial \hat{w}}{\partial \hat{y}} + \hat{w}\frac{\partial \hat{w}}{\partial \hat{z}} = -\frac{1}{\rho}\frac{\partial \hat{p}}{\partial \hat{z}} + \nu\left(\frac{\partial^2 \hat{w}}{\partial \hat{x}^2} + \frac{\partial^2 \hat{w}}{\partial \hat{y}^2} + \frac{\partial^2 \hat{w}}{\partial \hat{z}^2}\right), \qquad (1.2.1c)$$

$$\frac{\partial \hat{u}}{\partial \hat{x}} + \frac{\partial \hat{v}}{\partial \hat{y}} + \frac{\partial \hat{w}}{\partial \hat{z}} = 0. \qquad (1.2.1d)$$

These should be solved subject to the no-slip condition on the two plates:

$$\hat{u} = \hat{v} = \hat{w} = 0 \quad \text{at} \quad \hat{y} = h \text{ and } \hat{y} = -h. \qquad (1.2.2)$$

If the channel is sufficiently long, then in the majority of the flow (excluding the channel ends)

$$\frac{\partial \hat{u}}{\partial \hat{x}} = 0. \qquad (1.2.3)$$

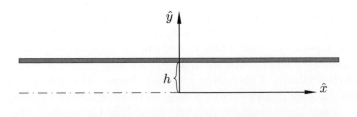

Fig. 1.7: The problem layout.

[2]Notice that in Part 1, slightly different notations were used.

Since the problem considered is invariant with respect to an arbitrary shift in the $\hat{z}$-direction, we can also claim that the derivative of any function with respect to $\hat{z}$ is zero, that is

$$\frac{\partial}{\partial \hat{z}} = 0. \tag{1.2.4}$$

With (1.2.3) and (1.2.4) the continuity equation (1.2.1d) reduces to

$$\frac{\partial \hat{v}}{\partial \hat{y}} = 0$$

which, being integrated with the boundary condition for $\hat{v}$ in (1.2.2), yields

$$\hat{v} \equiv 0 \tag{1.2.5}$$

in the entire flow field.

Using (1.2.3), (1.2.4), and (1.2.5) in the $\hat{x}$-momentum equation (1.2.1a), we find that

$$\frac{\partial^2 \hat{u}}{\partial \hat{y}^2} = \frac{1}{\mu} \frac{\partial \hat{p}}{\partial \hat{x}}. \tag{1.2.6}$$

Here $\mu = \rho \nu$ is the dynamic viscosity coefficient. Since the flow considered is incompressible, we can assume μ to be constant over the entire flow field.

Substitution of (1.2.5) into the $\hat{y}$-momentum equation (1.2.1b) yields

$$\frac{\partial \hat{p}}{\partial \hat{y}} = 0,$$

and we see that in the basic flow the pressure does not vary across the channel.

Differentiating (1.2.6) with respect to $\hat{x}$ we have

$$\frac{\partial}{\partial \hat{x}} \left(\frac{\partial \hat{p}}{\partial \hat{x}} \right) = \mu \frac{\partial^2}{\partial \hat{y}^2} \left(\frac{\partial \hat{u}}{\partial \hat{x}} \right),$$

which, in view of (1.2.3), is zero. Consequently, the pressure gradient remains constant over the entire flow field and may be calculated as

$$\frac{d\hat{p}}{d\hat{x}} = \frac{\Delta \hat{p}}{L}, \tag{1.2.7}$$

where $\Delta \hat{p}$ is the pressure difference between the channel ends, and L is the length of the channel.

Integrating (1.2.6) with the no-slip conditions on the two plates

$$\hat{u} \Big|_{\hat{y}=-h} = \hat{u} \Big|_{\hat{y}=h} = 0,$$

we find that the velocity profile is parabolic,

$$\hat{u} = -\frac{1}{2\mu} \frac{d\hat{p}}{d\hat{x}} (h^2 - \hat{y}^2). \tag{1.2.8}$$

The maximum velocity

$$\hat{u}_{\max} = \frac{h^2}{2\mu} \left| \frac{d\hat{p}}{d\hat{x}} \right| \tag{1.2.9}$$

is achieved in the middle of the channel ($\hat{y} = 0$).

This completes the analysis of the basic flow state. We now need to consider the perturbations. In order to non-dimensionalize the Navier–Stokes equations (1.1.1) we shall use, instead of (1.1.3), the following transformations

$$\left.\begin{aligned}
\hat{x} = hx, \qquad \hat{y} = hy, \qquad \hat{z} = hz, \qquad \hat{t} = \frac{h}{\hat{u}_{\max}}t, \\[2mm]
\hat{u} = \hat{u}_{\max}u, \qquad \hat{v} = \hat{u}_{\max}v, \qquad \hat{w} = \hat{u}_{\max}w, \qquad \hat{p} = p_0 + \rho\hat{u}_{\max}^2 p.
\end{aligned}\right\} \qquad (1.2.10)$$

Here the choice of p_0 is not important; one can choose, for example, p_0 to be the value of the pressure in the steady basic flow at the channel cross-section with $\hat{x} = 0$. The basic flow solution (1.2.8) is expressed in the non-dimensional variables as

$$U = 1 - y^2, \quad V = 0, \quad W = 0, \quad P = -\frac{2}{Re}x, \qquad (1.2.11)$$

with the Reynolds number given by

$$Re = \frac{\rho\hat{u}_{\max}h}{\mu}. \qquad (1.2.12)$$

Substitution of (1.2.10) into the Navier–Stokes equations (1.1.1) leads again to equations (1.1.4), with the Reynolds number now defined by (1.2.12). Decomposing the solution into the steady basic flow and small perturbations (1.1.7), and seeking the perturbation functions in the normal-mode form (1.1.12) leads to equations (1.1.13). In view of (1.2.11), they now assume the form:

$$\sigma\tilde{u} + U\frac{\partial\tilde{u}}{\partial x} = -\frac{\partial\tilde{p}}{\partial x} + \frac{1}{Re}\left(\frac{\partial^2\tilde{u}}{\partial x^2} + \frac{\partial^2\tilde{u}}{\partial y^2} + \frac{\partial^2\tilde{u}}{\partial z^2}\right), \qquad (1.2.13a)$$

$$\sigma\tilde{v} + U\frac{\partial\tilde{v}}{\partial x} = -\frac{\partial\tilde{p}}{\partial y} + \frac{1}{Re}\left(\frac{\partial^2\tilde{v}}{\partial x^2} + \frac{\partial^2\tilde{v}}{\partial y^2} + \frac{\partial^2\tilde{v}}{\partial z^2}\right), \qquad (1.2.13b)$$

$$\sigma\tilde{w} + U\frac{\partial\tilde{w}}{\partial x} = -\frac{\partial\tilde{p}}{\partial z} + \frac{1}{Re}\left(\frac{\partial^2\tilde{w}}{\partial x^2} + \frac{\partial^2\tilde{w}}{\partial y^2} + \frac{\partial^2\tilde{w}}{\partial z^2}\right), \qquad (1.2.13c)$$

$$\frac{\partial\tilde{u}}{\partial x} + \frac{\partial\tilde{v}}{\partial y} + \frac{\partial\tilde{w}}{\partial z} = 0. \qquad (1.2.13d)$$

These have to be solved subject to the no-slip conditions on the two walls of the channel:

$$\tilde{u} = \tilde{v} = \tilde{w} = 0 \quad \text{at} \quad y = -1 \text{ and } y = 1. \qquad (1.2.14)$$

It should be noticed that U is a function of y only. This means that all the coefficients in equations (1.2.13) are independent of x and z. Hence, if the initial perturbation introduced into the flow at $t = 0$ is 'localized' in the (x, z)-plane, that is, decays fast enough as $x^2 + z^2 \to \infty$, then the solution of (1.2.13), (1.2.14) may be constructed using Fourier transforms. To introduce the corresponding notations we shall consider, as an example, the function $\tilde{u}(x, y, z; \sigma)$. Its Fourier transform $\breve{u}$ is given by

$$\breve{u}(y; \sigma, \alpha, \beta) = \iint\limits_{-\infty}^{\infty} \tilde{u}(x, y, z, \sigma)e^{-i\alpha x - i\beta z}\,dx dz.$$

In order to recover the original function $\tilde{u}$, one needs to calculate the integral

$$\tilde{u}(x,y,z;\sigma) = \frac{1}{4\pi^2} \int\!\!\!\int_{-\infty}^{\infty} \breve{u}(y;\sigma,\alpha,\beta) e^{i\alpha x + i\beta z} \, d\alpha d\beta. \qquad (1.2.15)$$

It follows from (1.2.15) that the solution can be represented as a superposition of the *normal modes*:

$$\left.\begin{aligned}
\tilde{u} &= \breve{u}(y;\sigma,\alpha,\beta)\,e^{i\alpha x + i\beta z}, & \tilde{v} &= \breve{v}(y;\sigma,\alpha,\beta)\,e^{i\alpha x + i\beta z}, \\
\tilde{w} &= \breve{w}(y;\sigma,\alpha,\beta)\,e^{i\alpha x + i\beta z}, & \tilde{p} &= \breve{p}(y;\sigma,\alpha,\beta)\,e^{i\alpha x + i\beta z}.
\end{aligned}\right\} \qquad (1.2.16)$$

Substituting (1.2.16) into (1.2.13) and introducing, instead of σ, a new complex parameter ω such that

$$\sigma = -i\omega, \qquad (1.2.17)$$

we have

$$i\big(\alpha U - \omega\big)\breve{u} + \frac{dU}{dy}\breve{v} = -i\alpha\breve{p} + \frac{1}{Re}\left[\frac{d^2\breve{u}}{dy^2} - (\alpha^2 + \beta^2)\breve{u}\right], \qquad (1.2.18a)$$

$$i\big(\alpha U - \omega\big)\breve{v} = -\frac{d\breve{p}}{dy} + \frac{1}{Re}\left[\frac{d^2\breve{v}}{dy^2} - (\alpha^2 + \beta^2)\breve{v}\right], \qquad (1.2.18b)$$

$$i\big(\alpha U - \omega\big)\breve{w} = -i\beta\breve{p} + \frac{1}{Re}\left[\frac{d^2\breve{w}}{dy^2} - (\alpha^2 + \beta^2)\breve{w}\right], \qquad (1.2.18c)$$

$$i\alpha\breve{u} + \frac{d\breve{v}}{dy} + i\beta\breve{w} = 0, \qquad (1.2.18d)$$

while the boundary conditions (1.2.14) become

$$\breve{u} = \breve{v} = \breve{w} = 0 \quad \text{at} \quad y = -1 \ \text{and} \ y = 1. \qquad (1.2.19)$$

We have shown that if the basic flow is steady and parallel, then the set of partial differential equations (1.8) describing the behaviour of the perturbations, may be reduced to four ordinary differential equations (1.2.18). We shall simplify these equations further by eliminating $\breve{u}$, $\breve{w}$, and $\breve{p}$, and formulating a single equation for $\breve{v}$. We start by introducing the function

$$f = i\alpha\breve{u} + i\beta\breve{w},$$

which allows us to express the continuity equation (1.2.18d) in the form

$$f = -\frac{d\breve{v}}{dy}. \qquad (1.2.20)$$

We then multiply equations (1.2.18a) and (1.2.18c) by $i\alpha$ and $i\beta$, respectively, and add the results together. This leads to

$$i\big(\alpha U - \omega\big)f + i\alpha\frac{dU}{dy}\breve{v} = (\alpha^2 + \beta^2)\breve{p} + \frac{1}{Re}\left[\frac{d^2 f}{dy^2} - (\alpha^2 + \beta^2)f\right]. \qquad (1.2.21)$$

The function f can be eliminated from equation (1.2.21) with the help of the continuity equation (1.2.20). We have

$$-i(\alpha U - \omega)\frac{d\breve{v}}{dy} + i\alpha\frac{dU}{dy}\breve{v} = (\alpha^2 + \beta^2)\breve{p} - \frac{1}{Re}\left[\frac{d^3\breve{v}}{dy^3} - (\alpha^2 + \beta^2)\frac{d\breve{v}}{dy}\right]. \qquad (1.2.22)$$

To eliminate the pressure $\breve{p}$ we use equation (1.2.18b). Differentiating (1.2.22) with respect to y,

$$-i(\alpha U - \omega)\frac{d^2\breve{v}}{dy^2} + i\alpha\frac{d^2 U}{dy^2}\breve{v} = (\alpha^2 + \beta^2)\frac{d\breve{p}}{dy} - \frac{1}{Re}\left[\frac{d^4\breve{v}}{dy^4} - (\alpha^2 + \beta^2)\frac{d^2\breve{v}}{dy^2}\right], \qquad (1.2.23)$$

and combining (1.2.23) with (1.2.18b), we finally have

$$\frac{1}{iRe}\left[\frac{d^4\breve{v}}{dy^4} - 2(\alpha^2 + \beta^2)\frac{d^2\breve{v}}{dy^2} + (\alpha^2 + \beta^2)^2\breve{v}\right]$$
$$= (\alpha U - \omega)\left[\frac{d^2\breve{v}}{dy^2} - (\alpha^2 + \beta^2)\breve{v}\right] - \alpha\frac{d^2 U}{dy^2}\breve{v}, \qquad (1.2.24)$$

which is known as the *Orr–Sommerfeld equation*. It is a fourth order ordinary differential equation, and requires four boundary conditions for $\breve{v}$. Two of these,

$$\breve{v} = 0 \quad \text{at} \quad y = -1 \text{ and } y = 1,$$

follow directly from (1.2.19). To formulate the other two, we notice that with $\breve{u}$ and $\breve{w}$ being zero on the channel walls, the function f should also vanish, but then it follows from (1.2.20) that

$$\frac{d\breve{v}}{dy} = 0 \quad \text{at} \quad y = -1 \text{ and } y = 1.$$

Thus the full set of the boundary conditions that should be used with the Orr–Sommerfeld equation (1.2.24) are

$$\breve{v} = \frac{d\breve{v}}{dy} = 0 \quad \text{at} \quad y = -1 \text{ and } y = 1. \qquad (1.2.25)$$

Before analysing the boundary-value problem (1.2.24), (1.2.25) let us recall the procedure we used for reducing the perturbation equations (1.1.8) to the Orr–Sommerfeld equation (1.2.24). Firstly, assuming the basic flow steady, we found that the flow perturbation functions u', v', w', and p' may be expressed as a superposition of the normal modes (1.1.12). Secondly, if the basic flow is parallel, then equations (1.2.16) become applicable. Let us now substitute (1.2.16) back into (1.1.12), and then, using (1.2.17), we find that the lateral velocity component perturbation function may be represented in the *normal-mode form*

$$v'(t, x, y, z) = e^{i(\alpha x + \beta z - \omega t)}\breve{v}(y), \qquad (1.2.26)$$

with similar expressions for the rest of the fluid-dynamic functions, u', w', and p'.

The boundary-value problem (1.2.24), (1.2.25) is homogeneous, and since we are interested in non-trivial solutions, it should be treated as an eigen-value problem.

When dealing with plane Poiseuille flow, it is natural to assume that the Fourier parameters α and β are real. For a given Reynolds number Re, solving the problem (1.2.24), (1.2.25) leads to a set of eigen-values

$$\omega = \omega_r + i\omega_i \tag{1.2.27}$$

for each pair of α and β. Substitution of (1.2.27) into (1.2.26) yields

$$v'(t, x, y, z) = e^{\omega_i t} e^{i(\alpha x + \beta z - \omega_r t)} \breve{v}(y), \tag{1.2.28}$$

which represents a simple wave motion with α and β being the wave numbers in the x- and z-directions, respectively, ω_r is the frequency of oscillations, and ω_i the amplification rate. Instability of the flow develops when ω_i becomes positive.

The above analysis shows that parallel flows are subject to three-dimensional perturbations in the form of oblique waves (1.2.28). However, in order to study the stability of the flow, one does not need to solve the eigen-value problem (1.2.24), (1.2.25) for all possible combinations of the parameters Re, α, and β. The following theorem due to Squire (1933) shows that, when studying the instability of incompressible parallel flows, we can restrict our attention to two-dimensional perturbations ($\beta = 0$).

Theorem 1.1 *If a growing three-dimensional perturbation can be found in a parallel flow at Reynolds number Re, then a growing two-dimensional perturbation exists in this flow at a lower Reynolds number $Re_* < Re$, and it exhibits a higher growth rate than the corresponding three-dimensional perturbation.*

Proof   Given α and β, we define the 'two-dimensional wave number' as

$$\alpha_* = \sqrt{\alpha^2 + \beta^2}. \tag{1.2.29}$$

We then use the transformations

$$\omega = \frac{\alpha}{\alpha_*}\omega_*, \qquad Re = \frac{\alpha_*}{\alpha}Re_* \tag{1.2.30}$$

to define the two-dimensional frequency ω_* and Reynolds number Re_*. Keeping the functions $U(y)$ and $\breve{v}(y)$ unchanged, we substitute (1.2.29), (1.2.30) into the Orr–Sommerfeld equation (1.2.24). This leads to

$$\frac{1}{iRe_*}\left(\frac{d^4\breve{v}}{dy^4} - 2\alpha_*^2\frac{d^2\breve{v}}{dy^2} + \alpha_*^4\breve{v}\right) = (\alpha_* U - \omega_*)\left(\frac{d^2\breve{v}}{dy^2} - \alpha_*^2\breve{v}\right) - \alpha_*\frac{d^2U}{dy^2}\breve{v},$$

which is the two-dimensional version of the Orr–Sommerfeld equation. Since the function $\breve{v}(y)$ remains unchanged, the boundary conditions (1.2.25) are still satisfied. This proves that the transformations (1.2.29), (1.2.30) turn the three-dimensional problem (1.2.24), (1.2.25) into a two-dimensional one.

Now, to compare the characteristics of the three- and two-dimensional perturbations, we need to use equations (1.2.30). It follows from (1.2.29) that for any $\beta \neq 0$

$$\frac{\alpha_*}{\alpha} > 1,$$

and therefore,

$$\omega_{*i} > \omega_i \quad \text{and} \quad Re_* < Re,$$

in accordance with the statement of the theorem. $\square$

1.2.2 Analysis of two-dimensional perturbations

For two-dimensional perturbations the normal-mode representation (1.2.26), (1.2.28) is written as

$$v'(t, x, y, z) = e^{i(\alpha x - \omega t)} \breve{v}(y) = e^{\omega_i t} e^{i(\alpha x - \omega_r t)} \breve{v}(y). \tag{1.2.31}$$

If instead of the frequency ω we use the *phase velocity* c, such that

$$\omega = \alpha c,$$

then (1.2.31) becomes

$$v'(t, x, y, z) = e^{i\alpha(x - ct)} \breve{v}(y) = e^{\alpha c_i t} e^{i\alpha(x - c_r t)} \breve{v}(y), \tag{1.2.32}$$

and the Orr–Sommerfeld equation (1.2.24) assumes the form

$$\frac{1}{i\alpha Re} \left(\frac{d^4 \breve{v}}{dy^4} - 2\alpha^2 \frac{d^2 \breve{v}}{dy^2} + \alpha^4 \breve{v} \right) = (U - c) \left(\frac{d^2 \breve{v}}{dy^2} - \alpha^2 \breve{v} \right) - \frac{d^2 U}{dy^2} \breve{v}. \tag{1.2.33a}$$

For channel flow the associated boundary conditions are

$$\breve{v} = \frac{d\breve{v}}{dy} = 0 \quad \text{at} \quad y = -1 \text{ and } y = 1. \tag{1.2.33b}$$

Unfortunately, an analytical solution of the eigen-value problem (1.2.33) is not possible even for a basic flow as simple as (1.2.11). Instead the problem has to be tackled numerically. In the course of the calculations one can expect to find the value of the complex phase speed $c = c_r + ic_i$ for each pair of real Re and α. It can be demonstrated (see Problem 1 in Exercises 1) that without loss of generality only positive α need to be considered. Of course, the Reynolds number Re is positive by its nature, which is why the results of the calculations are normally presented in the first quadrant of the (Re, α)-plane. As has been already mentioned, to each point in this plane there corresponds a certain value of the phase speed $c = c_r + ic_i$. Of primary interest are points where $c_i = 0$, and the perturbation (1.2.32) represents a wave of constant amplitude propagating with velocity c_r. The locus of such points is referred to as the *neutral curve*. For plane Poiseuille flow it is shown in Figure 1.8(a). On this curve the perturbations neither grow nor decay. Inside the curve lies the region of instability $(c_i > 0)$; everywhere outside the neutral curve $c_i < 0$, and the perturbations decay. It is interesting to notice that there exists a *critical* (minimum) value of the Reynolds number on the neutral curve; we shall denote it as Re_c. If $Re < Re_c$, then the perturbations decay for any wave number α. If, on the other hand, $Re > Re_c$, then the perturbations are unstable for the band of wave numbers between the lower and upper branches of the neutral curve. For plane Poiseuille flow

$$Re_c \simeq 5772.2,$$

with the corresponding values of the wave number and the phase speed being

$$\alpha_c \simeq 1.02, \qquad c_{rc} \simeq 0.264.$$

Figure 1.8(b) shows an alternative way of plotting the neutral curve. Since at each point on the neutral curve in Figure 1.8(a) the imaginary part of the phase velocity

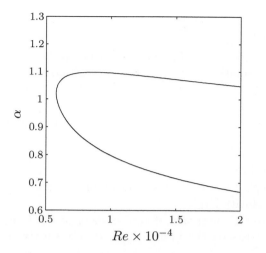
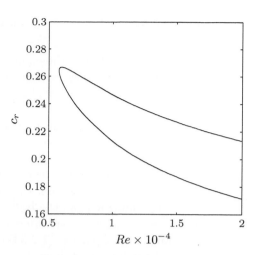

(a) Wave number α against Reynolds number Re.   (b) Phase speed c_r against Reynolds number Re.

Fig. 1.8: Neutral curves for plane Poiseuille flow, $U = 1 - y^2$.

$c_i = 0$, and the real part of the phase velocity c_r is known, we can use this information to produce a plot of the neutral curve in the (Re, c_r)-plane. Figure 1.8(b) shows such a plot for plane Poiseuille flow.

The following two comments are appropriate here. Firstly, the eigen-value problem (1.2.33) produces not one but a set of solutions for each pair of Re and α. However, only one of these, namely, the one discussed above, shows an instability for certain Re and α; the rest of the eigen-solutions are stable for all Re and α. Secondly, in plane Poiseuille flow the basic solution (1.2.11) is even with respect to the centreline of the channel, in which case all the solutions of the Orr–Sommerfeld equation may be subdivided into two categories. In the first one are the solutions with even $\breve{v}$ and odd $\breve{u}$; these are called *sinuous modes*. In the second category are the solutions where $\breve{v}$ is odd with respect to $y = 0$ and $\breve{u}$ is even; these are called *varicose modes*. The mode shown in Figure 1.8 is a sinuous mode, and it is the only one that exhibits the flow instability.

Of course, the approach described above may be applied to other parallel flows, including plane Couette flow between two parallel plates and Hagen–Poiseuille flow in a circular pipe.[3] Interestingly enough, both of these flows prove to be linearly stable for all values of the Reynolds number.

Exercises 1

1. Assume that for a pair of real wave numbers $\alpha = \alpha_1$ and $\beta = \beta_1$ the Orr–Sommerfeld equation (1.2.24) admits a solution with eigen-value $\omega = \omega_1$ and eigen-function $\breve{v} = \breve{v}_1$.

   Then, taking the complex conjugate on both sides of (1.2.24), show that for $\alpha_2 = -\alpha_1$ and $\beta_2 = -\beta_1$ the Orr–Sommerfeld equation (1.2.24) also admits a

---

[3]See Sections 2.1.1 and 2.1.3 in Part 1 of this book series.

solution with

$$\omega_2 = -\overline{\omega}_1, \qquad \check{v}_2 = \overline{\check{v}}_1,$$

where $\overline{\omega}_1$ and $\overline{\check{v}}_1$ denote the complex conjugates of ω_1 and $\check{v}_1$, respectively.

Write down the first and second solutions for v' in the normal-mode form (1.2.26). Add them together, and confirm that the resulting formula produces a real function.

2. (a) Explain the physical meaning of the transformation (1.2.30) of the Reynolds number. For this purpose determine the direction of propagation of the normal-mode disturbance (1.2.26) in the (x, z)-plane and find the projection of the basic flow velocity vector in this direction.

   (b) Modify Squire's theorem by keeping the Reynolds number unchanged, and transforming instead the basic velocity $U(y)$.

3. (a) Prove by direct substitution into (1.2.33a) that if the solution of the Orr–Sommerfeld equation (1.2.33a) exists on the interval $y \in [0, -1]$, satisfying the boundary conditions

$$\check{v} = \frac{d\check{v}}{dy} = 0 \quad \text{at} \quad y = -1 \quad \text{and} \quad \frac{d\check{v}}{dy} = \frac{d^3\check{v}}{dy^3} = 0 \quad \text{at} \quad y = 0,$$

then it may be extended into $y \in [0, 1]$ as

$$\check{v}(y) = \check{v}(-y)$$

forming a solution where $\check{v}$ is even with respect to $y = 0$. Show that this solution satisfies the boundary conditions (1.2.33b).

Is the corresponding solution for $\check{u}$ even with respect to $y = 0$ or odd?

   (b) Formulate a similar statement for the solution which is anti-symmetric for $\check{v}$.

4. Deduce the two-dimensional Orr–Sommerfeld equation (1.2.33a) starting from the two-dimensional version of the perturbation equations (1.1.8). Confirm that in the case of a parallel basic flow, equations (1.1.8) can be written as

$$\frac{\partial u'}{\partial t} + U\frac{\partial u'}{\partial x} + \frac{dU}{dy}v' = -\frac{\partial p'}{\partial x} + \frac{1}{Re}\left(\frac{\partial^2 u'}{\partial x^2} + \frac{\partial^2 u'}{\partial y^2}\right), \qquad (1.2.34\text{a})$$

$$\frac{\partial v'}{\partial t} + U\frac{\partial v'}{\partial x} = -\frac{\partial p'}{\partial y} + \frac{1}{Re}\left(\frac{\partial^2 v'}{\partial x^2} + \frac{\partial^2 v'}{\partial y^2}\right), \qquad (1.2.34\text{b})$$

$$\frac{\partial u'}{\partial x} + \frac{\partial v'}{\partial y} = 0. \qquad (1.2.34\text{c})$$

Exclude the pressure from this set of equations by cross-differentiation of (1.2.34a) and (1.2.34b). Then, based on the continuity equation (1.2.34c), introduce the stream function $\psi'(x, y)$ such that

$$u' = \frac{\partial \psi'}{\partial y}, \qquad v' = -\frac{\partial \psi'}{\partial x},$$

and formulate the equation for $\psi'(x, y)$.

Seek the solution of this equation in the form

$$\psi'(t, x, y) = e^{i\alpha(x-ct)}\phi(y),$$

and show that

$$\frac{1}{i\alpha Re}\left(\frac{d^4\phi}{dy^4} - 2\alpha^2\frac{d^2\phi}{dy^2} + \alpha^4\phi\right) = (U - c)\left(\frac{d^2\phi}{dy^2} - \alpha^2\phi\right) - \frac{d^2U}{dy^2}\phi. \qquad (1.2.35)$$

What are the boundary conditions for $\phi(y)$ in the case of plane Poiseuille flow?

5. Consider the Hagen–Poiseuille flow of an incompressible fluid through a tube of circular cross-section as shown in Figure 1.9.

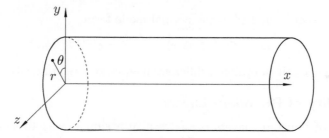

Fig. 1.9: The cylindrical geometry for Hagen–Poiseuille flow.

Assuming both the basic laminar flow and the perturbations to be axisymmetric, the Navier–Stokes equations may be written in dimensionless variables as

$$\frac{\partial u}{\partial t} + u\frac{\partial u}{\partial x} + v\frac{\partial u}{\partial r} = -\frac{\partial p}{\partial x} + \frac{1}{Re}\left(\frac{\partial^2 u}{\partial x^2} + \frac{\partial^2 u}{\partial r^2} + \frac{1}{r}\frac{\partial u}{\partial r}\right), \qquad (1.2.36a)$$

$$\frac{\partial v}{\partial t} + u\frac{\partial v}{\partial x} + v\frac{\partial v}{\partial r} = -\frac{\partial p}{\partial r} + \frac{1}{Re}\left(\frac{\partial^2 v}{\partial x^2} + \frac{\partial^2 v}{\partial r^2} + \frac{1}{r}\frac{\partial v}{\partial r} - \frac{v}{r^2}\right), \qquad (1.2.36b)$$

$$\frac{\partial u}{\partial x} + \frac{\partial v}{\partial r} + \frac{v}{r} = 0. \qquad (1.2.36c)$$

Here cylindrical coordinates are used with (u, v) being the velocity components in the axial and radial directions, respectively.

Accept without proof that the basic laminar flow is given by

$$U(r) = 1 - r^2, \qquad V = 0, \qquad P(x) = -\frac{4}{Re}x.$$

Perturb this solution by writing

$$u = U(r) + \varepsilon u'(t, x, r), \quad v = \varepsilon v'(t, x, r), \quad p = P(x) + \varepsilon p'(t, x, r). \qquad (1.2.37)$$

Assuming $\varepsilon \ll 1$, formulate the linearized equations for u', v' and p'.

Eliminate the pressure perturbation function p' through cross-differentiation of the two momentum equations. Show that the vorticity

$$\omega' = \frac{\partial v'}{\partial x} - \frac{\partial u'}{\partial r} \qquad (1.2.38)$$

satisfies the equation

$$\frac{\partial \omega'}{\partial t} + U \frac{\partial \omega'}{\partial x} - \left(\frac{d^2 U}{dr^2} - \frac{1}{r} \frac{dU}{dr} \right) v' = \frac{1}{Re} \left(\frac{\partial^2 \omega'}{\partial x^2} + \frac{\partial^2 \omega'}{\partial r^2} + \frac{1}{r} \frac{\partial \omega'}{\partial r} - \frac{\omega'}{r^2} \right).$$

Based on the continuity equation, introduce the stream function ψ' such that

$$u' = \frac{1}{r} \frac{\partial \psi'}{\partial r}, \qquad v' = -\frac{1}{r} \frac{\partial \psi'}{\partial x}. \qquad (1.2.39)$$

Then represent ω' and ψ' in the normal-mode form

$$\omega' = e^{i\alpha(x-ct)} \breve{\omega}(r), \qquad \psi' = e^{i\alpha(x-ct)} \phi(r),$$

and deduce a pair of ordinary differential equations relating $\breve{\omega}(r)$ and $\phi(r)$.

1.3   Stability of Boundary Layers

The stability of boundary layers is analysed in a similar way to that performed in Section 1.2 for plane Poiseuille flow. The first step is to derive the basic unperturbed flow. We shall consider, as an example, the Blasius boundary layer that forms on the surface of a flat plate aligned with the free-stream velocity vector; see Figure 1.10.

1.3.1   Basic flow

Here we shall give a brief description of Blasius flow.[4] We assume that the fluid is incompressible with its density ρ and the kinematic viscosity coefficient ν remaining constant throughout the flow field. We also assume that the flow is two-dimensional. We denote the plate length by L and introduce Cartesian coordinates $(\hat{x}, \hat{y})$ with $\hat{x}$ measured along the flat plate surface from the leading edge O, and $\hat{y}$ in the perpendicular direction, as shown in Figure 1.10. The corresponding velocity components are $\hat{u}$ and $\hat{v}$, respectively, while $\hat{p}$ is the pressure. In Blasius flow, the free-stream velocity V_∞ is directed parallel to the plate surface.

If the Reynolds number $Re = V_\infty L / \nu$ is large, then the laminar flow may be described within the framework of classical boundary-layer theory (see Prandtl, 1904). In this theory, the flow field is divided into two distinct regions: the inviscid region occupying almost the entire flow field and a thin viscous boundary layer that forms on the body surface. For the flow past a flat plate, the inviscid part of the flow remains unperturbed in the leading-order approximation:

$$\hat{u} = V_\infty, \quad \hat{v} = 0, \quad \hat{p} = p_\infty,$$

with p_∞ being the free-stream pressure.

---

[4]For a more detailed description of this flow, the reader is referred to Section 1.1 in Part 3 of this book series.

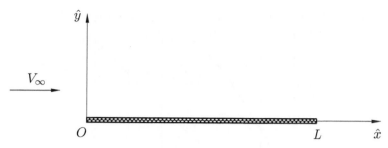

Fig. 1.10: Flow past a flat plate.

Turning to the boundary-layer flow, we recall that in the leading-order approximation the pressure does not change across the boundary layer, and therefore, it remains
constant, being given by $\hat{p} = p_\infty$. To determine the velocity field $(\hat{u}, \hat{v})$, one needs to
solve the Prandtl equations. These are written in dimensional form as

$$\hat{u}\frac{\partial \hat{u}}{\partial \hat{x}} + \hat{v}\frac{\partial \hat{u}}{\partial \hat{y}} = \nu\frac{\partial^2 \hat{u}}{\partial \hat{y}^2}, \tag{1.3.1a}$$

$$\frac{\partial \hat{u}}{\partial \hat{x}} + \frac{\partial \hat{v}}{\partial \hat{y}} = 0. \tag{1.3.1b}$$

They should be solved subject to the initial condition at the leading edge of the plate

$$\hat{u} = V_\infty \quad \text{at} \quad \hat{x} = 0, \;\; \hat{y} \in (0, \infty), \tag{1.3.1c}$$

the no-slip conditions on the plate surface

$$\hat{u} = \hat{v} = 0 \quad \text{at} \quad \hat{y} = 0, \;\; \hat{x} \in [0, L] \tag{1.3.1d}$$

and the matching condition with the solution in the external inviscid flow

$$\hat{u} = V_\infty \quad \text{at} \quad \hat{y} = \infty. \tag{1.3.1e}$$

The solution of the boundary-value problem (1.3.1) may be found in the following
self-similar form:

$$\hat{u} = V_\infty \varphi'(\eta), \qquad \hat{v} = \frac{1}{2}\sqrt{\frac{V_\infty \nu}{\hat{x}}}\left(\eta\varphi' - \varphi\right), \tag{1.3.2}$$

with

$$\eta = \sqrt{\frac{V_\infty}{\nu}}\frac{\hat{y}}{\sqrt{\hat{x}}}. \tag{1.3.3}$$

The function $\varphi(\eta)$ satisfies the *Blasius equation* (see Blasius, 1908):

$$\varphi''' + \frac{1}{2}\varphi\,\varphi'' = 0, \tag{1.3.4a}$$

which is solved subject to the boundary conditions

$$\varphi(0) = \varphi'(0) = 0, \quad \varphi'(\infty) = 1. \tag{1.3.4b}$$

The results of a numerical solution of the boundary-value problem (1.3.4) are displayed
in Figure 1.11.

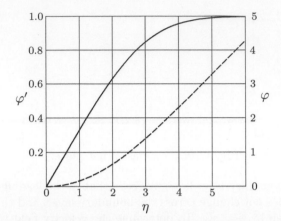

Fig. 1.11: Solution of the Blasius problem (1.3.4). The solid line shows the longitudinal velocity profile $\varphi'(\eta)$; the dashed line represents the function $\varphi(\eta)$.

We shall conclude the basic flow analysis by presenting asymptotic formulae that describe the behaviour of the solution close to the plate surface and at the outer edge of the boundary layer. For small η, the function $\varphi(\eta)$ may be written in the form of the Taylor expansion:

$$\varphi(\eta) = \varphi(0) + \varphi'(0)\eta + \frac{1}{2}\varphi''(0)\eta^2 + \cdots.$$

Since $\varphi(0) = \varphi'(0) = 0$, we have

$$\varphi(\eta) = \frac{1}{2}\lambda\eta^2 + \cdots \quad \text{as} \quad \eta \to 0, \tag{1.3.5}$$

where $\lambda = \varphi''(0)$. The numerical value of λ is found to be $\lambda \simeq 0.332057$. It follows from (1.3.2), (1.3.3), and (1.3.5) that the skin friction is proportional to λ, namely,

$$\hat{\tau}_w = \mu \frac{\partial \hat{u}}{\partial \hat{y}}\bigg|_{\hat{y}=0} = \rho V_\infty \sqrt{\frac{V_\infty \nu}{\hat{x}}}\lambda.$$

For this reason we shall call λ the *skin friction parameter*.

If η is large, then[5]

$$\varphi(\eta) = \eta - A' + \cdots \quad \text{as} \quad \eta \to \infty, \tag{1.3.6}$$

where constant $A' \simeq 1.7208$. Equation (1.3.6) may be used to calculate the *displacement thickness* of the Blasius boundary layer, $\delta^*(\hat{x})$. In the general case it is given by integral (1.2.37) on page 27 in Part 3 of this book series. Writing this integral in dimensional form we have

$$\delta^* = \int\limits_0^\infty \left(1 - \frac{\hat{u}}{V_\infty}\right) d\hat{y}. \tag{1.3.7}$$

---

[5]A derivation of (1.3.6) is given on pages 15, 16 in Part 3 of this book series.

Substituting the first equation in (1.3.2) together with (1.3.3) into (1.3.7) and using (1.3.6) we find that

$$\delta^* = A'\sqrt{\frac{\nu\hat{x}}{V_\infty}}. \tag{1.3.8}$$

1.3.2   The parallel flow approximation

Now we turn to the analysis of small perturbations to the basic flow. It should be noted that the normal-mode presentation (1.2.26) of the perturbations is applicable to a rather narrow class of parallel flows, where the longitudinal velocity component U in the basic flow is a function of the lateral coordinate y only, and $V = W = 0$. Hence, strictly speaking, the Orr–Sommerfeld equation (1.2.24) is not applicable to Blasius flow (1.3.2), (1.3.3) where the longitudinal velocity component $\hat{u}$ depends on both $\hat{x}$ and $\hat{y}$, and the lateral velocity component $\hat{v}$ is not zero as is evident from (1.3.2). Fortunately, if the Reynolds number is large enough, then the variation of the longitudinal velocity $\hat{u}$ in the $\hat{x}$-direction appears to be much slower than that in the $\hat{y}$-direction. In fact, it follows from (1.3.2) and (1.3.3) that

$$\frac{\partial\hat{u}}{\partial\hat{x}} = -\frac{1}{2}V_\infty\eta\,\varphi''(\eta)\frac{1}{\hat{x}}, \qquad \frac{\partial\hat{u}}{\partial\hat{y}} = V_\infty\varphi''(\eta)\sqrt{\frac{V_\infty}{\nu\hat{x}}},$$

and therefore,

$$\frac{\partial\hat{u}}{\partial\hat{x}}\Big/\frac{\partial\hat{u}}{\partial\hat{y}} = -\frac{1}{2}\eta\sqrt{\frac{\nu}{V_\infty\hat{x}}} \sim Re_{\hat{x}}^{-1/2},$$

where the Reynolds number

$$Re_{\hat{x}} = \frac{V_\infty\hat{x}}{\nu}$$

is assumed large.

   Guided by the results of the stability analysis of plane Poiseuille flow (see Section 1.2) one can expect the wavelength ℓ of the normal-mode perturbation to be comparable with the characteristic length scale across the boundary layer. The latter is represented by the displacement thickness of the boundary layer (1.3.8). We therefore have

$$\frac{\ell}{\hat{x}} \sim \frac{\delta^*}{\hat{x}} = A'\sqrt{\frac{\nu}{V_\infty\hat{x}}} \sim Re_{\hat{x}}^{-1/2},$$

which shows that the distance ℓ, over which the perturbation performs a complete oscillation cycle, is too small for the basic flow to experience any noticeable change. Thus, when analysing the stability of the boundary layer at a position $\hat{x}$ along the body surface, we can 'freeze' the basic velocity profile $\hat{u}$ and consider it as a function of $\hat{y}$ only. We can also set the lateral velocity $\hat{v}$ in the basic flow to zero. This is due to the fact that in the laminar boundary layer, $\hat{v}$ is much smaller than $\hat{u}$. Indeed, it follows from (1.3.2) that

$$\frac{\hat{v}}{\hat{u}} = \frac{\eta\,\varphi' - \varphi}{2\varphi'}\sqrt{\frac{\nu}{V_\infty\hat{x}}} \sim Re_{\hat{x}}^{-1/2}.$$

1.3.3 Stability analysis

In view of Squire's theorem (Theorem 1.1 on page 19), we shall consider the two-dimensional version of the Navier–Stokes equations (1.1.1):

$$
\left.
\begin{aligned}
\frac{\partial \hat{u}}{\partial \hat{t}} + \hat{u}\frac{\partial \hat{u}}{\partial \hat{x}} + \hat{v}\frac{\partial \hat{u}}{\partial \hat{y}} &= -\frac{1}{\rho}\frac{\partial \hat{p}}{\partial \hat{x}} + \nu\left(\frac{\partial^2 \hat{u}}{\partial \hat{x}^2} + \frac{\partial^2 \hat{u}}{\partial \hat{y}^2}\right), \\
\frac{\partial \hat{v}}{\partial \hat{t}} + \hat{u}\frac{\partial \hat{v}}{\partial \hat{x}} + \hat{v}\frac{\partial \hat{v}}{\partial \hat{y}} &= -\frac{1}{\rho}\frac{\partial \hat{p}}{\partial \hat{y}} + \nu\left(\frac{\partial^2 \hat{v}}{\partial \hat{x}^2} + \frac{\partial^2 \hat{v}}{\partial \hat{y}^2}\right), \\
\frac{\partial \hat{u}}{\partial \hat{x}} + \frac{\partial \hat{v}}{\partial \hat{y}} &= 0.
\end{aligned}
\right\}
\tag{1.3.9}
$$

We choose a position $\hat{x} = \hat{x}_0$ on the plate surface, calculate the corresponding value of the displacement thickness (1.3.8), and assuming δ^* constant, introduce the non-dimensional variables as follows:

$$
\left.
\begin{aligned}
\hat{x} &= \hat{x}_0 + \delta^* x, & \hat{y} &= \delta^* y, & \hat{t} &= \frac{\delta^*}{V_\infty}t, \\
\hat{u} &= V_\infty u, & \hat{v} &= V_\infty v, & \hat{p} &= p_\infty + \rho V_\infty^2 p.
\end{aligned}
\right\}
\tag{1.3.10}
$$

This renders the Navier–Stokes equations (1.3.9) in the form:

$$
\frac{\partial u}{\partial t} + u\frac{\partial u}{\partial x} + v\frac{\partial u}{\partial y} = -\frac{\partial p}{\partial x} + \frac{1}{Re_*}\left(\frac{\partial^2 u}{\partial x^2} + \frac{\partial^2 u}{\partial y^2}\right),
\tag{1.3.11a}
$$

$$
\frac{\partial v}{\partial t} + u\frac{\partial v}{\partial x} + v\frac{\partial v}{\partial y} = -\frac{\partial p}{\partial y} + \frac{1}{Re_*}\left(\frac{\partial^2 v}{\partial x^2} + \frac{\partial^2 v}{\partial y^2}\right),
\tag{1.3.11b}
$$

$$
\frac{\partial u}{\partial x} + \frac{\partial v}{\partial y} = 0,
\tag{1.3.11c}
$$

where

$$
Re_* = \frac{V_\infty \delta^*}{\nu}.
\tag{1.3.12}
$$

Before performing the stability analysis, we need to cast the basic laminar solution (1.3.2), (1.3.3) in terms of dimensionless variables. We start with the similarity variable η. Substituting $\hat{x}$ and $\hat{y}$ from (1.3.10) into (1.3.3) and taking into account that $\delta^* \ll \hat{x}_0$, we have

$$
\eta = \sqrt{\frac{V_\infty}{\nu}}\,\frac{\delta^* y}{\sqrt{\hat{x}_0}}.
\tag{1.3.13}
$$

At the position $\hat{x} = \hat{x}_0$ the displacement thickness of the boundary layer (1.3.8) is calculated as

$$
\delta^* = A'\sqrt{\frac{\nu \hat{x}_0}{V_\infty}}.
\tag{1.3.14}
$$

Substituting (1.3.14) into (1.3.13), we find that

$$
\eta = A'y.
$$

Hence, the longitudinal velocity in the Blasius solution (1.3.2) is written in terms of the non-dimensional variables as

$$ u = \frac{d\varphi}{d\eta}\bigg|_{\eta=A'y} . $$

Remember that the lateral velocity component $\hat{v}$ is small, as are the pressure perturbations with respect to the free-stream value p_∞. Hence, the basic flow solution is given by

$$ U(y) = \frac{d\varphi}{d\eta}\bigg|_{\eta=A'y} , \qquad V = 0, \qquad P = 0. \tag{1.3.15} $$

The stability analysis of this flow is conducted in the same way as it was done for plane Poiseuille flow. We add to (1.3.15) small amplitude perturbations by writing

$$ u = U(y) + \varepsilon u'(t,x,y), \qquad v = \varepsilon v'(t,x,y), \qquad p = \varepsilon p'(t,x,y), \tag{1.3.16} $$

and seek the perturbation functions u', v', and p' in the normal-mode form

$$ u' = e^{i(\alpha x - \omega t)}\breve{u}(y), \qquad v' = e^{i(\alpha x - \omega t)}\breve{v}(y), \qquad p' = e^{i(\alpha x - \omega t)}\breve{p}(y). \tag{1.3.17} $$

Now we need to substitute (1.3.17) into (1.3.16) and then into the Navier–Stokes equations (1.3.11). Working with the $O(\varepsilon)$ terms, we find that

$$ i(\alpha U - \omega)\breve{u} + \frac{dU}{dy}\breve{v} = -i\alpha\breve{p} + \frac{1}{Re_*}\left(\frac{d^2\breve{u}}{dy^2} - \alpha^2\breve{u}\right), \tag{1.3.18a} $$

$$ i(\alpha U - \omega)\breve{v} = -\frac{d\breve{p}}{dy} + \frac{1}{Re_*}\left(\frac{d^2\breve{v}}{dy^2} - \alpha^2\breve{v}\right), \tag{1.3.18b} $$

$$ i\alpha\breve{u} + \frac{d\breve{v}}{dy} = 0. \tag{1.3.18c} $$

Elimination of $\breve{u}$ and $\breve{p}$ from (1.3.18) leads to the Orr–Sommerfeld equation

$$ \frac{1}{i\alpha Re_*}\left(\frac{d^4\breve{v}}{dy^4} - 2\alpha^2\frac{d^2\breve{v}}{dy^2} + \alpha^4\breve{v}\right) = (U - c)\left(\frac{d^2\breve{v}}{dy^2} - \alpha^2\breve{v}\right) - \frac{d^2U}{dy^2}\breve{v}, \tag{1.3.19} $$

where $c = \omega/\alpha$.

Two boundary conditions for equation (1.3.19) are the no-slip conditions on the plate surface:

$$ \breve{v} = \frac{d\breve{v}}{dy} = 0 \quad \text{at} \quad y = 0. \tag{1.3.20} $$

The other two should be formulated at the outer edge of the boundary layer. Since

$$ U \to 1 \quad \text{as} \quad y \to \infty, \tag{1.3.21} $$

the coefficients in (1.3.19) become constant in this limit, and therefore, for large values of y, the complementary solutions to (1.3.19) may be sought in the form

$$ \breve{v} = Ce^{\lambda y}. \tag{1.3.22} $$

Substitution of (1.3.22) and (1.3.21) into (1.3.19) results in the following equation for λ:

$$\lambda^4 - 2\alpha^2\lambda^2 + \alpha^4 = i\alpha Re_*(1-c)(\lambda^2 - \alpha^2).$$

The four solutions of this equation are

$$\lambda_{1,2} = \pm\alpha, \qquad \lambda_{3,4} = \pm\sqrt{i\alpha Re_*(1-c) + \alpha^2}.$$

Two of these have positive real part, and give rise to exponentially growing perturbations (1.3.22). The other two are exponentially decaying with y. Our task is to identify natural oscillations of the boundary layer, not the oscillations caused by large amplitude disturbances outside the boundary layer. Therefore, we have to suppress the two growing solutions. This is achieved by imposing the boundary condition

$$\check{v} \to 0 \quad \text{as} \quad y \to \infty. \tag{1.3.23}$$

The Orr–Sommerfeld equation (1.3.19) with the boundary conditions (1.3.20) and (1.3.23) constitute an eigen-value problem that should be treated in the same way as the corresponding problem (1.2.33) for plane Poiseuille flow, namely, for each pair of real Re_* and α it allows us to find complex $c = c_r + ic_i$. Representing the perturbations (1.3.17) in the form

$$v' = e^{i(\alpha x - \omega t)}\check{v}(y) = e^{i\alpha(x - ct)}\check{v}(y) = e^{\alpha c_i t}e^{i\alpha(x - c_r t)}\check{v}(y),$$

we see that c_r represents the phase speed, and c_i the amplification rate. In Figure 1.12 the *neutral curves*, where $c_i = 0$, are shown in the (α, Re_*)- and (c, Re_*)-planes. The perturbations decay ($c_i < 0$) everywhere outside these curve, and the region inside the neutral curves is the instability region ($c_i > 0$). The minimal value of the Reynolds number on the neutral curve is termed the *critical Reynolds number*. For the Blasius boundary layer it is found to be

$$Re_{*c} \simeq 518.0, \tag{1.3.24}$$

with the corresponding wave number $\alpha_c \simeq 0.303$.

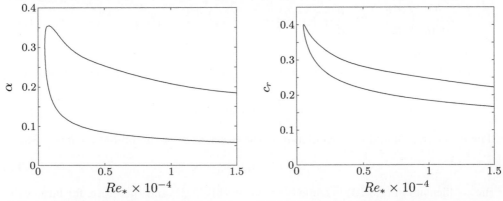

(a) Wave number α against Reynolds number Re_*.

(b) Phase speed c_r against Reynolds number Re_*.

Fig. 1.12: Neutral curves for the Blasius boundary layer.

1.3.4   Temporal and spatial instabilities

Up to now, we have been treating the flow instability as a *temporal instability*, Using this approach, when adopting the normal-mode representation

$$v' = e^{i(\alpha x - \omega t)} \breve{v}(y), \tag{1.3.25}$$

we assume α real, and ω complex, namely

$$\omega = \omega_r + i\omega_i,$$

which turns (1.3.25) into

$$v' = e^{\omega_i t} e^{i(\alpha x - \omega_r t)} \breve{v}(y).$$

The perturbation appears to be periodic in the spatial variable x, and can grow or decay with time t depending on the sign of ω_i. This approach is natural for Poiseuille flow, but not for boundary layers, where the perturbations are observed to grow in the downstream direction; see Figure I.4. This situation may be described by assuming that ω is real, while α is complex:

$$\alpha = \alpha_r + i\alpha_i.$$

This turns (1.3.25) into

$$v' = e^{-\alpha_i x} e^{i(\alpha_r x - \omega t)} \breve{v}(y),$$

leading to the concept of *spatial instability*. The flow is stable if $\alpha_i > 0$, and unstable if $\alpha_i < 0$. Of course, for neutrally stable modes, the spatial and temporal instabilities are indistinguishable from one another, and the critical value (1.3.24) of the Reynolds number proves to be the same for both forms of instability.

Substitution of (1.3.8) into (1.3.12) allows us to express the Reynolds number as

$$Re_* = A' \sqrt{\frac{V_\infty \hat{x}}{\nu}},$$

which shows that there is a 'critical position' $\hat{x} = \hat{x}_c$ in the boundary layer that corresponds to (1.3.24). Upstream of this position, $Re_* < Re_{*c}$, and the boundary layer is expected to remain laminar. As an observer moves downstream, Re_* increases, and as soon as it reaches the critical value Re_{*c}, neutral oscillations become possible. Further downstream lies a region of growing perturbations. These are known as the *Tollmien–Schlichting waves*.

Tollmien–Schlichting waves were first described theoretically (see Tollmien, 1929; Schlichting, 1933), but it took some time before their existence was confirmed experimentally. This became possible when so-called *low turbulence wind tunnels* were built. The level of turbulence is measured as the ratio of the velocity pulsations u' to the average velocity V_∞ in the test section. In conventional wind tunnels, used for routine aerodynamic tests, the level of turbulence is about 1%. In a low turbulence wind tunnel it is reduced to 0.05%.

The results of early experiments in low turbulence wind tunnels were presented by Dryden (1947), Schubauer and Skramstad (1948), and Klebanoff and Tidstrom (1959).

Fig. 1.13: Generation of a Tollmien–Schlichting wave by a vibrating ribbon.

One of the techniques these authors used to study the behaviour of the Tollmien–Schlichting waves, was to generate such a wave artificially by means of a vibrating ribbon installed a small distance above the plate surface (see Figure 1.13), and then follow the development of the wave in the boundary layer. Their apparatus allowed the ribbon to perform harmonic oscillations with the frequency ω chosen by the experimentalist. The perturbations downstream of the ribbon were detected using a 'hot wire'. The measurements showed that the initial amplitude of the Tollmien–Schlichting wave was proportional to the amplitude of the ribbon oscillations. Also it was observed that the frequency ω of the Tollmien–Schlichting wave remained unchanged as the wave travelled downstream in the boundary layer, and was the same as the frequency of the ribbon. This is explained by the fact that the basic Blasius flow (1.3.2) is independent of time.

When the ribbon was installed at a point x_r upstream of the lower branch of the neutral curve (see Figure 1.14), the amplitude of the Tollmien–Schlichting wave was observed to decay on the interval $[x_r, x_l]$, where x_l corresponds to the lower branch of the neutral curve for given frequency ω. The amplitude of the Tollmien–Schlichting wave then started to grow on the interval $[x_l, x_u]$ between the lower and upper branches

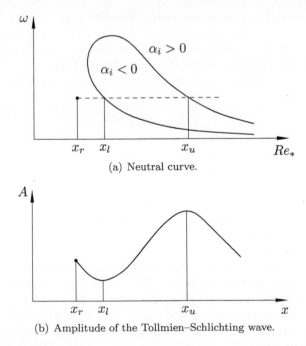

(a) Neutral curve.

(b) Amplitude of the Tollmien–Schlichting wave.

Fig. 1.14: Development of a Tollmien–Schlichting wave in the boundary layer.

of the neutral curve, where the boundary layer was unstable. After crossing the upper branch at the point x_u the Tollmien–Schlichting wave decayed again. Repeating the experiment for other values of the frequency ω, a neutral curve was obtained. It appeared to be in good agreement with the theoretical curve obtained by solving the Orr–Sommerfeld equation (1.3.19).

Of course, if the amplitude of the Tollmien–Schlichting wave becomes large enough to cause nonlinear effects, the boundary layer turns turbulent before the upper branch of the neutral curve is reached.

Exercises 2

1. Given that $Re_* = V_\infty \delta^*/\nu$, where $\delta^* = A'\sqrt{\nu \hat{x}/V_\infty}$, show that

$$Re_* = A' Re^{1/2}, \tag{1.3.26}$$

where $Re = V_\infty \hat{x}/\nu$.

2. The *Eckhaus equation*

$$\frac{1}{R}\left(\frac{\partial^2 u}{\partial y^2} + \frac{\partial^2 u}{\partial x^2}\right) - \frac{\partial^4 u}{\partial x^4} - \frac{\partial u}{\partial t} = \frac{\partial u}{\partial y}\frac{\partial^2 u}{\partial x^2} \tag{1.3.27}$$

is often used as a simplified version of the Navier–Stokes equations. It should be solved for $u(t, x, y)$ in the infinite strip $x \in (-\infty, \infty)$, $y \in (0, 1)$. The role of the Reynolds number is played by the parameter R which is real and positive. The boundary conditions for (1.3.27) are

$$u\Big|_{y=0} = 0, \qquad u\Big|_{y=1} = 1 \tag{1.3.28}$$

Perform the following tasks:
(a) By considering the linear stability of the basic state, $U = y$, with respect to normal-mode perturbations in the form

$$u' = e^{\sigma t + i\alpha x}\breve{u}(y),$$

show that the amplification rate

$$\sigma = \alpha^2 - \alpha^4 - \frac{n^2\pi^2 + \alpha^2}{R}, \tag{1.3.29}$$

where n is a positive integer, $n = 1, 2, 3, \ldots$
(b) Set $n = 1$, and show that the 'flow' becomes unstable for

$$R > R_c = \frac{\pi^2 + \alpha_c^2}{\alpha_c^2 - \alpha_c^4},$$

where

$$\alpha_c^2 = \pi^2\big[\sqrt{1 + \pi^{-2}} - 1\big].$$

1.4   Inviscid Stability Theory

The stability analysis of two-dimensional parallel flows of 'inviscid fluids' is based on the *Rayleigh equation*

$$(U - c)\left(\frac{d^2\phi}{dy^2} - \alpha^2\phi\right) - \frac{d^2U}{dy^2}\phi = 0, \tag{1.4.1}$$

that can be deduced from the Orr–Sommerfeld equation (1.2.35) by setting[6]

$$Re \to \infty, \quad \alpha = O(1), \quad c = O(1). \tag{1.4.2}$$

We shall see that the limit (1.4.2), in which the wave number α and the phase speed c are presumed finite as $Re \to \infty$, is too restrictive for some flows. Still the Rayleigh equation proves to be useful in identifying various important forms of instability, especially, for flows which are capable of supporting unsteady perturbations with high amplification rates.

Since the viscous terms are disregarded in (1.4.1), the boundary conditions for $\phi(y)$ should be modified, accordingly. In particular, for Poiseuille flow between two flat plates positioned at $y = -1$ and $y = 1$, we shall use the impermeability condition on both plates, namely,

$$\phi = 0 \quad \text{at} \quad y = -1 \text{ and } y = 1. \tag{1.4.3a}$$

When dealing with the boundary layer on a rigid body surface, we shall use the conditions

$$\phi = 0 \quad \text{at} \quad y = 0 \quad \text{and} \quad \phi \to 0 \quad \text{as} \quad y \to \infty. \tag{1.4.3b}$$

The first of these is the impermeability condition on the body surface; the second serves to ensure that the perturbations considered represent the natural oscillations of the boundary layer, rather than the oscillations forced by external perturbations.

1.4.1   Properties of the Rayleigh equation

We start with a simple observation that if for a given real α the Rayleigh eigen-value problem (1.4.1), (1.4.3) admits a solution with the eigen-value c and eigen-function $\phi(y)$, then it also admits a solution with $\bar{c}$ and $\bar{\phi}(y)$, which are complex conjugates of c and $\phi(y)$, respectively. This can be seen by taking complex conjugates of all the terms in the equation (1.4.1), leading to

$$(U - \bar{c})\left(\frac{d^2\bar{\phi}}{dy^2} - \alpha^2\bar{\phi}\right) - \frac{d^2U}{dy^2}\bar{\phi} = 0,$$

while the boundary conditions (1.4.3a) and (1.4.3b) turn into

$$\bar{\phi} = 0 \quad \text{at} \quad y = -1 \text{ and } y = 1,$$

and

$$\bar{\phi} = 0 \quad \text{at} \quad y = 0 \quad \text{and} \quad \bar{\phi} \to 0 \quad \text{as} \quad y \to \infty.$$

---

[6]In fact, Rayleigh deduced this equation as early as 1880, well before the Orr–Sommerfeld equation became known.

It is therefore indeed the case that $\bar{c}$, $\bar{\phi}(y)$ also represent a solution of the Rayleigh problem (1.4.1), (1.4.3).

Consequently, if there exists a stable solution with $c_i < 0$, then there also exists an unstable solution with $c_i > 0$, and *vice versa*. This implies, on one hand, that the only way an inviscid flow may be stable is that *all* modes are neutrally stable with real eigen-values c; but this is not surprising since an inviscid flow is a non-dissipative system. On the other hand, the flow becomes unstable if there exists an eigen solution with complex c, because one of the pair c and $\bar{c}$ will have an imaginary part greater than zero.

The following important theorem due to Rayleigh (1880), known as the *Inflexion Point Theorem*, helps to narrow down the class of flows which are inviscidly unstable.

Theorem 1.2 *Suppose that in a parallel flow the basic velocity profile $U(y)$ is continuous in the open interval $y \in (y_1, y_2)$ together with its first and second derivatives, $U'(y)$ and $U''(y)$. Then a necessary condition for inviscid instability is that $U(y)$ possesses an inflexion point $U'' = 0$ somewhere between y_1 and y_2.*

Here we assume that $\phi = 0$ on both ends of the interval $y \in (y_1, y_2)$ to satisfy the boundary conditions (1.4.3). We shall allow y_2 to be finite or infinite to model the channel or boundary-layer flows, respectively.

Proof Suppose that the flow at hand is unstable, which means that there exists a solution to (1.4.1), (1.4.3) with $c_i \neq 0$. For such a solution $U - c \neq 0$, and all the terms in the Rayleigh equation (1.4.1) may be divided by $U - c$, leading to

$$\frac{d^2\phi}{dy^2} = \alpha^2\phi + \frac{U''(y)}{U - c}\phi. \tag{1.4.4}$$

Let us multiply both sides in (1.4.4) by the complex conjugate solution $\bar{\phi}$, leading to

$$\frac{d^2\phi}{dy^2}\bar{\phi} = \alpha^2|\phi|^2 + \frac{U''(y)}{U - c}|\phi|^2.$$

Now we integrate the above equation over the interval $y \in (y_1, y_2)$:

$$\int_{y_1}^{y_2} \frac{d^2\phi}{dy^2}\bar{\phi}\,dy = \int_{y_1}^{y_2} \alpha^2|\phi|^2\,dy + \int_{y_1}^{y_2} \frac{U''(y)}{U - c}|\phi|^2\,dy.$$

Performing integration by parts on the left hand side of this equation, we have

$$\bar{\phi}\frac{d\phi}{dy}\Big|_{y_1}^{y_2} - \int_{y_1}^{y_2} \left|\frac{d\phi}{dy}\right|^2\,dy = \int_{y_1}^{y_2} \alpha^2|\phi|^2\,dy + \int_{y_1}^{y_2} \frac{U''(y)}{U - c}|\phi|^2\,dy. \tag{1.4.5}$$

Since $\phi(y_1) = \phi(y_2) = 0$, it follows that $\bar{\phi}(y_1) = \bar{\phi}(y_2) = 0$, and therefore, the first term on the left-hand side disappears, reducing (1.4.5) to

$$-\int_{y_1}^{y_2} \left|\frac{d\phi}{dy}\right|^2\,dy = \int_{y_1}^{y_2} \alpha^2|\phi|^2\,dy + \int_{y_1}^{y_2} \frac{U''(y)}{U - c}|\phi|^2\,dy. \tag{1.4.6}$$

Let us now consider the integrand in the second term on the right-hand side of (1.4.6). We have

$$\frac{1}{U-c} = \frac{1}{U-c_r - ic_i} = \frac{U-c_r + ic_i}{(U-c_r)^2 + c_i^2} = \frac{U-c_r}{|U-c|^2} + i\frac{c_i}{|U-c|^2},$$

which allows to express equation (1.4.6) in the form

$$-\int_{y_1}^{y_2} \left|\frac{d\phi}{dy}\right|^2 dy = \int_{y_1}^{y_2} \alpha^2 |\phi|^2 \, dy$$

$$+ \int_{y_1}^{y_2} \frac{U-c_r}{|U-c|^2} U''(y)|\phi|^2 dy + ic_i \int_{y_1}^{y_2} \frac{U''(y)}{|U-c|^2}|\phi|^2 \, dy. \tag{1.4.7}$$

The imaginary part of this equation is

$$c_i \int_{y_1}^{y_2} \frac{U''(y)}{|U-c|^2}|\phi|^2 dy = 0. \tag{1.4.8}$$

We have assumed that the flow is unstable, that is, $c_i \neq 0$. Therefore,

$$\int_{y_1}^{y_2} \frac{U''(y)}{|U-c|^2}|\phi|^2 dy = 0. \tag{1.4.9}$$

Equation (1.4.9) can only be satisfied if $U''(y)$ changes sign within the interval $y \in (y_1, y_2)$. Since $U''(y)$ is continuous, there should be at least one inflexion point of $U(y)$ between y_1 and y_2.   □

Interestingly enough, neither plane Poiseuille flow nor flat plate Blasius flow have inflexion points. Indeed, it is easily seen from the first equation in (1.2.11) that in Poiseuille flow $U'' = -2$. In the Blasius boundary layer U'' is also negative throughout the flow field (see Problem 1 in Exercises 3). Consequently, by Theorem 1.2 we have to conclude that both flows are stable. This, however, contradicts the results of the solution of the Orr–Sommerfeld equation, which show that Poiseuille flow (see Figures 1.8) as well as the Blasius boundary layer (see Figure 1.12) prove to be unstable for all $Re > Re_c$, no matter how large.

To resolve this 'contradiction' one needs to examine the neutral curve more closely. The asymptotic analysis of the Orr–Sommerfeld equation at large values of the Reynolds number (see Sections 2.3 and 2.4) shows that both the upper and lower branches of the neutral curve approach asymptotically the abscissa in the (Re, α)-plane, as shown schematically in Figure 1.15. The limit (1.4.2) is represented in this figure by a straight line that corresponds to a fixed value of the wave number, say, $\alpha = \alpha'$. It is clear that for any $\alpha' > 0$ this line leads, as Re increases, into the region of decaying perturbations. An important (some might find it 'paradoxical') conclusion may be drawn from

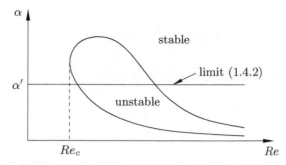

Fig. 1.15: Neutral curve for a flow without inflexion point.

the above discussion. While the action of internal viscosity is normally expected to be a dissipative process, it also proves to be capable of destabilizing various flows.

If instead of the Blasius boundary layer we consider a more general case when the boundary layer is exposed to a non-zero pressure gradient, then equation (1.3.1a) should be written as

$$\hat{u}\frac{\partial \hat{u}}{\partial \hat{x}} + \hat{v}\frac{\partial \hat{u}}{\partial \hat{y}} = -\frac{1}{\rho}\frac{d\hat{p}}{d\hat{x}} + \nu\frac{\partial^2 \hat{u}}{\partial \hat{y}^2}.$$

Setting $\hat{y} = 0$ in this equation and using the no-slip conditions (1.3.1d), we have

$$\left.\frac{\partial^2 \hat{u}}{\partial \hat{y}^2}\right|_{\hat{y}=0} = \frac{1}{\rho\nu}\frac{d\hat{p}}{d\hat{x}}.$$

Hence, if the pressure gradient is 'favourable' (i.e. $d\hat{p}/d\hat{x} < 0$), then at the wall $U''(0) < 0$. If, on the other hand, the pressure gradient is 'adverse' ($d\hat{p}/d\hat{x} > 0$), then $U''(0)$ becomes positive. Near the outer edge of the boundary layer U'' is always negative. This means that an adverse pressure gradient leads to a formation of an inflexion point inside the boundary layer, making it susceptible to inviscid instability. The latter shows a higher amplification rate, and hence, a more rapid transition to turbulence as compared to boundary layers with a favourable pressure gradient.

The following theorem due to Fjortoft (1950) gives an additional necessary condition for inviscid instability.

Theorem 1.3 *Suppose that in a parallel flow the basic velocity profile $U(y)$ has an inflexion point y_0 such that $U''(y_0) = 0$. Then a necessary condition for inviscid instability is that $(U - U_0)U'' < 0$ on at least a part of the interval $y \in (y_1, y_2)$. Here U_0 is the value of the basic velocity at the inflexion point, i.e. $U_0 = U(y_0)$.*

Proof The real part of equation (1.4.7) is written as

$$\int\limits_{y_1}^{y_2} \frac{U - c_r}{|U - c|^2}U''(y)|\phi|^2 \, dy = -\int\limits_{y_1}^{y_2}\left|\frac{d\phi}{dy}\right|^2 dy - \int\limits_{y_1}^{y_2} \alpha^2|\phi|^2 \, dy. \tag{1.4.10}$$

Multiplying equation (1.4.9) by $c_r - U_0$, we have

$$\int\limits_{y_1}^{y_2} \frac{c_r - U_0}{|U - c|^2} U''(y)|\phi|^2 dy = 0. \qquad (1.4.11)$$

Adding (1.4.10) and (1.4.11) together, yields

$$\int\limits_{y_1}^{y_2} \frac{(U - U_0)U''(y)}{|U - c|^2}|\phi|^2 dy = -\int\limits_{y_1}^{y_2} \left|\frac{d\phi}{dy}\right|^2 dy - \int\limits_{y_1}^{y_2} \alpha^2|\phi|^2 dy. \qquad (1.4.12)$$

Since the right-hand side of (1.4.12) is negative, the integrand on the left-hand side must be negative on some part of the interval $y \in (y_1, y_2)$. $\qquad \square$

The following results, concerning necessary conditions for the existence of a neutral solution to Rayleigh's equation, are due to Foote and Lin (1950).

Theorem 1.4 *Let $U(y)$ have the same properties as given in the statement of Theorem 1.2 and suppose there is a solution $\phi(y)$ of the Rayleigh equation over $[y_1, y_2]$, vanishing at the end points, with $c_i = 0$.*

(i) If $U(y)$ is a strictly increasing or decreasing function over the range $y_1 \le y \le y_2$, then at the location y_c where $U(y_c) = c$ we must have $U''(y_c) = 0$.

(ii) If $U(y)$ is not monotonic and $\alpha \neq 0$, there exists at least one location $y_I \in (y_1, y_2)$ where $U''(y_I) = 0$.

Proof   To prove Part (i) of the theorem it is convenient to start with the equation (1.4.8), which may be written as

$$c_i \int\limits_{y_1}^{y_2} \frac{U''(y)|\phi(y)|^2}{(U(y) - c_r)^2 + c_i^2} dy = 0. \qquad (1.4.13)$$

Since $U(y)$ is strictly increasing or decreasing, there exists a unique value of y_c such that $U(y_c) = c_r$. We then introduce a new variable Y such that $y = y_c + c_i Y$, so that (1.4.13) can be rewritten as

$$c_i^2 \int\limits_{-(y_c - y_1)/c_i}^{(y_2 - y_c)/c_i} \frac{U''(y_c + c_i Y)|\phi(y_c + c_i Y)|^2}{\left[U(y_c + c_i Y) - c_r\right]^2 + c_i^2} dY = 0.$$

Letting $c_i \to 0$, and using Taylor series expansions, we obtain

$$U''(y_c)|\phi(y_c)|^2 \int\limits_{-\infty}^{\infty} \frac{dY}{\left[U'(y_c)\right]^2 Y^2 + 1} + O(c_i) = 0.$$

Evaluating the integral in the above equation, we can further write

$$\frac{U''(y_c)|\phi(y_c)|^2}{|U'(y_c)|} \left[\tan^{-1}\left(|U'(y_c)|Y\right)\right]_{-\infty}^{\infty} + O(c_i) = 0,$$

implying that

$$\frac{\pi U''(y_c)|\phi(y_c)|^2}{|U'(y_c)|} = 0. \qquad (1.4.14)$$

The quantity $\phi(y_c)$ can be taken to be equal to unity as a normalization condition on the Rayleigh eigen-value problem. We therefore conclude that for (1.4.14) to hold, $U''(y_c)$ must be equal to zero, which proves part (i) of the theorem.

(ii) If the profile $U(y)$ is not monotonic, the situation is a little more complicated. Then there will be multiple critical locations $y_c^{(n)}$, $n = 1, \dots, N$, say, where $c_r = U(y_c^{(n)})$.

First we note that if $\alpha = 0$ it may be possible to satisfy the boundary conditions by choosing $\phi = U - c$, since this is an exact solution of the Rayleigh equation in this case. We exclude this possibility by henceforth assuming $\alpha > 0$.

If we split the range of integration in (1.4.13) into regions where $U(y)$ is monotonic, then each interval will contain either no critical point (in which case the contribution to (1.4.13) will tend to zero as $c_i \to 0$) or one critical point. To the latter intervals we then employ the transformation $y = y_c^{(n)} + c_i Y$, obtaining in place of (1.4.14):

$$\sum_{n=1}^{N} \frac{\pi U''(y_c^{(n)})|\phi(y_c^{(n)})|^2}{|U'(y_c^{(n)})|} = 0. \qquad (1.4.15)$$

Although equation (1.4.15) can be satisfied if U'' vanishes at each of the critical locations this is not necessary to ensure a neutral solution: in general the contributions from each critical location just need to sum to zero, which requires U'' to vanish at least once in the interval (y_1, y_2). This completes the proof of Part (ii). □

A consequence of Part (i) of this result is that if, for a monotonic $U(y)$, we have a neutral solution of the Orr–Sommerfeld equation with $\alpha = O(1)$, $c = O(1)$ as $Re \to \infty$, it must be the case that $U'' = 0$ at the location where $U(y) = c$. This is because the Rayleigh equation is the leading-order approximation to the Orr–Sommerfeld equation in this limit, as remarked upon at the beginning of this section.

All of the above theorems provide necessary conditions under which unstable or neutral solutions to Rayleigh's equation exist. It is difficult to establish sufficient conditions for the Rayleigh problem that are not overly restrictive, too complicated to implement or provide little further physical insight. However, several authors have established relatively simple necessary and sufficient conditions for instability for profiles which are monotonic. For example, if the monotonic profile possesses a single inflexion point, Rosenbluth and Simon (1964), building upon results from Lin (1955), show that a necessary and sufficient condition for an instability to occur with $\alpha > 0$ is that such an instability exists at zero wave number. For a monotonic profile with multiple inflexion points, Balmforth and Morrison (1999) have obtained necessary and sufficient conditions for instability which involve the solution of a related integral equation, with this numerical procedure considerably less computationally intensive than the original Rayleigh eigen-value problem.

1.4.2 Inviscid instability of a laminar jet

As an example of inviscid instability, we shall consider here the instability of the laminar jet that was discussed in detail in Section 1.5 in Part 3 of this book series. The jet is produced by forcing a fluid through a narrow slit in a flat barrier OO'. As it emerges from the slit, the jet penetrates into a semi-infinite region filled with a fluid at rest; see Figure 1.16. We assume for simplicity that the fluid in the jet is identical to that in the surrounding region. Due to the action of viscous forces between the jet and the surrounding fluid, the latter is also expected to start moving.

To describe this flow it is convenient to use Cartesian coordinates $(\hat{x}, \hat{y})$ with $\hat{x}$ measured from the centre of the slit normal to the barrier OO'. Here, as usual, 'hat' stands for dimensional variables. We assume the flow two-dimensional. We also assume that the jet is strong enough for boundary-layer theory to be applicable. According to this theory, the pressure does not vary across the jet, and since the fluid is motionless above and below the jet, the pressure proves to be constant throughout the flow field. This allows us to write the boundary-layer equations as

$$\hat{u}\frac{\partial \hat{u}}{\partial \hat{x}} + \hat{v}\frac{\partial \hat{u}}{\partial \hat{y}} = \nu\frac{\partial^2 \hat{u}}{\partial \hat{y}^2}, \tag{1.4.16}$$

$$\frac{\partial \hat{u}}{\partial \hat{x}} + \frac{\partial \hat{v}}{\partial \hat{y}} = 0. \tag{1.4.17}$$

These equations have to be solved subject to the following boundary conditions. Firstly, on both sides of the jet the longitudinal velocity should tend to zero,

$$\hat{u} = 0 \quad \text{at} \quad \hat{y} = \pm\infty, \quad \hat{x} > 0, \tag{1.4.18}$$

to satisfy the matching condition with the motionless fluid surrounding the jet. Secondly, due to the symmetry of the flow with respect to the $\hat{x}$-axis, we have

$$\hat{v} = 0 \quad \text{at} \quad \hat{y} = 0, \quad \hat{x} > 0. \tag{1.4.19}$$

In addition, to avoid a trivial solution, the condition of conservation of the total momentum flux through a plane perpendicular to the jet axis is imposed

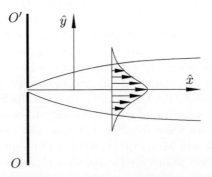

Fig. 1.16: A laminar jet.

$$J_0 = \rho \int_{-\infty}^{\infty} \hat{u}^2 d\hat{y}. \tag{1.4.20}$$

When discussing this problem in Part 3 of this book series, we introduced the stream function $\hat{\psi}(\hat{x}, \hat{y})$ such that

$$\frac{\partial \hat{\psi}}{\partial \hat{x}} = -\hat{v}, \qquad \frac{\partial \hat{\psi}}{\partial \hat{y}} = \hat{u}, \tag{1.4.21}$$

and we found that the solution to (1.4.16)–(1.4.20) may be written in the self-similar form

$$\hat{\psi}(\hat{x}, \hat{y}) = \sqrt[3]{\frac{J_0 \nu}{\rho}} \hat{x}^{1/3} \varphi(\eta), \qquad \eta = \sqrt[3]{\frac{J_0}{\rho \nu^2}} \frac{\hat{y}}{\hat{x}^{2/3}}, \tag{1.4.22}$$

where

$$\varphi(\eta) = \sqrt{6 C_1} \frac{1 - \exp\left(-\sqrt{\frac{2}{3}C_1}\,\eta\right)}{1 + \exp\left(-\sqrt{\frac{2}{3}C_1}\,\eta\right)}, \tag{1.4.23}$$

with

$$C_1 = \frac{1}{2}\sqrt[3]{\frac{3}{4}}.$$

For the purpose of the stability analysis, we need to determine the longitudinal velocity $\hat{u}$ of the basic unperturbed flow. Substituting (1.4.23) into (1.4.22) and then into the second equation in (1.4.21), we find that

$$\hat{u} = \sqrt[3]{\frac{3J_0^2}{32\rho^2\nu\hat{x}}} \operatorname{sech}^2\left(\sqrt{\frac{C_1}{6}}\eta\right), \qquad \sqrt{\frac{C_1}{6}}\eta = \sqrt[3]{\frac{J_0}{48\rho\nu^2}} \frac{\hat{y}}{\hat{x}^{2/3}}. \tag{1.4.24}$$

Let us now consider a certain position $(\hat{x} = \hat{x}_0)$ along the jet. The maximum velocity at this position is obtained by setting $\eta = 0$ in the first equation in (1.4.24). We have

$$\hat{u}_{\max} = \sqrt[3]{\frac{3J_0^2}{32\rho^2\nu\hat{x}_0}}.$$

Also, guided by the second equation in (1.4.24), we define the thickness δ of the jet as

$$\delta = \sqrt[3]{\frac{48\rho\nu^2}{J_0}} \hat{x}_0^{2/3}.$$

Using these quantities, we introduce non-dimensional variables as follows:

$$\hat{u} = \hat{u}_{\max}u, \qquad \hat{y} = \delta y.$$

As a result, the basic velocity profile (1.4.24) assumes the form

$$U(y) = \operatorname{sech}^2(y). \tag{1.4.25}$$

It should be noted that this flow has an inflexion point and so it is a candidate for inviscid instability according to Theorem 1.2; see Problem 4 in Exercises 3.

We are ready now to study the stability of the jet, for which purpose we shall use the Rayleigh equation

$$(U - c)\left(\frac{d^2\phi}{dy^2} - \alpha^2\phi\right) - \frac{d^2U}{dy^2}\phi = 0. \qquad (1.4.26)$$

The boundary conditions for (1.4.26) are

$$\phi = 0 \quad \text{at} \quad y = \pm\infty. \qquad (1.4.27)$$

Given the symmetry of the velocity profile (1.4.25), there are two distinct types of eigen-solutions. In the first one, the perturbations of the longitudinal velocity, being proportional to $\phi'(y)$, are taken to be anti-symmetric with respect to the centreline $(y = 0)$. This is known as the *sinuous mode*. It is studied by imposing the following conditions at $y = 0$:

$$\phi(0) = 1, \quad \phi'(0) = 0. \qquad (1.4.28)$$

Here it is taken into account that the solution of the eigen-value problem (1.4.26), (1.4.27) is defined to within an arbitrary constant. To be definite, we choose this constant according to the first condition in (1.4.28).

In the second, the longitudinal velocity perturbations are symmetric with respect to the centreline giving rise to the boundary conditions

$$\phi(0) = 0, \quad \phi'(0) = 1. \qquad (1.4.29)$$

This is referred to as the *varicose mode*.

The numerical solution of the Rayleigh equation (1.4.26) with either (1.4.28) or (1.4.29) is constructed by marching from $y = 0$ with a suitable guess for the phase velocity c for a given wave number α. The value of c is iterated upon to ensure the attenuation condition $\phi(\infty) = 0$ is satisfied. The results of the calculations are displayed in Figure 1.17, where we show the perturbations' amplification rate αc_i for the sinuous mode (marked by S) and for the varicose mode (marked by V). In both cases,

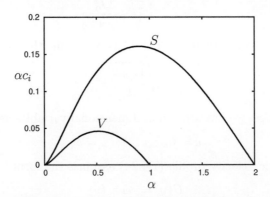

Fig. 1.17: Perturbation growth rate αc_i as a function of the wave number α; the sinuous and varicose modes are marked by S and V, respectively.

the instability starts at $\alpha = 0$ and extends over a finite interval $\alpha \in (0, \alpha_c)$. The latter is twice as long for the sinuous mode. Also, the sinuous mode exhibits a significantly larger amplification rate as compared to the varicose mode.

To conclude we note that if the neutral curve (see Figure 1.15) was calculated for finite values of the Reynolds number, then the lower branch would still tend to $\alpha = 0$ as $Re \rightarrow \infty$, but the upper branch would now be 'lifted' with the asymptote being $\alpha = \alpha_c$. An example of such behaviour will be discussed in Section 1.6 which is devoted to the cross-flow instability; see Figure 1.30.

Exercises 3

1. Consider Blasius equation (1.3.4a)

$$\varphi''' + \frac{1}{2}\varphi\,\varphi'' = 0 \qquad (1.4.30)$$

describing the unperturbed steady flow in the boundary layer on a flat plate. Your task is to show that the second derivative of the basic velocity $U(y)$ is negative everywhere in the boundary layer except on the plate surface $(y = 0)$ and at the outer edge of the boundary layer $(y = \infty)$. You may assume without proof that $\varphi(\eta) > 0$ for all $\eta > 0$.

You may perform this task in the following steps:

(a) Write equation (1.4.30) in the form

$$\frac{\varphi'''}{\varphi''} = -\frac{1}{2}\varphi. \qquad (1.4.31)$$

Integrating (1.4.31) with the initial condition $\varphi''(0) = \lambda$, show that

$$\varphi'' = \lambda e^{-\frac{1}{2}\int_0^\eta \varphi(\eta')\,d\eta'}.$$

Hence, conclude that $\varphi'' > 0$ everywhere except at $\eta = \infty$.

(b) Now return to equation (1.4.30) and argue that $\varphi''' < 0$ everywhere in the boundary layer except at $\eta = 0$ and $\eta = \infty$.

(c) Finally, recall that according to (1.3.15)

$$U(y) = \varphi'(\eta), \qquad \eta = A'y, \qquad (1.4.32)$$

where A' is a positive constant. Differentiate (1.4.32) two times, and conclude that $U'' < 0$.

2. Consider four incompressible, inviscid, parallel flows with the basic velocity profiles shown in Figure 1.18. Which two of these are stable by the Rayleigh theorem? Which of the remaining is stable by the Fjortoft theorem?

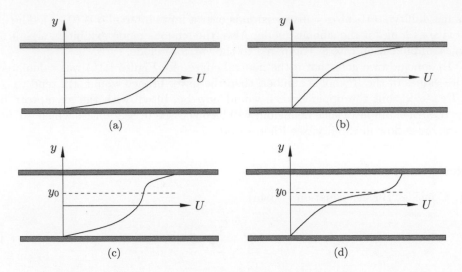

Fig. 1.18: Basic velocity profiles. In (c) and (d), y_0 is a point of inflexion.

3. Consider a flow that is inviscidly unstable, that is there exists a solution to (1.4.1), (1.4.3) with $c_i \neq 0$. Perform the following tasks:

(a) Show that the Rayleigh equation

$$\frac{d^2\phi}{dy^2} - \alpha^2\phi = \frac{U''}{U-c}\phi$$

may be written in terms of the function

$$f(y) = \frac{\phi(y)}{U-c}$$

as

$$\frac{d}{dy}\left[(U-c)\frac{df}{dy}\right] + U'\frac{df}{dy} - \alpha^2(U-c)f = 0. \qquad (1.4.33)$$

What boundary conditions should be imposed on the function $f(y)$?

(b) Multiply all terms of the equation (1.4.33) by $(U-c)\overline{f}$, where $\overline{f}$ is the complex conjugate of f, and integrate between the flow boundaries, $y = y_1$ and $y = y_2$. After applying integration by parts, show that

$$\int_{y_1}^{y_2} (U-c)^2 Q\, dy = 0, \qquad (1.4.34)$$

where $Q = |f'|^2 + \alpha^2|f|^2$.

(c) Equating the imaginary parts of (1.4.34) show that

$$\int_{y_1}^{y_2} UQ\, dy = c_r \int_{y_1}^{y_2} Q\, dy. \qquad (1.4.35)$$

(d) By equating the real parts of (1.4.34) and using (1.4.35) show that

$$\int_{y_1}^{y_2} U^2 Q \, dy = (c_r^2 + c_i^2) \int_{y_1}^{y_2} Q \, dy. \tag{1.4.36}$$

(e) Let $U_{\min}$ and $U_{\max}$ be the minimum and maximum velocities respectively of the basic flow. Then clearly

$$\int_{y_1}^{y_2} (U - U_{\min})(U - U_{\max}) Q \, dy \leq 0. \tag{1.4.37}$$

Multiplying out the brackets in (1.4.37) and substituting in (1.4.35) and (1.4.36) show that

$$\left[c_r^2 + c_i^2 - (U_{\min} + U_{\max}) c_r + U_{\min} U_{\max} \right] \int_{y_1}^{y_2} Q \, dy \leq 0.$$

Hence, conclude that

$$\left[c_r^2 + c_i^2 - (U_{\min} + U_{\max}) c_r + U_{\min} U_{\max} \right] \leq 0, \tag{1.4.38}$$

(f) Show that (1.4.38) can be arranged as

$$\left[c_r - \frac{1}{2}(U_{\min} + U_{\max}) \right]^2 + c_i^2 \leq \left[(U_{\max} - U_{\min}) \right]^2. \tag{1.4.39}$$

Equation (1.4.39) proves *Howard's semicircle theorem* (see Howard, 1961):

Theorem 1.5 *For an unstable wave, the complex phase velocity c lies inside, or on, the semi-circle centred on $(U_{\max} + U_{\min})/2$ with radius $(U_{\max} - U_{\min})/2$; see Figure 1.19.*

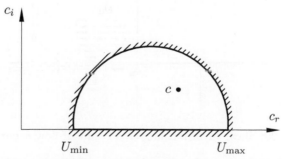

Fig. 1.19: Graphical illustration of Howard's semicircle theorem.

4. Consider the laminar jet velocity profile

$$U(y) = \text{sech}^2(y).$$

Show that it has inflexion points at

$$y_0 = \pm \tanh^{-1}(1/\sqrt{3}).$$

Verify that the boundary-value problem (1.4.26), (1.4.27) admits eigen-solutions corresponding to taking $c = U(y_0) = 2/3$ with

$$\phi(y) = \text{sech}^2(y), \quad \alpha = 2,$$

for the sinuous mode and

$$\phi = \text{sech}(y)\tanh(y), \quad \alpha = 1,$$

for the varicose mode.

1.5   Kelvin–Helmholtz Instability

A classical example of a flow with an inflexion point is the shear layer that forms when a flow separates from a body surface. In Figure 1.20 we present a sketch of the flow separating from a corner O of a rigid body contour.[7] Before separation, the fluid moves along the front face AO of the body with velocity U_0. The separation brings this flow in contact with the stagnant fluid behind the corner. Of course, the internal viscosity will smooth out the velocity profile as shown in Figure 1.20. However, if one is interested in perturbations with wavelength much larger than the characteristic thickness δ of the shear layer, then the internal structure of the shear layer can be disregarded. In such circumstances the shear layer can simply be considered as a contact line with the tangential velocity displaying a jump across this line; see Figure 1.21(a).

If the density ρ of the fluid is the same on both sides of the contact line, then it is convenient to use the coordinate frame that moves along the contact line with speed $\frac{1}{2}U_0$. In this frame (see Figure 1.21b) the flow becomes symmetrical with respect to the $\hat{x}$-axis, and we can expect the phase velocity of the perturbations to be zero.

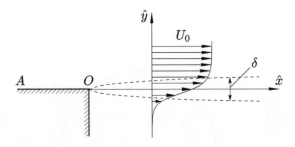

Fig. 1.20: Shear layer forming as a result of flow separation.

[7]A detailed analysis of the steady shear layer flow is given in Section 1.4.1 in Part 3 of this book series.

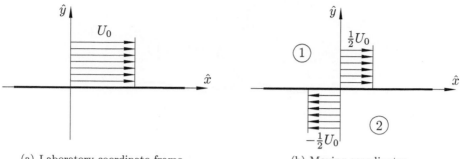

(a) Laboratory coordinate frame.   (b) Moving coordinates.

Fig. 1.21: Galilean transformation.

Since the basic velocity profile is discontinuous

$$\hat{u} = \begin{cases} \frac{1}{2}U_0 & \text{if} \quad \hat{y} > 0, \\ -\frac{1}{2}U_0 & \text{if} \quad \hat{y} < 0, \end{cases}$$

the applicability of the Rayleigh equation (1.4.1) becomes questionable. We shall, therefore, use as our starting point the Euler equations:[8]

$$\frac{\partial \hat{u}}{\partial \hat{t}} + \hat{u}\frac{\partial \hat{u}}{\partial \hat{x}} + \hat{v}\frac{\partial \hat{u}}{\partial \hat{y}} = -\frac{1}{\rho}\frac{\partial \hat{p}}{\partial \hat{x}}, \tag{1.5.1a}$$

$$\frac{\partial \hat{v}}{\partial \hat{t}} + \hat{u}\frac{\partial \hat{v}}{\partial \hat{x}} + \hat{v}\frac{\partial \hat{v}}{\partial \hat{y}} = -\frac{1}{\rho}\frac{\partial \hat{p}}{\partial \hat{y}}, \tag{1.5.1b}$$

$$\frac{\partial \hat{u}}{\partial \hat{x}} + \frac{\partial \hat{v}}{\partial \hat{y}} = 0. \tag{1.5.1c}$$

Here $\hat{t}$ is time, and $(\hat{x}, \hat{y})$ are the Cartesian coordinates, with $\hat{x}$ measured along the discontinuity line in its undisturbed state; see Figure 1.21. The fluid is assumed incompressible with the density ρ being a known constant. All the variables in (1.5.1) are dimensional.

Equations (1.5.1) are applicable both above the discontinuity line (region 1) and below it (region 2). The solutions in regions 1 and 2 are linked to one another through the kinematic and dynamic conditions on the contact line. In order to formulate these conditions, let us assume that the contact line is perturbed, and its shape is given by the equation (see Figure 1.22),

$$\hat{y} = \eta(\hat{t}, \hat{x}). \tag{1.5.2}$$

The kinematic condition states that a fluid particle residing on the line (1.5.2) must remain there for a finite period of time. Therefore, we can relate the coordinates

$$\hat{x} = \hat{x}_p(\hat{t}), \qquad \hat{y} = \hat{y}_p(\hat{t})$$

---

[8]In fact, Kelvin (1871) solved this problem before the Rayleigh equation became known.

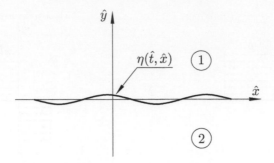

Fig. 1.22: Disturbed discontinuity line.

of such a fluid particle by equation (1.5.2) as follows:

$$\hat{y}_p(\hat{t}) = \eta\big[\hat{t}, \hat{x}_p(\hat{t})\big]. \tag{1.5.3}$$

Differentiation of (1.5.3) with respect to time $\hat{t}$ yields

$$\frac{d\hat{y}_p}{d\hat{t}} = \frac{\partial \eta}{\partial \hat{t}} + \frac{\partial \eta}{\partial \hat{x}}\frac{d\hat{x}_p}{d\hat{t}}.$$

The derivatives $d\hat{x}_p/d\hat{t}$ and $d\hat{y}_p/d\hat{t}$ coincide with the fluid velocity components,

$$\hat{u} = \frac{d\hat{x}_p}{d\hat{t}}, \qquad \hat{v} = \frac{d\hat{y}_p}{d\hat{t}},$$

and therefore, on the contact line we have

$$\hat{v} = \frac{\partial \eta}{\partial \hat{t}} + \hat{u}\frac{\partial \eta}{\partial \hat{x}} \quad \text{at} \quad \hat{y} = \eta(\hat{t}, \hat{x}). \tag{1.5.4}$$

Equation (1.5.4) represents the required kinematic condition. It is applicable to the solution in region 1 as well as to the solution in region 2. The dynamic condition is simply the requirement that the pressure immediately above the contact line should coincide with the pressure immediately below this line:

$$\hat{p}\Big|_{\hat{y}=\eta(\hat{t},\hat{x})+0} = \hat{p}\Big|_{\hat{y}=\eta(\hat{t},\hat{x})-0}. \tag{1.5.5}$$

Let us now linearize the governing equations (1.5.1) and boundary conditions (1.5.4), (1.5.5). We assume that the perturbations to the shape of the contact line are weak, i.e.

$$\eta(\hat{t}, \hat{x}) = \varepsilon\eta'(\hat{t}, \hat{x}), \tag{1.5.6}$$

where ε is a small parameter. Corresponding to (1.5.6), the velocity components and the pressure are represented in region 1 as

$$\hat{u} = \frac{1}{2}U_0 + \varepsilon u_1'(\hat{t}, \hat{x}, \hat{y}), \quad \hat{v} = \varepsilon v_1'(\hat{t}, \hat{x}, \hat{y}), \quad p = P_0 + \varepsilon p_1'(\hat{t}, \hat{x}, \hat{y}), \tag{1.5.7}$$

with P_0 denoting the unperturbed value of pressure both above and below the contact line. Substituting (1.5.7) into (1.5.1) and working with the $O(\varepsilon)$ terms, we arrive at the linearized set of Euler equations:

$$\frac{\partial u'_1}{\partial \hat{t}} + \frac{U_0}{2}\frac{\partial u'_1}{\partial \hat{x}} = -\frac{1}{\rho}\frac{\partial p'_1}{\partial \hat{x}}, \tag{1.5.8a}$$

$$\frac{\partial v'_1}{\partial \hat{t}} + \frac{U_0}{2}\frac{\partial v'_1}{\partial \hat{x}} = -\frac{1}{\rho}\frac{\partial p'_1}{\partial \hat{y}}, \tag{1.5.8b}$$

$$\frac{\partial u'_1}{\partial \hat{x}} + \frac{\partial v'_1}{\partial \hat{y}} = 0. \tag{1.5.8c}$$

Next, we consider the kinematic condition (1.5.4). With (1.5.7) and (1.5.6), the left-hand side of (1.5.4) is written as

$$\hat{v} = \varepsilon v'_1\big[\hat{t}, \hat{x}, \varepsilon\eta'(\hat{t}, \hat{x})\big].$$

Since the third argument of the function v'_1 is small, we can use the Taylor expansion

$$v'_1\big[\hat{t}, \hat{x}, \varepsilon\eta'(\hat{t}, \hat{x})\big] = v'_1(\hat{t}, \hat{x}, 0) + \frac{\partial v'_1}{\partial \hat{y}}(\hat{t}, \hat{x}, 0)\,\varepsilon\eta'(\hat{t}, \hat{x}) + \cdots.$$

Dealing in the same way with the function $\hat{u}$ on the right-hand side of (1.5.4), the kinematic condition becomes

$$\varepsilon v'_1(\hat{t}, \hat{x}, 0) + \varepsilon^2\eta'(\hat{t}, \hat{x})\frac{\partial v'_1}{\partial \hat{y}}(\hat{t}, \hat{x}, 0) + \cdots = \varepsilon\frac{\partial \eta'}{\partial \hat{t}} + \left[\frac{1}{2}U_0 + \varepsilon u'_1(\hat{t}, \hat{x}, 0) + \cdots\right]\varepsilon\frac{\partial \eta'}{\partial \hat{x}}.$$

Restricting our attention to the $O(\varepsilon)$ terms, we have the kinematic condition for (1.5.8) in the form

$$v'_1(\hat{t}, \hat{x}, 0) = \frac{\partial \eta'}{\partial \hat{t}} + \frac{U_0}{2}\frac{\partial \eta'}{\partial \hat{x}}. \tag{1.5.9}$$

Following the conventional routine of hydrodynamic stability theory, we use the normal-mode representation of the contact line function

$$\eta'(\hat{t}, \hat{x}) = \eta_0 e^{\sigma\hat{t} + i\alpha\hat{x}}, \tag{1.5.10}$$

where η_0 is a constant representing the amplitude of the perturbations. Correspondingly the velocity components and the pressure are sought in region 1 in the form

$$u'_1 = e^{\sigma\hat{t} + i\alpha\hat{x}}\breve{u}_1(\hat{y}), \quad v'_1 = e^{\sigma\hat{t} + i\alpha\hat{x}}\breve{v}_1(\hat{y}), \quad p'_1 = e^{\sigma\hat{t} + i\alpha\hat{x}}\breve{p}_1(\hat{y}). \tag{1.5.11}$$

Substitution of (1.5.11) into (1.5.8) leads to the following set of ordinary differential equations:

$$\left(\sigma + \frac{i\alpha}{2}U_0\right)\breve{u}_1 = -\frac{i\alpha}{\rho}\breve{p}_1, \tag{1.5.12a}$$

$$\left(\sigma + \frac{i\alpha}{2}U_0\right)\breve{v}_1 = -\frac{1}{\rho}\frac{d\breve{p}_1}{d\hat{y}}, \tag{1.5.12b}$$

$$i\alpha\breve{u}_1 + \frac{d\breve{v}_1}{d\hat{y}} = 0. \tag{1.5.12c}$$

We shall now eliminate $\breve{u}_1$ and $\breve{v}_1$ from (1.5.12), and formulate a single equation for $\breve{p}_1$. Solving equation (1.5.12c) for $\breve{u}_1$,

$$\breve{u}_1 = \frac{i}{\alpha} \frac{d\breve{v}_1}{d\hat{y}}, \tag{1.5.13}$$

and substituting (1.5.13) into (1.5.12a), we have

$$\left(\sigma + \frac{i\alpha}{2} U_0 \right) \frac{d\breve{v}_1}{d\hat{y}} = -\frac{\alpha^2}{\rho} \breve{p}_1. \tag{1.5.14}$$

Next, we differentiate equation (1.5.12b) with respect to $\hat{y}$. This leads to

$$\left(\sigma + \frac{i\alpha}{2} U_0 \right) \frac{d\breve{v}_1}{d\hat{y}} = -\frac{1}{\rho} \frac{d^2 \breve{p}_1}{d\hat{y}^2}. \tag{1.5.15}$$

Comparing (1.5.15) with (1.5.14), we see that the function $\breve{p}_1$ satisfies the following equation

$$\frac{d^2 \breve{p}_1}{d\hat{y}^2} - \alpha^2 \breve{p}_1 = 0.$$

The general solution of this equation is written as

$$\breve{p}_1 = C_1 e^{\alpha \hat{y}} + C_2 e^{-\alpha \hat{y}}. \tag{1.5.16}$$

Assuming α positive, we have to set $C_1 = 0$ to satisfy the condition of attenuation of the perturbations away from the contact line ($\hat{y} \to \infty$). To find the constant C_2, the kinematic condition (1.5.9) is used. Substituting (1.5.10) together with the second of equations (1.5.11) into (1.5.9), we have

$$\breve{v}_1 \Big|_{\hat{y}=0} = \eta_0 \left(\sigma + \frac{i\alpha}{2} U_0 \right).$$

This condition is easily converted into the corresponding boundary condition for the pressure. Indeed, setting $\hat{y} = 0$ in (1.5.12b), we find that

$$\frac{d\breve{p}_1}{d\hat{y}} \Big|_{\hat{y}=0} = -\rho \left(\sigma + \frac{i\alpha}{2} U_0 \right) \breve{v}_1 \Big|_{\hat{y}=0} = -\eta_0 \rho \left(\sigma + \frac{i\alpha}{2} U_0 \right)^2. \tag{1.5.17}$$

If we now substitute (1.5.16) into (1.5.17), then we will see that

$$C_2 = \frac{\eta_0 \rho}{\alpha} \left(\sigma + \frac{i\alpha}{2} U_0 \right)^2.$$

Hence, the pressure distribution in region 1 is given by

$$\breve{p}_1 = \frac{\eta_0 \rho}{\alpha} \left(\sigma + \frac{i\alpha}{2} U_0 \right)^2 e^{-\alpha \hat{y}}. \tag{1.5.18}$$

The analysis of the flow in region 2 below the discontinuity line is conducted in a similar way. However, instead of (1.5.7) we now have to write

$$\hat{u} = -\frac{1}{2}U_0 + \varepsilon u_2'(\hat{t}, \hat{x}, \hat{y}), \qquad \hat{v} = \varepsilon v_2'(\hat{t}, \hat{x}, \hat{y}), \qquad p = P_0 + \varepsilon p_2'(\hat{t}, \hat{x}, \hat{y}). \qquad (1.5.19)$$

The perturbation functions are again sought in the normal-mode form

$$u_2' = e^{\sigma \hat{t} + i\alpha \hat{x}} \breve{u}_2(\hat{y}), \qquad v_2' = e^{\sigma \hat{t} + i\alpha \hat{x}} \breve{v}_2(\hat{y}), \qquad p_2' = e^{\sigma \hat{t} + i\alpha \hat{x}} \breve{p}_2(\hat{y}), \qquad (1.5.20)$$

and it is found that the pressure in region 2 is given by

$$\breve{p}_2 = -\frac{\eta_0 \rho}{\alpha} \left(\sigma - \frac{i\alpha}{2} U_0 \right)^2 e^{\alpha \hat{y}}. \qquad (1.5.21)$$

To complete the flow analysis, one needs to satisfy the dynamic condition (1.5.5) on the contact line. Using the representations of $\hat{p}$ in (1.5.7) and (1.5.19) on the left- and right-hand sides of (1.5.5), respectively, we have

$$P_0 + \varepsilon p_1' \big[\hat{t}, \hat{x}, \varepsilon \eta'(\hat{t}, \hat{x})\big] = P_0 + \varepsilon p_2' \big[\hat{t}, \hat{x}, \varepsilon \eta'(\hat{t}, \hat{x})\big]. \qquad (1.5.22)$$

We now take advantage of the fact that ε is small and express p_1' and p_2' in terms of their Taylor expansions:

$$\left.\begin{aligned}
p_1' \big[\hat{t}, \hat{x}, \varepsilon \eta'(\hat{t}, \hat{x})\big] &= p_1'(\hat{t}, \hat{x}, 0) + \frac{\partial p_1'}{\partial \hat{y}}(\hat{t}, \hat{x}, 0)\varepsilon \eta'(\hat{t}, \hat{x}) + \cdots, \\
p_2' \big[\hat{t}, \hat{x}, \varepsilon \eta'(\hat{t}, \hat{x})\big] &= p_2'(\hat{t}, \hat{x}, 0) + \frac{\partial p_2'}{\partial \hat{y}}(\hat{t}, \hat{x}, 0)\varepsilon \eta'(\hat{t}, \hat{x}) + \cdots.
\end{aligned}\right\} \qquad (1.5.23)$$

Substituting (1.5.23) into (1.5.22), and working with the $O(\varepsilon)$ terms, we arrive at the conclusion that

$$p_1' \Big|_{\hat{y}=0+} = p_2' \Big|_{\hat{y}=0-}. \qquad (1.5.24)$$

Using the normal-mode representations of p_1' and p_2', given by (1.5.11) and (1.5.20), we can express condition (1.5.24) in the form

$$\breve{p}_1 \Big|_{\hat{y}=0+} = \breve{p}_2 \Big|_{\hat{y}=0-}. \qquad (1.5.25)$$

It remains to substitute (1.5.18) and (1.5.21) into (1.5.25), which leads to the following *dispersion relation*

$$\left(\sigma + \frac{i\alpha}{2} U_0 \right)^2 = -\left(\sigma - \frac{i\alpha}{2} U_0 \right)^2. \qquad (1.5.20)$$

The two solutions of (1.5.26) are

$$\sigma_1 = \frac{1}{2} U_0 \alpha, \qquad \sigma_2 = -\frac{1}{2} U_0 \alpha. \qquad (1.5.27)$$

The first represents growing perturbations, signifying that the flow considered is unstable.

Fig. 1.23: Spiral vortices with Kelvin–Helmholtz instability modes created by the flow separation from a tip of a body moving upwards; see Pierce (1961).

In Figure 1.23 we reproduce a visualization by Pierce (1961) of the flow separation from a sharp tip. One can clearly see that the flow is subject to a Kelvin–Helmholtz instability. It should be noted that according to (1.5.27), the growth rate σ_1 of the unstable mode is proportional to the wave number α, which means that the perturbations from a short wavelength spectrum are the 'most dangerous'. Another important observation is that the theory presented in this section is applicable not only to shear layers with a straight centre-line, as in Figure 1.20, but also for the case of a curved centre-line, as in Figure 1.23, provided that the perturbation wavelength $2\pi/\alpha$ is small compared to the radius of curvature of the centre-line.

Exercises 4

1. Consider a parallel flow in a channel ($-1 < y < 1$) with the basic velocity profile

$$U(y) = \sin(ry) + \sin(r), \qquad (1.5.28)$$

where r is a positive parameter.
   Perform the following tasks:
   (a) Sketch the velocity profile (1.5.28), and show that if $0 < r < \pi$, then $U(y)$ has a single inflexion point.
   (b) Show further that with a suitably chosen phase velocity c, the function

$$\phi = \sin\left[\tfrac{1}{2}\pi(y+1)\right]$$

   satisfies Rayleigh's equation

$$\frac{d^2\phi}{dy^2} - \left(\alpha^2 + \frac{U''}{U-c}\right)\phi = 0$$

   as well as the impermeability conditions on the channel walls:

$$\phi = 0 \quad \text{at} \quad y = \pm 1,$$

provided that the wave number

$$\alpha = \sqrt{r^2 - \frac{\pi^2}{4}}.$$

2. Consider a planar jet of width $2h$ propagating through a fluid at rest with a uniform velocity U_0; see Figure 1.24. The density of the fluid in the jet is ρ_1, and the density of the surrounding fluid is ρ_2.

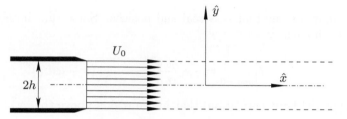

Fig. 1.24: Flow layout.

Investigate the stability of the jet using the Euler equations:

$$\left.\begin{aligned}
\frac{\partial \hat{u}}{\partial \hat{t}} + \hat{u}\frac{\partial \hat{u}}{\partial \hat{x}} + \hat{v}\frac{\partial \hat{u}}{\partial \hat{y}} &= -\frac{1}{\rho}\frac{\partial \hat{p}}{\partial \hat{x}}, \\
\frac{\partial \hat{v}}{\partial \hat{t}} + \hat{u}\frac{\partial \hat{v}}{\partial \hat{x}} + \hat{v}\frac{\partial \hat{v}}{\partial \hat{y}} &= -\frac{1}{\rho}\frac{\partial \hat{p}}{\partial \hat{y}}, \\
\frac{\partial \hat{u}}{\partial \hat{x}} + \frac{\partial \hat{v}}{\partial \hat{y}} &= 0.
\end{aligned}\right\} \qquad (1.5.29)$$

Restrict your attention to varicose modes where the perturbations of $\hat{u}$ are symmetric with respect to the $\hat{x}$-axis, while the perturbations of $\hat{v}$ are antisymmetric. Thanks to this symmetry, only half of the flow, say the upper half, needs to be considered.

Express the 'perturbed' upper boundary of the jet by the equation $\hat{y} = h + \eta(\hat{t}, \hat{x})$, and use (without proof) the following kinematic condition

$$\hat{v} = \frac{\partial \eta}{\partial \hat{t}} + \hat{u}\frac{\partial \eta}{\partial \hat{x}} \quad \text{at} \quad \hat{y} = h + \eta(\hat{t}, \hat{x}) \qquad (1.5.30)$$

on the boundary of the jet. Also take into account that in the case of the varicose mode, the boundary condition on the jet axis is written as

$$v = 0 \quad \text{at} \quad y = 0. \qquad (1.5.31)$$

Perform your analysis in the following steps:

(a) Assume that $\eta = \varepsilon\eta'(\hat{t}, \hat{x})$, where $\varepsilon \ll 1$, and represent the solution inside the jet in the form

$$\hat{u} = U_0 + \varepsilon u'(\hat{t}, \hat{x}, \hat{y}), \quad \hat{v} = \varepsilon v'(\hat{t}, \hat{x}, \hat{y}), \quad \hat{p} = P_0 + \varepsilon p'(\hat{t}, \hat{x}, \hat{y}).$$

Deduce the linearized equations of motion. Also linearize the boundary conditions (1.5.30) and (1.5.31).

(b) Seek the solution of the linear boundary-value problem in the normal-mode form

$$\eta' = \eta_0 e^{\sigma\hat{t} + i\alpha\hat{x}}, \quad u' = \breve{u}(\hat{y})e^{\sigma\hat{t} + i\alpha\hat{x}}, \quad v' = \breve{v}(\hat{y})e^{\sigma\hat{t} + i\alpha\hat{x}}, \quad p' = \breve{p}(\hat{y})e^{\sigma\hat{t} + i\alpha\hat{x}}.$$

You may assume that α is real and positive. Show that immediately below the jet boundary

$$\breve{p}\Big|_{\hat{y} = h - 0} = -\rho_1 \frac{\eta_0}{\alpha} \frac{e^{\alpha h} + e^{-\alpha h}}{e^{\alpha h} - e^{-\alpha h}} (\sigma + i\alpha U_0)^2.$$

(c) Accept (without derivation) that in the region above the jet

$$\breve{p} = \frac{\rho_2 \sigma^2 \eta_0}{\alpha e^{-\alpha h}} e^{-\alpha\hat{y}}, \tag{1.5.32}$$

and show that the dispersion equation relating σ and α may be written as

$$(\sigma + i\alpha U_0)^2 = -A\sigma^2, \quad \text{where} \quad A = \frac{\rho_2}{\rho_1}\tanh(\alpha h) > 0.$$

Solve this equation and decide if the flow is stable or unstable.

1.6 Cross-Flow Vortices

For the flow past a swept wing, the Tollmien–Schlichting waves come into competition with another form of instability, which can be seen in the form of *cross-flow vortices*. The nature of the laminar-turbulent transition is found to be dependent on the *sweep angle* Λ. The latter is defined as the angle between the free-stream velocity vector and the normal to the leading edge of the wing. According to experiments, if $\Lambda \lesssim 35°$, then transition is caused by the growth of Tollmien–Schlichting waves. If, on the other hand, $\Lambda \gtrsim 35°$, then cross-flow vortices dominate in the transition process. For Λ close to $35°$, both instability mechanisms play a role in the boundary-layer transition. Figure 1.25 reproduces a visualization of cross-flow vortices from experiments performed by Saeed *et al.* (2016). Here the boundary-layer instability reveals itself as a pattern of regularly spaced stationary vortices inclined at a certain angle to the velocity vector at the outer edge of the boundary layer. Notice a row of small roughnesses at the leading edge of the wing. These are installed to trigger the cross-flow vortices in a controlled manner. The process of generation of the vortices is termed the boundary-layer *receptivity*. For a detailed discussion of this process the reader is referred to Chapter 3.

The reason for this form of instability may be explained as follows. Let us consider a streamline AA' that lies just above the boundary layer as shown in Figure 1.26(a).

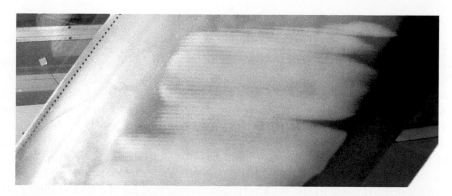

Fig. 1.25: Cross-flow vortices over a swept wing from experiments by Saeed *et al.* (2016). Figure courtesy of Prof. J. F. Morrison.

We choose a point on the wing surface directly below AA', and use it as the coordinate origin O of a Cartesian coordinate system with the $\hat{x}$-axis drawn tangentially to the wing surface and parallel to the streamline AA'. The $\hat{y}$-axis is directed perpendicular to the wing surface, and the $\hat{z}$-axis is perpendicular to the $(\hat{x}, \hat{y})$-plane. In the three-dimensional boundary layer on the wing surface, in addition to the longitudinal velocity component $\hat{u}$, the velocity vector also has a non-zero spanwise velocity component $\hat{w}$. The latter is produced by pressure variations in the $\hat{z}$-direction, which lead to a deflection of the streamline AA' at the outer edge of the boundary layer, but an even stronger deflection inside the boundary layer, where the slower moving fluid is more susceptible to the pressure variations. This is illustrated in Figure 1.26(b), where curve 1 shows a typical distribution of the $\hat{x}$-component $\hat{u}$ of the velocity vector along the $\hat{y}$-axis, with curve 2 showing the $\hat{z}$-component $\hat{w}$ of the velocity vector. Notice that the latter has a 'jet-like' profile. We also show (curve 3) the projection $\hat{u}_S$ of the velocity vector onto the surface S that makes an angle χ with the $\hat{x}$-axis (see Figure 1.26a).

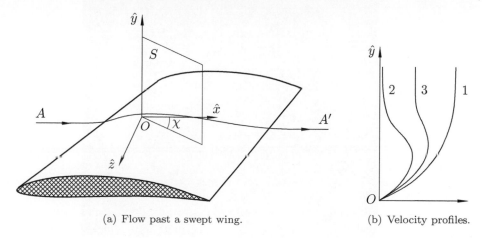

(a) Flow past a swept wing.

(b) Velocity profiles.

Fig. 1.26: Formation of the *secondary flow* in the boundary layer.

It is calculated as

$$\hat{u}_S = \hat{u}\cos\chi + \hat{w}\sin\chi,$$

and shows an inflexional behaviour, which is believed to be the reason for the cross-flow instability.

Another canonical example of a flow that is susceptible to cross-flow instability is the flow produced by a thin disk rotating in its plane inside a large tank of otherwise stagnant fluid. The laminar state of this flow is described by the Kármán solution, which is one of the exact solutions of the Navier–Stokes equations. A detailed discussion of the properties of this flow may be found in Section 2.1.7 of Part 1 of this book series. When performing the laminar flow analysis in Part 1, an inertial coordinate frame, motionless with respect to the tank, was used. However, for the flow stability studies, it is more convenient to use a coordinate frame rotating with the disk. We shall see that the Kármán flow allows for both stationary and moving vortices to exist. Still, it is more often one observes the cross-flow vortices that are produced by wall roughness; these remain motionless with respect to the disk surface. Similarly, the cross-flow vortices on a swept wing are usually produced by the wing surface roughness, and remain stationary as shown in Figure 1.25.

When dealing with the rotating disk problem, it is convenient to use cylindrical polar coordinates with the $\hat{z}$-axis aligned with the axis of rotation of the disk; see Figure 1.28. Remember that in these coordinates, the position of a point M is defined by (i) the distance $\hat{r}$ from the $O\hat{z}$-axis, (ii) the circumferential angle ϕ, and (iii) the altitude $\hat{z}$ of the point M above the disk. Correspondingly, the unit vector triad $(\mathbf{e}_1, \mathbf{e}_2, \mathbf{e}_3)$ is constructed at point M in the following way. Vector $\mathbf{e}_1$ points in the radial direction, vector $\mathbf{e}_2$ is tangent to the circle C that lies in the plane drawn through point M perpendicular to the $O\hat{z}$-axis and has its centre on this axis, and, finally,

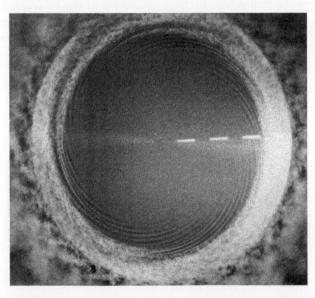

Fig. 1.27: Spiral vortices on a spinning disk; visualization by Kobayashi *et al.* (1980).

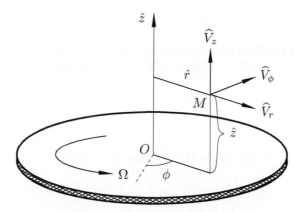

Fig. 1.28: Kármán flow.

vector $\mathbf{e}_3$ is parallel to the $O\hat{z}$-axis. Using this triad, the fluid velocity vector $\widehat{\mathbf{V}}$ is decomposed as

$$\widehat{\mathbf{V}} = \widehat{V}_r \mathbf{e}_1 + \widehat{V}_\phi \mathbf{e}_2 + \widehat{V}_z \mathbf{e}_3,$$

and the Navier–Stokes equations have the form:[9]

$$\frac{\partial \widehat{V}_r}{\partial \hat{t}} + \widehat{V}_r \frac{\partial \widehat{V}_r}{\partial \hat{r}} + \frac{\widehat{V}_\phi}{\hat{r}} \frac{\partial \widehat{V}_r}{\partial \phi} + \widehat{V}_z \frac{\partial \widehat{V}_r}{\partial \hat{z}} - \frac{\widehat{V}_\phi^2}{\hat{r}} = \hat{f}_r - \frac{1}{\rho} \frac{\partial \hat{p}}{\partial \hat{r}}$$

$$+ \nu \left(\frac{\partial^2 \widehat{V}_r}{\partial \hat{z}^2} + \frac{1}{\hat{r}^2} \frac{\partial^2 \widehat{V}_r}{\partial \phi^2} - \frac{2}{\hat{r}^2} \frac{\partial \widehat{V}_\phi}{\partial \phi} + \frac{\partial^2 \widehat{V}_r}{\partial \hat{r}^2} + \frac{1}{\hat{r}} \frac{\partial \widehat{V}_r}{\partial \hat{r}} - \frac{\widehat{V}_r}{\hat{r}^2} \right), \qquad (1.6.1a)$$

$$\frac{\partial \widehat{V}_\phi}{\partial \hat{t}} + \widehat{V}_r \frac{\partial \widehat{V}_\phi}{\partial \hat{r}} + \frac{\widehat{V}_\phi}{\hat{r}} \frac{\partial \widehat{V}_\phi}{\partial \phi} + \widehat{V}_z \frac{\partial \widehat{V}_\phi}{\partial \hat{z}} + \frac{\widehat{V}_r \widehat{V}_\phi}{\hat{r}} = \hat{f}_\phi - \frac{1}{\rho \hat{r}} \frac{\partial \hat{p}}{\partial \phi}$$

$$+ \nu \left(\frac{\partial^2 \widehat{V}_\phi}{\partial \hat{z}^2} + \frac{1}{\hat{r}^2} \frac{\partial^2 \widehat{V}_\phi}{\partial \phi^2} + \frac{2}{\hat{r}^2} \frac{\partial \widehat{V}_r}{\partial \phi} + \frac{\partial^2 \widehat{V}_\phi}{\partial \hat{r}^2} + \frac{1}{\hat{r}} \frac{\partial \widehat{V}_\phi}{\partial \hat{r}} - \frac{\widehat{V}_\phi}{\hat{r}^2} \right), \qquad (1.6.1b)$$

$$\frac{\partial \widehat{V}_z}{\partial \hat{t}} + \widehat{V}_r \frac{\partial \widehat{V}_z}{\partial \hat{r}} + \frac{\widehat{V}_\phi}{\hat{r}} \frac{\partial \widehat{V}_z}{\partial \phi} + \widehat{V}_z \frac{\partial \widehat{V}_z}{\partial \hat{z}} = \hat{f}_z - \frac{1}{\rho} \frac{\partial \hat{p}}{\partial \hat{z}}$$

$$+ \nu \left(\frac{\partial^2 \widehat{V}_z}{\partial \hat{z}^2} + \frac{1}{\hat{r}^2} \frac{\partial^2 \widehat{V}_z}{\partial \phi^2} + \frac{\partial^2 \widehat{V}_z}{\partial \hat{r}^2} + \frac{1}{\hat{r}} \frac{\partial \widehat{V}_z}{\partial \hat{r}} \right), \qquad (1.6.1c)$$

$$\frac{\partial \widehat{V}_r}{\partial \hat{r}} + \frac{1}{\hat{r}} \frac{\partial \widehat{V}_\phi}{\partial \phi} + \frac{\partial \widehat{V}_z}{\partial \hat{z}} + \frac{\widehat{V}_r}{\hat{r}} = 0, \qquad (1.6.1d)$$

with $\hat{f}_r$, $\hat{f}_\phi$, and $\hat{f}_z$ being the components of the inertial body force $\hat{\mathbf{f}}$. Assuming that the coordinate frame rotates with the disk with angular velocity Ω, the body force can

---

[9]See equations (1.8.45) on page 84 in Part 1 of this book series.

be calculated as[10]

$$\hat{\mathbf{f}} = (\widehat{\mathbf{\Omega}} \times \hat{\mathbf{r}}) \times \widehat{\mathbf{\Omega}} + 2(\widehat{\mathbf{V}} \times \widehat{\mathbf{\Omega}}). \tag{1.6.2}$$

Here $\widehat{\mathbf{\Omega}} = \Omega \mathbf{e}_3$ and $\hat{\mathbf{r}}$ is the position vector that connects the coordinate origin O with the point of observation M; see Figure 1.28. It is easily seen that

$$(\widehat{\mathbf{\Omega}} \times \hat{\mathbf{r}}) = \mathbf{e}_2 \Omega \hat{r},$$

and therefore,

$$(\widehat{\mathbf{\Omega}} \times \hat{\mathbf{r}}) \times \widehat{\mathbf{\Omega}} = \begin{vmatrix} \mathbf{e}_1 & \mathbf{e}_2 & \mathbf{e}_3 \\ 0 & \Omega \hat{r} & 0 \\ 0 & 0 & \Omega \end{vmatrix} = \mathbf{e}_1 \Omega^2 \hat{r}. \tag{1.6.3}$$

We further have

$$(\widehat{\mathbf{V}} \times \widehat{\mathbf{\Omega}}) = \begin{vmatrix} \mathbf{e}_1 & \mathbf{e}_2 & \mathbf{e}_3 \\ \widehat{V}_r & \widehat{V}_\phi & \widehat{V}_z \\ 0 & 0 & \Omega \end{vmatrix} = \mathbf{e}_1 \widehat{V}_\phi \Omega + \mathbf{e}_2 (-\widehat{V}_r \Omega). \tag{1.6.4}$$

Substitution of (1.6.3) and (1.6.4) into (1.6.2) yields

$$\hat{\mathbf{f}} = \mathbf{e}_1 (\Omega^2 \hat{r} + 2\widehat{V}_\phi \Omega) + \mathbf{e}_2 (-2\widehat{V}_r \Omega),$$

and we can conclude that the components of the inertial body force are given by

$$\hat{f}_r = \Omega^2 \hat{r} + 2\widehat{V}_\phi \Omega, \qquad \hat{f}_\phi = -2\widehat{V}_r \Omega, \qquad \hat{f}_z = 0. \tag{1.6.5}$$

1.6.1 Basic flow

We start with the basic laminar flow. It is represented by the time-independent solution of the Navier–Stokes equations (1.6.1) subject to the no-slip conditions

$$\widehat{V}_r = \widehat{V}_\phi = \widehat{V}_z = 0 \quad \text{at} \quad \hat{z} = 0, \tag{1.6.6}$$

on the disk surface and the following condition far away from the disk:

$$\widehat{V}_r = 0, \quad \widehat{V}_\phi = -\Omega \hat{r}, \quad \hat{p} = \hat{p}_0 \quad \text{at} \quad \hat{z} = \infty, \tag{1.6.7}$$

where $\hat{p}_0$ denotes the pressure inside the tank.

Following the arguments presented in Section 2.1.7 in Part 1 of this book series, we seek the fluid-dynamic functions in the form

$$\widehat{V}_r = \Omega \hat{r} F(z), \quad \widehat{V}_\phi = \Omega \hat{r} G(z), \quad \widehat{V}_z = \sqrt{\Omega \nu} H(z), \quad \hat{p} = \hat{p}_0 + \rho \Omega \nu P(z), \tag{1.6.8}$$

where

$$z = \sqrt{\frac{\Omega}{\nu}} \hat{z}. \tag{1.6.9}$$

---

[10]See Section 1.2 in Part 1 of this book series.

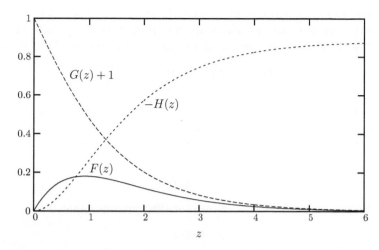

Fig. 1.29: Radial $F(z)$, circumferential $G(z)$, and axial $H(z)$ velocity profiles for rotating-disk flow.

Substitution of (1.6.8) into the Navier–Stokes equations (1.6.1) results in the following set of ordinary differential equations:

$$F^2 + HF' - (G+1)^2 = F'', \qquad (1.6.10\text{a})$$

$$2F(G+1) + HG' = G'', \qquad (1.6.10\text{b})$$

$$HH' = -P' + H'', \qquad (1.6.10\text{c})$$

$$2F + H' = 0. \qquad (1.6.10\text{d})$$

As expected, this reduces to the corresponding set of equations (2.1.87) in Part 1 through a simple substitution of G with $G-1$ which accounts for the rotation of the coordinate frame.

One can see that equations (1.6.10a,b,d) do not involve the pressure $P(z)$, and may be solved separately from equation (1.6.10c). Considered together, they constitute a set of ordinary differential equations of fifth order, and require five boundary conditions. These may be obtained by substituting (1.6.8) into (1.6.6) and (1.6.7). We have

$$\left.\begin{aligned} F = 0, \quad G = 0, \quad H = 0 \quad \text{at} \quad z = 0, \\ F = 0, \quad G = -1 \qquad\qquad \text{at} \quad z = \infty. \end{aligned}\right\} \qquad (1.6.11)$$

The results of a numerical solution of (1.6.10), (1.6.11) are displayed in Figure 1.29.

1.6.2 Linear stability analysis

Turning to the stability analysis we notice that in the basic flow (1.6.8), the radial and circumferential velocity components depend not only on the vertical coordinate $\hat{z}$, measured across the viscous layer, but also on the radial coordinate $\hat{r}$. Therefore, in order to use the normal-mode representation of the perturbations we need to rely on the parallel flow approximation as in Section 1.3 where the stability of boundary layers was discussed. Here, the parallel flow approximation is only applicable to situations where the variations of $\widehat{V}_r$ and $\widehat{V}_\phi$ with respect to $\hat{r}$ are weak compared to the variations with respect to $\hat{z}$. This requirement may be expressed in terms of the inequalities

$$\frac{\partial \widehat{V}_r}{\partial \hat{r}} \ll \frac{\partial \widehat{V}_r}{\partial \hat{z}}, \qquad \frac{\partial \widehat{V}_\phi}{\partial \hat{r}} \ll \frac{\partial \widehat{V}_\phi}{\partial \hat{z}}. \tag{1.6.12}$$

In what follows we will be interested in examining the flow stability near a radial location, say $\hat{r} = a$. It follows from (1.6.9) that the characteristic thickness of the layer of fluid set into motion by the rotating disk may be estimated as

$$\delta = \sqrt{\frac{\nu}{\Omega}}. \tag{1.6.13}$$

Obviously, both conditions in (1.6.12) are satisfied provided that

$$\delta \ll a. \tag{1.6.14}$$

Using (1.6.13) in (1.6.14), we have

$$\frac{1}{a}\sqrt{\frac{\nu}{\Omega}} \ll 1. \tag{1.6.15}$$

Condition (1.6.15) is more conveniently expressed in terms of the Reynolds number. We shall define the latter using the characteristic fluid velocity Ωa and the thickness (1.6.13) of the moving fluid layer as

$$Re_* = \frac{\Omega a \cdot \delta}{\nu} = a\sqrt{\frac{\Omega}{\nu}}. \tag{1.6.16}$$

Clearly, condition (1.6.15) is satisfied provided that $Re_* \gg 1$.

By analogy with the boundary-layer flows (see Section 1.3), we can expect the perturbations to have radial and circumferential wavelengths comparable with δ, and a period of oscillation estimated as

$$\hat{t} \sim \frac{\delta}{\Omega a} \sim \frac{1}{\Omega} Re_*^{-1}.$$

Correspondingly, we define the dimensionless independent variables t, r, φ, and z as

$$\hat{t} = \frac{Re_*^{-1}}{\Omega}t, \quad \hat{r} = a(1 + Re_*^{-1}r), \quad \phi = Re_*^{-1}\varphi, \quad \hat{z} = aRe_*^{-1}z, \tag{1.6.17}$$

and seek the fluid-dynamic functions in the form

$$\left.\begin{aligned}
\widehat{V}_r &= \Omega a\big[(1 + Re_*^{-1}r)F(z) + \varepsilon u'(t,r,\varphi,z)\big], \\
\widehat{V}_\phi &= \Omega a\big[(1 + Re_*^{-1}r)G(z) + \varepsilon v'(t,r,\varphi,z)\big], \\
\widehat{V}_z &= \Omega a\big[Re_*^{-1}H(z) + \varepsilon w'(t,r,\varphi,z)\big], \\
\hat{p} &= \hat{p}_0 + \rho(\Omega a)^2\big[Re_*^{-2}P(z) + \varepsilon p'(t,r,\varphi,z)\big],
\end{aligned}\right\} \tag{1.6.18}$$

where ε is the amplitude of the perturbations, presumed small.

We now need to substitute (1.6.18) together with (1.6.17) into the Navier–Stokes equations (1.6.1). We shall demonstrate the procedure using, as an example, the second term on the left-hand side of the radial momentum equation (1.6.1a). We have

$$\widehat{V}_r \frac{\partial \widehat{V}_r}{\partial \hat{r}} = \Omega^2 a \left[(1 + Re_*^{-1} r) F(z) + \varepsilon u'(t, r, \varphi, z) \right] \left[F(z) + \varepsilon Re_* \frac{\partial u'}{\partial r} \right].$$

Our interest is in the $O(\varepsilon)$ terms, which are

$$\widehat{V}_r \frac{\partial \widehat{V}_r}{\partial \hat{r}} = \Omega^2 a \left[\cdots + \underbrace{\varepsilon Re_* F(z) \frac{\partial u'}{\partial r}}_{1} + \underbrace{\varepsilon F(z) u'}_{2} + \underbrace{\varepsilon r F(z) \frac{\partial u'}{\partial r}}_{3} + \cdots \right]. \qquad (1.6.19)$$

Clearly, for large values of Re_*, term 1 is dominant. In fact, this is the only term that was taken from $u \partial u / \partial x$ in (1.3.11a) when deriving the Orr–Sommerfeld equation (1.3.19). Here, we shall also keep the smaller term 2. Unlike term 3, it does not prevent the normal-mode representation

$$u'(t, r, \varphi, z) = e^{i(\alpha r + \beta \varphi - \omega t)} \breve{u}(z) \qquad (1.6.20)$$

from being used. Disregarding term 3 and using (1.6.20) for terms 1 and 2, we have[11]

$$\widehat{V}_r \frac{\partial \widehat{V}_r}{\partial \hat{r}} = \Omega^2 a \left[\cdots + \varepsilon \left(i \alpha Re_* F \breve{u} + F \breve{u} \right) e^{i(\alpha r + \beta \varphi - \omega t)} + \cdots \right].$$

Treating the rest of the terms in (1.6.1a) in the same manner and keeping the $O(Re_*)$ and $O(1)$ terms but disregarding the the $O(Re_*^{-1})$ and smaller terms we find that

$$\frac{d^2 \breve{u}}{dz^2} - H \frac{d\breve{u}}{dz} - \left[i Re_* (\alpha F + \beta G - \omega) + \alpha^2 + \beta^2 + F \right] \breve{u}$$

$$+ 2(G + 1) \breve{v} - Re_* \frac{dF}{dz} \breve{w} - i \alpha Re_* \breve{p} = 0. \qquad (1.6.21a)$$

Here $\breve{v}(z)$, $\breve{w}(z)$, and $\breve{p}(z)$ are the amplitude functions in the normal-mode representation of v', w', and p', respectively, that is

$$v' = e^{i(\alpha r + \beta \varphi - \omega t)} \breve{v}(z), \qquad w' = e^{i(\alpha r + \beta \varphi - \omega t)} \breve{w}(z), \qquad p' = e^{i(\alpha r + \beta \varphi - \omega t)} \breve{p}(z).$$

A similar treatment of equations (1.6.1b)–(1.6.1d) reduces them to

---

[11] The practice of inclusion of smaller terms is used by researchers in an attempt to account for the non-parallelism in the basic flow. However, it should be noted that there is no rational justification for an approach where some smaller terms are retained but others are discarded despite being the same order of magnitude.

$$\frac{d^2\breve{v}}{dz^2} - H\frac{d\breve{v}}{dz} - \left[iRe_*(\alpha F + \beta G - \omega) + \alpha^2 + \beta^2 + F\right]\breve{v}$$
$$- 2(G+1)\breve{u} - Re_*\frac{dG}{dz}\breve{w} - i\beta Re_*\breve{p} = 0, \quad (1.6.21b)$$

$$\frac{d^2\breve{w}}{dz^2} - H\frac{d\breve{w}}{dz} - \left[iRe_*(\alpha F + \beta G - \omega) + \alpha^2 + \beta^2 + \frac{dH}{dz}\right]\breve{w} - Re_*\frac{d\breve{p}}{dz} = 0, \quad (1.6.21c)$$

$$\left(i\alpha + \frac{1}{Re_*}\right)\breve{u} + i\beta\breve{v} + \frac{d\breve{w}}{dz} = 0. \quad (1.6.21d)$$

Equations (1.6.21) have to be solved subject to the no-slip conditions on the disk surface

$$\breve{u} = \breve{v} = \breve{w} = 0 \quad \text{at} \quad z = 0, \quad (1.6.22)$$

and the conditions of attenuation of the perturbations far from the disk

$$\breve{u} = \breve{v} = \breve{w} = \breve{p} = 0 \quad \text{at} \quad z = \infty. \quad (1.6.23)$$

The boundary-value problem (1.6.21)–(1.6.23) is homogeneous, and since we are interested in non-trivial solutions, it should be treated as an eigen-value problem. As such, it allows us to find

$$\alpha = \alpha_r + i\alpha_i$$

for each set of

$$(Re_*, \omega, \beta). \quad (1.6.24)$$

In particular, we can assume β in (1.6.24) to be real. If we also want α to be real, then for fixed values of Re_* and ω, we can make α_i zero by an appropriate choice of β. Thus for real wave numbers (α, β), the solution of the eigen-value problem (1.6.21)–(1.6.23) defines α and β as functions of Re_* and ω.

The results of the numerical solution of the eigen-value problem (1.6.21)–(1.6.23) are shown in Figure 1.30. We performed the calculations for *stationary vortices* ($\omega = 0$) as well as for *travelling vortices* with $\omega = -5$ and $\omega = 7.9$. Figure 1.30(a) displays the corresponding neutral curves in the (Re_*, α)-plane, and Figure 1.30(b) in the (Re_*, β)-plane. As usual, inside the neutral curves the disturbances are unstable. For each frequency ω there is a critical (minimal) value of the Reynolds number. In particular for stationary vortices, it is

$$Re_* \simeq 286.05. \quad (1.6.25)$$

If Re_* is smaller than the critical value (1.6.25) then all the perturbations decay, which explains why the flow remains laminar (see Figure 1.27) within a circle of radius[12]

$$a_* = 286.05\sqrt{\frac{\nu}{\Omega}}. \quad (1.6.26)$$

The numerical solution of the eigen-value problem (1.6.21)–(1.6.23) also provides us with the values of α and β at the critical point. For the stationary vortices, they

---

[12]Equation (1.6.26) is easily obtained by substituting the value of Re_* in (1.6.16) from (1.6.25).

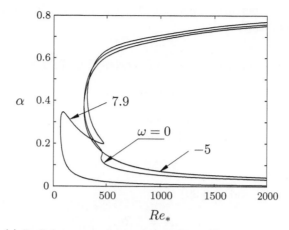

(a) Radial wave number α against Reynolds number Re_*.

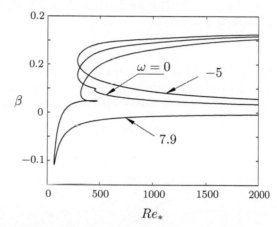

(b) Circumferential wave number β against Reynolds number Re_*.

Fig. 1.30: Neutral curves for the cross-flow instability of rotating disk flow for different frequencies $\omega = 0$, -5, and 7.9.

are $\alpha_* \simeq 0.3829$, $\beta_* \simeq 0.0773$. Using these values one can calculate the angle ϵ of deflection of the vortices from the direction of the fluid motion at the outer edge of the boundary layer:[13]

$$\epsilon = \arctan\left(\beta_*/\alpha_*\right) \simeq 11.41°.$$

This is reasonably close to the experimentally observed angle $\epsilon \approx 14°$.

The theory is less successful at predicting the number of vortices that form on the disk surface. Theoretically it can be calculated as follows. In the normal-mode

---

[13]See Problem 1 in Exercises 5.

representation of the perturbations (1.6.20), the angular interval $\Delta\varphi$ between two maxima (or minima) is given by

$$\beta\Delta\varphi = 2\pi. \tag{1.6.27}$$

Using the scaling for ϕ in (1.6.17) we can express (1.6.27) in the form

$$\beta Re_*\Delta\phi = 2\pi. \tag{1.6.28}$$

On the other hand, if we denote the number of vortices by $\mathcal{N}$, then to cover the entire circumference of the disk, we should have

$$\mathcal{N}\Delta\phi = 2\pi. \tag{1.6.29}$$

It follows from (1.6.28) and (1.6.29) that

$$\mathcal{N} = \beta Re_*.$$

Using for β its critical value $\beta_* = 0.0773$, we find that $\mathcal{N} \approx 22$, while the experimentally observed number of vortices is about 30.

Exercises 5

1. Let an observation point M be situated on a circle C whose radius $\hat{r} = a$; see Figure 1.31. Consider the vortex that crosses circle C at point M. Your task is to show that the angle ϵ the vortex makes with the circle is calculated as

$$\epsilon = \arctan{(\beta/\alpha)}.$$

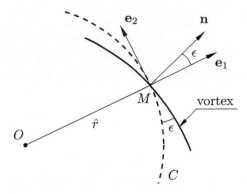

Fig. 1.31: Calculation of the angle between circle C and a vortex.

Suggestions: Using the normal-mode representation of the perturbations (1.6.20) argue that the equation for the vortex axis may be written as

$$\alpha r + \beta\varphi = c_1, \tag{1.6.30}$$

where c_1 is a constant.

Return to the dimensional variables (1.6.17), and cast (1.6.30) in the form

$$\Psi(\hat{r}, \phi) = \frac{\alpha}{a}\hat{r} + \beta\phi = c_2,$$

with c_2 being another constant.

Finally, show that the vector **n** normal to the vortex is

$$\mathbf{n}\Big|_{\hat{r}=a} = \frac{\alpha}{a}\mathbf{e}_1 + \frac{\beta}{a}\mathbf{e}_2.$$

You may use without derivation the fact that in cylindrical polar coordinates, the gradient of $\Psi(\hat{r}, \phi)$ is calculated as

$$\nabla\Psi = \frac{\partial\Psi}{\partial\hat{r}}\mathbf{e}_1 + \frac{1}{\hat{r}}\frac{\partial\Psi}{\partial\phi}\mathbf{e}_2 + \frac{\partial\Psi}{\partial\hat{z}}\mathbf{e}_3.$$

2. Notice that for large values of the Reynolds number, the neutral curves (see Figure 1.30) remain open implying inviscid instability. Your task is to show that in the limit $Re_* \to \infty$, equations (1.6.21) may be reduced to the Rayleigh equation, which allows us to judge the stability of the flow with the help of the inflexion point theorem. You can perform this task in the following steps:

(a) Assume $Re_* \to \infty$, and show that in this limit, equations (1.6.21) reduce to

$$i(\alpha F + \beta G - \omega)\breve{u} + \frac{dF}{dz}\breve{w} = -i\alpha\breve{p}, \tag{1.6.31a}$$

$$i(\alpha F + \beta G - \omega)\breve{v} + \frac{dG}{dz}\breve{w} = -i\beta\breve{p}, \tag{1.6.31b}$$

$$i(\alpha F + \beta G - \omega)\breve{w} = -\frac{d\breve{p}}{dz}, \tag{1.6.31c}$$

$$i\alpha\breve{u} + i\beta\breve{v} + \frac{d\breve{w}}{dz} = 0. \tag{1.6.31d}$$

(b) Consider the plane S that is drawn through the point of observation M perpendicular to the disk surface at an angle χ to the radial direction; see Figure 1.32.

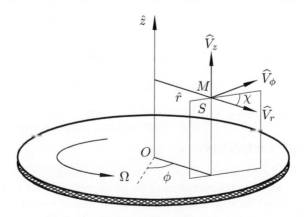

Fig. 1.32: Kármán flow.

Argue that the projection of the velocity vector onto the plane S is represented by

$$U = F \cos \chi + G \sin \chi.$$

(c) Consider a class of perturbations for which the wave front is perpendicular to the plane S. Show that for such perturbations, the radial and circumferential wave numbers are

$$\alpha = \gamma \cos \chi, \qquad \beta = \gamma \sin \chi,$$

where $\gamma = \sqrt{\alpha^2 + \beta^2}$.

(d) Multiply equations (1.6.31a) and (1.6.31b) by $\cos \chi$ and $\sin \chi$, respectively, and add them together. Then use the continuity equation (1.6.31d) to eliminate $\breve{u}$ and $\breve{v}$. Show that this leads to the equation

$$(U - c)\frac{d\breve{w}}{dz} - \frac{dU}{dz}\breve{w} = i\gamma\breve{p}, \qquad (1.6.32)$$

where $c = \omega/\gamma$ is the phase velocity.

(e) Finally, eliminate the pressure $\breve{p}$ from (1.6.32) and (1.6.31c). You should arrive at the Rayleigh equation

$$\frac{d^2\breve{w}}{dz^2} - \left(\frac{U''}{U - c} + \gamma^2\right)\breve{w} = 0. \qquad (1.6.33)$$

What are the boundary conditions for this equation?

3. Consider Rayleigh's equation (1.6.33) for a model velocity profile

$$U(z) = \left(e^{-z} - e^{-2z}\right)\cos \chi + \left(1 - e^{-z}\right)\sin \chi,$$

which is closely related to the asymptotic suction profile (see Gregory *et al.*, 1955). Perform the following tasks:

(a) Show that U is zero at a point of inflexion if $\tan \chi = -3$ or $\tan \chi = -1/3$.

(b) Deduce that for the case $\tan \chi = -1/3$ the function

$$\breve{w} = e^{-\frac{3}{2}z} - e^{-\frac{5}{2}z}$$

satisfies equation (1.6.33) as well as the boundary conditions

$$\breve{w}(0) = 0, \qquad \breve{w}(\infty) = 0,$$

for stationary modes, provided that $\gamma = 3/2$.

1.7 Centrifugal Instability

A classical example of centrifugal instability is shown in Figure 1.33. Here the fluid is contained between two coaxial cylinders. The inner cylinder rotates around its axis, while the outer cylinder is kept motionless, and is made of a transparent material to facilitate the flow visualization. The observations show that provided the angular velocity of the inner cylinder stays below a critical value, the fluid motion between

 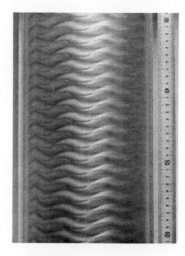

(a) Taylor vortices. (b) Wavy Taylor vortices.

Fig. 1.33: Instability of the flow between two coaxial cylinders, visualized by Burkhalter and Koschmieder (1974).

the cylinders assumes a rather simple form, which is described subsequently in Section 1.7.1. In this motion the fluid particles do not migrate from one plane perpendicular to the cylinders' axis to another. Instead, they follow circular trajectories that lie in these planes. The velocity distribution in the gap between the cylinders proves to be independent of the coordinate $\hat{z}$ measured along the cylinders' axis, except near the top and bottom ends of the apparatus.

However, once the angular velocity of the inner cylinder exceeds a critical value, the flow develops a periodic structure clearly visible in Figure 1.33(a). Each element of the structure represents a vortical motion of the fluid, where in addition to the main azimuthal motion, the fluid particles are involved in helical motion around the centre-line of each vortex. A detailed experimental and theoretical investigation of the process of formation of the vortical structure was conducted by Taylor (1923); hence the name *Taylor vortices.*

Interestingly enough, there exists a second critical value of the velocity of rotation of the inner cylinder, when the Taylor vortices themselves become unstable, and develop a wavy wave motion, shown in 1.33(b). In this presentation we shall restrict our attention to the first stage of the instability, namely, the formation of the Taylor vortices as in Figure 1.33(a).

When analysing the flow it is convenient to use cylindrical polar coordinates (see Figure 1.28) with the $\hat{z}$-axis aligned with the common axis of the two cylinders. Here we shall work in an inertial coordinate frame. Therefore, the body force $\hat{\mathbf{f}} = (\hat{f}_r, \hat{f}_\phi, \hat{f}_z)$ will be omitted in the Navier–Stokes equations (1.6.1). According to the observations both the basic laminar flow and the flow with Taylor vortices, shown in Figure 1.33(a), are axisymmetric. This allows us to disregard the derivatives with respect to ϕ in (1.6.1), reducing the Navier–Stokes equations to the form

$$\frac{\partial \widehat{V}_\phi}{\partial \hat{t}} + \widehat{V}_r \frac{\partial \widehat{V}_\phi}{\partial \hat{r}} + \widehat{V}_z \frac{\partial \widehat{V}_\phi}{\partial \hat{z}} + \frac{\widehat{V}_r \widehat{V}_\phi}{\hat{r}} = \nu \left(\frac{\partial^2 \widehat{V}_\phi}{\partial \hat{z}^2} + \frac{\partial^2 \widehat{V}_\phi}{\partial \hat{r}^2} + \frac{1}{\hat{r}} \frac{\partial \widehat{V}_\phi}{\partial \hat{r}} - \frac{\widehat{V}_\phi}{\hat{r}^2} \right), \qquad (1.7.1a)$$

$$\frac{\partial \widehat{V}_r}{\partial \hat{t}} + \widehat{V}_r \frac{\partial \widehat{V}_r}{\partial \hat{r}} + \widehat{V}_z \frac{\partial \widehat{V}_r}{\partial \hat{z}} - \frac{\widehat{V}_\phi^2}{\hat{r}} = -\frac{1}{\rho} \frac{\partial \hat{p}}{\partial \hat{r}}$$

$$+ \nu \left(\frac{\partial^2 \widehat{V}_r}{\partial \hat{z}^2} + \frac{\partial^2 \widehat{V}_r}{\partial \hat{r}^2} + \frac{1}{\hat{r}} \frac{\partial \widehat{V}_r}{\partial \hat{r}} - \frac{\widehat{V}_r}{\hat{r}^2} \right), \quad (1.7.1b)$$

$$\frac{\partial \widehat{V}_z}{\partial \hat{t}} + \widehat{V}_r \frac{\partial \widehat{V}_z}{\partial \hat{r}} + \widehat{V}_z \frac{\partial \widehat{V}_z}{\partial \hat{z}} = -\frac{1}{\rho} \frac{\partial \hat{p}}{\partial \hat{z}} + \nu \left(\frac{\partial^2 \widehat{V}_z}{\partial \hat{z}^2} + \frac{\partial^2 \widehat{V}_z}{\partial \hat{r}^2} + \frac{1}{\hat{r}} \frac{\partial \widehat{V}_z}{\partial \hat{r}} \right), \quad (1.7.1c)$$

$$\frac{\partial \widehat{V}_r}{\partial \hat{r}} + \frac{\partial \widehat{V}_z}{\partial \hat{z}} + \frac{\widehat{V}_r}{\hat{r}} = 0. \qquad (1.7.1d)$$

The basic laminar flow

The solution for the laminar flow between two cylinders is obtained by setting

$$\widehat{V}_r = \widehat{V}_z = 0, \qquad (1.7.2)$$

and disregarding the derivatives with respect to $\hat{z}$ and $\hat{t}$ in (1.7.1). For a detailed description of this procedure the reader is referred to Section 2.1.4 in Part 1 of this book series. Denoting the circumferential velocity $\widehat{V}_\phi$ and the pressure $\hat{p}$ in the laminar flow by $\widehat{U}$ and $\widehat{P}$, respectively, we have

$$\widehat{U} = C_1 \hat{r} + \frac{C_2}{\hat{r}}, \qquad (1.7.3a)$$

$$\widehat{P} = \hat{p}_0 + \tfrac{1}{2} \rho C_1^2 \hat{r}^2 + 2\rho C_1 C_2 \ln \hat{r} - \frac{\rho C_2^2}{2\hat{r}^2}. \qquad (1.7.3b)$$

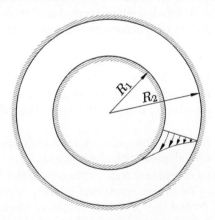

Fig. 1.34: The flow between two coaxial cylinders; cross-section view.

Here C_1 and C_2 are constants given by

$$C_1 = \frac{\Omega_1 R_1^2 - \Omega_2 R_2^2}{R_1^2 - R_2^2}, \qquad C_2 = (\Omega_1 - \Omega_2)\frac{R_1^2 R_2^2}{R_2^2 - R_1^2}, \qquad (1.7.4)$$

with Ω_1 and Ω_2 being the angular velocities of the inner and outer cylinders; R_1 and R_2 denote their radii; see Figure 1.34. In this presentation we shall concentrate our attention on the case where the outer cylinder is kept at rest, $\Omega_2 = 0$, so that

$$C_1 = -\Omega_1 \frac{R_1^2}{R_2^2 - R_1^2}, \qquad C_2 = \Omega_1 \frac{R_1^2 R_2^2}{R_2^2 - R_1^2}. \qquad (1.7.5)$$

The value of the constant $\hat{p}_0$ in the solution for the pressure in (1.7.3) has no significance for the analysis that follows.

1.7.1 Taylor vortices

To study the stability of the laminar flow (1.7.2), (1.7.3) we represent the solution of equations (1.7.1) in the form

$$\left. \begin{array}{ll} \widehat{V}_\phi = \widehat{U}(\hat{r}) + \varepsilon \hat{u}'(\hat{t}, \hat{r}, \hat{z}), & \widehat{V}_r = \varepsilon \hat{v}'(\hat{t}, \hat{r}, \hat{z}), \\[2mm] \widehat{V}_z = \varepsilon \hat{w}'(\hat{t}, \hat{r}, \hat{z}), & \hat{p} = \widehat{P}(\hat{r}) + \varepsilon \hat{p}'(\hat{t}, \hat{r}, \hat{z}). \end{array} \right\} \qquad (1.7.6)$$

Substituting (1.7.6) into (1.7.1) and assuming that the amplitude of the perturbations ε is small, we arrive at the conclusion that the perturbations are governed by the equations

$$\frac{\partial \hat{u}'}{\partial \hat{t}} + \frac{d\widehat{U}}{d\hat{r}}\hat{v}' + \frac{\widehat{U}}{\hat{r}}\hat{v}' = \nu\left(\frac{\partial^2 \hat{u}'}{\partial \hat{z}^2} + \frac{\partial^2 \hat{u}'}{\partial \hat{r}^2} + \frac{1}{\hat{r}}\frac{\partial \hat{u}'}{\partial \hat{r}} - \frac{\hat{u}'}{\hat{r}^2}\right), \qquad (1.7.7\text{a})$$

$$\frac{\partial \hat{v}'}{\partial \hat{t}} - 2\frac{\widehat{U}}{\hat{r}}\hat{u}' = -\frac{1}{\rho}\frac{\partial \hat{p}'}{\partial \hat{r}} + \nu\left(\frac{\partial^2 \hat{v}'}{\partial \hat{z}^2} + \frac{\partial^2 \hat{v}'}{\partial \hat{r}^2} + \frac{1}{\hat{r}}\frac{\partial \hat{v}'}{\partial \hat{r}} - \frac{\hat{v}'}{\hat{r}^2}\right), \qquad (1.7.7\text{b})$$

$$\frac{\partial \hat{w}'}{\partial \hat{t}} = -\frac{1}{\rho}\frac{\partial \hat{p}'}{\partial \hat{z}} + \nu\left(\frac{\partial^2 \hat{w}'}{\partial \hat{z}^2} + \frac{\partial^2 \hat{w}'}{\partial \hat{r}^2} + \frac{1}{\hat{r}}\frac{\partial \hat{w}'}{\partial \hat{r}}\right), \qquad (1.7.7\text{c})$$

$$\frac{\partial \hat{v}'}{\partial \hat{r}} + \frac{\partial \hat{w}'}{\partial \hat{z}} + \frac{\hat{v}'}{\hat{r}} = 0. \qquad (1.7.7\text{d})$$

Narrow gap case

To simplify equations (1.7.7) further we shall assume that the gap between the cylinders, $h = R_2 - R_1$, is small as compared with the radius of the inner cylinder, namely,

$$\frac{h}{R_1} \ll 1. \qquad (1.7.8)$$

It should be noted that in addition to facilitating the theoretical analysis of the flow, assumption (1.7.8) makes the theory directly applicable to the practical task of predicting laminar-turbulent transition in gas bearings.

We shall perform the flow analysis using non-dimensional variables. We start with the basic flow, for which the non-dimensional circumferential velocity U and pressure P are introduced as

$$\widehat{U} = \Omega_1 R_1 U(y), \qquad \widehat{P} = \widehat{P}_0 + \rho \Omega_1^2 h R_1 P(y), \tag{1.7.9}$$

with y being a scaled dimensionless radial coordinate defined by

$$\hat{r} = R_1 + hy. \tag{1.7.10}$$

It may be shown that in the case of a narrow gap[14]

$$U = 1 - y, \qquad P = -\tfrac{1}{3}(1 - y)^3.$$

Now we turn to the perturbation functions. To find an appropriate way to make these functions dimensionless we shall perform an *inspection analysis* of the flow, which is aimed at identifying the terms in equations (1.7.7) 'responsible' for the onset of centrifugal instability.[15] We start by noticing that the unsteady terms $\partial \hat{u}'/\partial \hat{t}$, $\partial \hat{v}'/\partial \hat{t}$, and $\partial \hat{w}'/\partial \hat{t}$ in (1.7.1a)–(1.7.1c) play a key role in determining the stability of the flow. These terms should always be retained in the governing equations. In order to identify other important terms, one needs to look closely at the physical processes leading to the formation of the Taylor vortices. The main reason why a centrifugal instability occurs is that the fluid particles near the inner cylinder carry higher azimuthal momentum. Due to the centrifugal effect they tend to migrate towards the outer cylinder. This migration is described by the second term in the azimuthal momentum equation (1.7.7a). Balancing this against the unsteady term we have

$$\frac{\partial \hat{u}'}{\partial \hat{t}} \sim \frac{d\widehat{U}}{d\hat{r}} \hat{v}'. \tag{1.7.11}$$

Taking into account that in the basic flow $\widehat{U} \sim \Omega_1 R_1$, we can estimate the derivative as

$$\frac{d\widehat{U}}{d\hat{r}} \sim \frac{\Omega_1 R_1}{h}. \tag{1.7.12}$$

Similarly, approximating the derivative on the left-hand side of (1.7.11) by finite differences and using (1.7.12) on the right-hand side, we have

$$\frac{\hat{u}'}{\Delta \hat{t}} \sim \frac{\Omega_1 R_1}{h} \hat{v}'. \tag{1.7.13}$$

It should be further noticed that when the Taylor vortices are formed, the fluid particles acquire, in addition to the azimuthal velocity, $\hat{V}_\phi$, also radial and axial velocities, $\hat{V}_r$ and $\hat{V}_z$. The latter is induced by the pressure gradient $\partial \hat{p}'/\partial \hat{z}$. Indeed, if

---

[14]See Problem 1 in Exercises 6.

[15]Inspection analysis has been used in this book series in a number of places, most notably, in Section 2.2.3 in Part 3, where, through physical arguments, it allowed us to identify the structure of the triple-deck flow.

$\partial \hat{p}'/\partial \hat{z}$ were negligible, then the solution of the axial momentum equation (1.7.7c), satisfying the no-slip conditions

$$\hat{w}'\Big|_{\hat{r}=R_1} = \hat{w}'\Big|_{\hat{r}=R_2} = 0,$$

would be $\hat{w}' \equiv 0$. This suggests that in (1.7.7c) the following balance should hold:

$$\frac{\partial \hat{w}'}{\partial \hat{t}} \sim \frac{1}{\rho}\frac{\partial \hat{p}'}{\partial \hat{z}}. \tag{1.7.14}$$

A non-zero axial pressure gradient $\partial \hat{p}'/\partial \hat{z}$, in its turn, is caused by the centrifugal effect. This is expressed by the following balance in the radial momentum equation (1.7.7b):

$$2\frac{\widehat{U}}{\hat{r}}\hat{u}' \sim \frac{1}{\rho}\frac{\partial \hat{p}'}{\partial \hat{r}}. \tag{1.7.15}$$

Since the unsteady term has to be retained in (1.7.7b), we can also write

$$\frac{\partial \hat{v}'}{\partial \hat{t}} \sim 2\frac{\widehat{U}}{\hat{r}}\hat{u}'. \tag{1.7.16}$$

Finally, in the continuity equation (1.7.7d) the principal balance is between the terms

$$\frac{\partial \hat{v}'}{\partial \hat{r}} \sim \frac{\partial \hat{w}'}{\partial \hat{z}}. \tag{1.7.17}$$

Expressing (1.7.16) in terms of finite differences,[16]

$$\frac{\hat{v}'}{\Delta \hat{t}} \sim \Omega_1 \hat{u}', \tag{1.7.18}$$

and solving (1.7.18) together with (1.7.13) for $\Delta\hat{t}$ and $\hat{v}'$, we find that

$$\Delta\hat{t} \sim \frac{1}{\Omega_1}\sqrt{\frac{h}{R_1}}, \qquad \hat{v}' \sim \sqrt{\frac{h}{R_1}}\,\hat{u}'. \tag{1.7.19}$$

Equations (1.7.14), (1.7.15), and (1.7.17) are written in finite-difference form as

$$\frac{\hat{w}'}{\Delta\hat{t}} \sim \frac{1}{\rho}\frac{\hat{p}'}{\Delta\hat{z}}, \qquad \Omega_1\hat{u}' \sim \frac{1}{\rho}\frac{\hat{p}'}{h}, \qquad \frac{\hat{v}'}{h} \sim \frac{\hat{w}'}{\Delta\hat{z}}.$$

Solving these equations for $\Delta\hat{z}$, $\hat{p}'$, and $\hat{w}'$, we have

$$\Delta\hat{z} \sim h, \qquad \hat{p}' \sim \rho\Omega_1 h\hat{u}', \qquad \hat{w}' \sim \sqrt{\frac{h}{R_1}}\,\hat{u}'. \tag{1.7.20}$$

This concludes the inspection analysis. One can see that $\hat{u}'$ remains arbitrary while the rest of the perturbation functions, $\hat{v}'$ in (1.7.19) as well as $\hat{p}'$ and $\hat{w}'$ in (1.7.20),

---

[16]Here we again use the fact that $\widehat{U} \sim \Omega_1 R_1$.

are proportional to $\hat{u}'$. This is due to the fact that equations (1.7.7) are homogeneous and do not allow us to determine the amplitude of the perturbations. For the purpose of rendering equations (1.7.7) dimensionless, it is convenient to scale $\hat{u}'$ with $\Omega_1 R_1$ as we did for $\widehat{U}$ in the basic flow solution (1.7.9). Then, guided by (1.7.19) and (1.7.20), we introduce the non-dimensional variables as follows:

$$
\left.
\begin{aligned}
&\hat{u}' = \Omega_1 R_1 \, u', \qquad &&\hat{v}' = \Omega_1 \sqrt{R_1 h}\, v', \qquad &&\hat{w}' = \Omega_1 \sqrt{R_1 h}\, w', \\
&\hat{p}' = \rho \Omega_1^2 R_1 h \, p', \qquad &&\hat{t} = \frac{1}{\Omega_1}\sqrt{\frac{h}{R_1}}, \qquad &&\hat{z} = hz.
\end{aligned}
\right\}
\tag{1.7.21}
$$

Substitution of (1.7.21) together with (1.7.9) and (1.7.9) into the equations for the perturbations (1.7.7) transforms them to

$$
\frac{\partial u'}{\partial t} + \frac{dU}{dy} v' + \frac{h}{R_1}\frac{Uv'}{1 + \frac{h}{R_1}y} =
$$
$$
= \frac{1}{Re}\left[\frac{\partial^2 u'}{\partial y^2} + \frac{\partial^2 u'}{\partial z^2} + \frac{h}{R_1}\frac{1}{1 + \frac{h}{R_1}y}\frac{\partial u'}{\partial y} - \frac{h^2}{R_1^2}\frac{u'}{\left(1 + \frac{h}{R_1}y\right)^2} \right],
\tag{1.7.22a}
$$

$$
\frac{\partial v'}{\partial t} - \frac{2Uu'}{1 + \frac{h}{R_1}y} = -\frac{\partial p'}{\partial y} +
$$
$$
+ \frac{1}{Re}\left[\frac{\partial^2 v'}{\partial y^2} + \frac{\partial^2 v'}{\partial z^2} + \frac{h}{R_1}\frac{1}{1 + \frac{h}{R_1}y}\frac{\partial v'}{\partial y} - \frac{h^2}{R_1^2}\frac{v'}{\left(1 + \frac{h}{R_1}y\right)^2} \right],
\tag{1.7.22b}
$$

$$
\frac{\partial w'}{\partial t} = -\frac{\partial p'}{\partial z} + \frac{1}{Re}\left[\frac{\partial^2 w'}{\partial y^2} + \frac{\partial^2 w'}{\partial z^2} + \frac{h}{R_1}\frac{1}{1 + \frac{h}{R_1}y}\frac{\partial w'}{\partial y} \right],
\tag{1.7.22c}
$$

$$
\frac{\partial v'}{\partial y} + \frac{\partial w'}{\partial z} + \frac{h}{R_1}\frac{v'}{1 + \frac{h}{R_1}y} = 0.
\tag{1.7.22d}
$$

Here Re is the Reynolds number given by

$$
Re = \frac{\Omega_1 h \sqrt{R_1 h}}{\nu}.
$$

It is easily seen that in the narrow gap limit $(h/R_1 \to 0)$ equations (1.7.22) reduce to

$$
\frac{\partial u'}{\partial t} + \frac{dU}{dy} v' = \frac{1}{Re}\left(\frac{\partial^2 u'}{\partial y^2} + \frac{\partial^2 u'}{\partial z^2} \right),
\tag{1.7.23a}
$$

$$
\frac{\partial v'}{\partial t} - 2Uu' = -\frac{\partial p'}{\partial y} + \frac{1}{Re}\left(\frac{\partial^2 v'}{\partial y^2} + \frac{\partial^2 v'}{\partial z^2} \right),
\tag{1.7.23b}
$$

$$
\frac{\partial w'}{\partial t} = -\frac{\partial p'}{\partial z} + \frac{1}{Re}\left(\frac{\partial^2 w'}{\partial y^2} + \frac{\partial^2 w'}{\partial z^2} \right),
\tag{1.7.23c}
$$

$$
\frac{\partial v'}{\partial y} + \frac{\partial w'}{\partial z} = 0.
\tag{1.7.23d}
$$

These equations are to be solved subject to the no-slip conditions on the two cylinders,

$$u' = v' = w' = 0 \quad \text{at} \quad y = 0 \text{ and } y = 1. \tag{1.7.24}$$

The normal-mode representation of the perturbations

$$u' = e^{\sigma t + i\beta z}\,\breve{u}(y), \quad v' = e^{\sigma t + i\beta z}\,\breve{v}(y), \quad w' = e^{\sigma t + i\beta z}\,\breve{w}(y), \quad p' = e^{\sigma t + i\beta z}\,\breve{p}(y)$$

turns equations (1.7.23) into

$$\sigma\breve{u} + \frac{dU}{dy}\breve{v} = \frac{1}{Re}\left(\frac{d^2\breve{u}}{dy^2} - \beta^2\breve{u}\right), \tag{1.7.25a}$$

$$\sigma\breve{v} - 2U\breve{u} = -\frac{d\breve{p}}{dy} + \frac{1}{Re}\left(\frac{d^2\breve{v}}{dy^2} - \beta^2\breve{v}\right), \tag{1.7.25b}$$

$$\sigma\breve{w} = -i\beta\breve{p} + \frac{1}{Re}\left(\frac{d^2\breve{w}}{dy^2} - \beta^2\breve{w}\right), \tag{1.7.25c}$$

$$\frac{d\breve{v}}{dy} + i\beta\breve{w} = 0, \tag{1.7.25d}$$

and the boundary conditions (1.7.24) assume the form

$$\breve{u} = \breve{v} = \breve{w} = 0 \quad \text{at} \quad y = 0 \text{ and } y = 1. \tag{1.7.26}$$

The pressure $\breve{p}$ may be eliminated from the set of equations (1.7.25). Indeed, if we multiply (1.7.25b) by $i\beta$ and differentiate (1.7.25c) with respect to y, we will have the following pair of equations:

$$i\beta\sigma\breve{v} - 2i\beta U\breve{u} = -i\beta\frac{d\breve{p}}{dy} + \frac{1}{Re}\left(i\beta\frac{d^2\breve{v}}{dy^2} - i\beta^3\breve{v}\right), \tag{1.7.27}$$

$$\sigma\frac{d\breve{w}}{dy} = -i\beta\frac{d\breve{p}}{dy} + \frac{1}{Re}\left(\frac{d^3\breve{w}}{dy^3} - \beta^2\frac{d\breve{w}}{dy}\right). \tag{1.7.28}$$

Subtracting (1.7.27) from (1.7.28), leads to

$$\sigma\frac{d\breve{w}}{dy} - i\beta\sigma\breve{v} + 2i\beta U\breve{u} = \frac{1}{Re}\left(\frac{d^3\breve{w}}{dy^3} - \beta^2\frac{d\breve{w}}{dy} - i\beta\frac{d^2\breve{v}}{dy^2} + i\beta^3\breve{v}\right). \tag{1.7.29}$$

Now we can eliminate $\breve{w}$ from (1.7.29) with the help of the continuity equation (1.7.25d). Solving (1.7.25d) for $\breve{w}$,

$$\breve{w} = \frac{i}{\beta}\frac{d\breve{v}}{dy}, \tag{1.7.30}$$

and substituting (1.7.30) into (1.7.29), results in

$$\sigma\frac{d^2\breve{v}}{dy^2} - \sigma\beta^2\breve{v} + 2\beta^2 U\breve{u} = \frac{1}{Re}\left(\frac{d^4\breve{v}}{dy^4} - 2\beta^2\frac{d^2\breve{v}}{dy^2} + \beta^4\breve{v}\right). \tag{1.7.31}$$

Equation (1.7.31) involves only two unknowns, $\breve{u}$ and $\breve{v}$, as does equation (1.7.25a). Thus, being considered together

$$\left.\begin{aligned}
\sigma\frac{d^2\breve{v}}{dy^2} - \sigma\beta^2\breve{v} + 2\beta^2 U\breve{u} &= \frac{1}{Re}\left(\frac{d^4\breve{v}}{dy^4} - 2\beta^2\frac{d^2\breve{v}}{dy^2} + \beta^4\breve{v}\right), \\
\sigma\breve{u} + \frac{dU}{dy}\breve{v} &= \frac{1}{Re}\left(\frac{d^2\breve{u}}{dy^2} - \beta^2\breve{u}\right),
\end{aligned}\right\} \tag{1.7.32}$$

they constitute a closed set of equations for $\breve{u}$ and $\breve{v}$. It follows from (1.7.26) and (1.7.30) that the boundary conditions for (1.7.32) may be written as

$$\breve{u} = \breve{v} = \frac{d\breve{v}}{dy} = 0 \quad \text{at} \quad y = 0 \text{ and } y = 1. \tag{1.7.33}$$

The boundary-value problem (1.7.32), (1.7.33) is homogeneous, and admits a trivial solution with $\breve{u} = \breve{v} = 0$ everywhere between the two cylinders. Hence, it should be treated as an eigen-value problem. Given the Reynolds number Re and the wave number β, this boundary-value problem allows us to determine the amplification rate σ. This is a numerical task, and the equations are already written in a convenient way for carrying out calculations. Still, traditionally, one more transformation,

$$\breve{v} = \frac{\tilde{v}}{Re}, \qquad \sigma = \frac{\tilde{\sigma}}{Re},$$

is usually performed to convert (1.7.32), (1.7.33) into

$$\frac{d^4\tilde{v}}{dy^4} - (\tilde{\sigma} + 2\beta^2)\frac{d^2\tilde{v}}{dy^2} + \beta^2(\tilde{\sigma} + \beta^2)\tilde{v} = \beta^2 TU\breve{u}, \tag{1.7.34a}$$

$$\frac{d^2\breve{u}}{dy^2} - (\tilde{\sigma} + \beta^2)\breve{u} = \frac{dU}{dy}\tilde{v}, \tag{1.7.34b}$$

$$\breve{u} = \tilde{v} = \frac{d\tilde{v}}{dy} = 0 \quad \text{at} \quad y = 0 \quad \text{and} \quad y = 1. \tag{1.7.34c}$$

Here

$$T = 2Re^2 = \frac{2\Omega_1^2 R_1 h^3}{\nu^2}$$

is referred to as the *Taylor number*.

The results of the numerical solution of (1.7.34) are depicted in Figure 1.35, where the neutral curve is shown in the (β, T)-plane. Everywhere below this curve, $\sigma < 0$, which means that for any $T < T_c$ the perturbations, introduced into the flow, will decay with time, and the flow will return to its basic state. When the Taylor number reaches the critical value $T = T_c \simeq 3389.9$ with $\beta_c \simeq 3.13$, Taylor vortices of wave length $2\pi/\beta_c$ become possible. In practice, they are triggered by any imperfection (say, wall roughness) in the apparatus. For each $T > T_c$ a band of wave numbers, that lie between the left- and right-hand side branches of the neutral curve, are unstable ($\sigma > 0$). The corresponding perturbations are expected to grow in time, according to the linear theory. However, as the amplitude of the Taylor vortices increases, non-linear effects come into play. They stop growing once they reach a certain amplitude, and then the Taylor vortices appear to remain steady to an observer. For a detailed discussion of this process the reader is referred to Chapter 3.

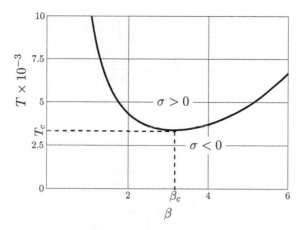

Fig. 1.35: Neutral curve for the flow between two coaxial cylinders. Here $T_c \simeq 3389.9$ and $\beta_c \simeq 3.13$.

Inviscid analysis

In this section we will not restrict ourselves to the narrow gap approximation. Instead we shall consider the general case where h is comparable with the cylinders' radii. We shall also allow both cylinders to rotate. In the context of this more general formulation we shall now exploit the simplifications that arise from the assumption of inviscid disturbances.

Restricting, as before, our attention to the axisymmetric Taylor vortices (see Figure 1.33a), we seek a solution of the form (1.7.6) leading to the disturbance equations (1.7.7) but this time we consider the perturbations to be predominantly inviscid. We take $\nu = 0$ in (1.7.7) so that

$$\frac{\partial \hat{u}'}{\partial \hat{t}} + \left(\frac{d\widehat{U}}{d\hat{r}} + \frac{\widehat{U}}{\hat{r}} \right) \hat{v}' = 0, \tag{1.7.35a}$$

$$\frac{\partial \hat{v}'}{\partial \hat{t}} - \frac{2\widehat{U}}{\hat{r}} \hat{u}' = -\frac{1}{\rho} \frac{\partial \hat{p}'}{\partial \hat{r}}, \tag{1.7.35b}$$

$$\frac{\partial \hat{w}'}{\partial \hat{t}} = -\frac{1}{\rho} \frac{\partial \hat{p}'}{\partial \hat{z}}, \tag{1.7.35c}$$

$$\frac{\partial \hat{v}'}{\partial \hat{r}} + \frac{\partial \hat{w}'}{\partial \hat{z}} + \frac{\hat{v}'}{\hat{r}} = 0. \tag{1.7.35d}$$

The normal-mode representation of the sought functions

$$\left. \begin{array}{ll} \hat{u}' = e^{\sigma \hat{t} + i\beta \hat{z}} \breve{u}(\hat{r}), & \hat{v}' = e^{\sigma \hat{t} + i\beta \hat{z}} \breve{v}(\hat{r}), \\ \hat{w}' = e^{\sigma \hat{t} + i\beta \hat{z}} \breve{w}(\hat{r}), & \hat{p}' = e^{\sigma \hat{t} + i\beta \hat{z}} \breve{p}(\hat{r}) \end{array} \right\} \tag{1.7.36}$$

converts (1.7.35) into

$$\sigma \breve{u} + \left(\frac{d\widehat{U}}{d\hat{r}} + \frac{\widehat{U}}{\hat{r}} \right) \breve{v} = 0, \tag{1.7.37a}$$

$$\sigma \breve{v} - \frac{2\widehat{U}}{\hat{r}} \breve{u} = -\frac{1}{\rho} \frac{d\breve{p}}{d\hat{r}}, \tag{1.7.37b}$$

$$\sigma \breve{w} = -\frac{i\beta}{\rho} \breve{p}, \tag{1.7.37c}$$

$$\frac{d\breve{v}}{d\hat{r}} + \frac{\breve{v}}{\hat{r}} + i\beta \breve{w} = 0. \tag{1.7.37d}$$

The set of equations (1.7.37) may be reduced to a single equation for $\breve{v}$. Indeed, solving (1.7.37d) for $\breve{w}$, and substituting the resulting equation into (1.7.37c), we find that

$$\breve{p} = -\rho \frac{\sigma}{\beta^2} \left(\frac{d\breve{v}}{d\hat{r}} + \frac{\breve{v}}{\hat{r}} \right). \tag{1.7.38}$$

Substitution of (1.7.38) into (1.7.37b) leads to the equation

$$\sigma \breve{v} - \frac{2\widehat{U}}{\hat{r}} \breve{u} = \frac{\sigma}{\beta^2} \left(\frac{d^2\breve{v}}{d\hat{r}^2} + \frac{1}{\hat{r}} \frac{d\breve{v}}{d\hat{r}} - \frac{\breve{v}}{\hat{r}^2} \right). \tag{1.7.39}$$

Finally, eliminating $\breve{u}$ from (1.7.39) and (1.7.37a), we arrive at the conclusion that

$$\frac{d^2\breve{v}}{d\hat{r}^2} + \frac{1}{\hat{r}} \frac{d\breve{v}}{d\hat{r}} - \left(\beta^2 + \frac{1}{\hat{r}^2} \right) \breve{v} = \lambda \Phi \breve{v}, \tag{1.7.40}$$

where

$$\Phi = \frac{1}{\hat{r}^3} \frac{d}{d\hat{r}} \left[(\hat{r}\widehat{U})^2 \right], \qquad \lambda = \frac{\beta^2}{\sigma^2}.$$

Since we are dealing with an inviscid disturbance, the boundary conditions for the equation (1.7.40) have to be deduced from the impermeability condition on the two cylinders,

$$\widehat{V}_r \Big|_{\hat{r}=R_1} = \widehat{V}_r \Big|_{\hat{r}=R_2} = 0. \tag{1.7.41}$$

According to (1.7.6) and (1.7.36),

$$\widehat{V}_r = \varepsilon e^{\sigma \hat{t} + i\beta \hat{z}} \breve{v}(\hat{r}),$$

which being substituted into (1.7.41), yields

$$\breve{v} \Big|_{\hat{r}=R_1} = \breve{v} \Big|_{\hat{r}=R_2} = 0. \tag{1.7.42}$$

Equation (1.7.40) subject to boundary conditions (1.7.42) constitute an eigen-value problem of the Sturm–Liouville type. Its properties are described by the following theorem:

Theorem 1.6 *The eigen-values, λ, of the problem (1.7.40), (1.7.42) are all negative if $\Phi > 0$ in the entire interval $\hat{r} \in [R_1, R_2]$, and they are all positive if $\Phi < 0$ throughout the interval $[R_1, R_2]$.*

A way to prove this theorem is outlined in Problem 2, Exercises 6.

When applying Theorem 1.6 to the flow between two cylinders, one needs to calculate the function Φ using the basic flow solution (1.7.3a), (1.7.4). Assuming that the outer cylinder is kept at rest ($\Omega_2 = 0$), and substituting (1.7.5) into (1.7.3a), we find that

$$\widehat{U} = \Omega_1 \frac{R_1^2 R_2}{R_2^2 - R_1^2} \left(\frac{R_2}{\hat{r}} - \frac{\hat{r}}{R_2} \right). \tag{1.7.43}$$

Using (1.7.43), the function Φ is calculated as

$$\Phi = \frac{1}{\hat{r}^3} \frac{d}{d\hat{r}} \left[(\hat{r}\widehat{U})^2 \right] = 2\frac{\widehat{U}}{\hat{r}^2} \frac{d}{d\hat{r}} (\hat{r}\widehat{U}) = -4\Omega_1^2 \frac{R_1^4}{(R_2^2 - R_1^2)^2} \left(\frac{R_2}{\hat{r}} - \frac{\hat{r}}{R_2} \right) \frac{R_2}{\hat{r}}.$$

Since everywhere between the cylinders, $R_2/\hat{r} > 1$, it follows that the function Φ is negative, and hence the eigen-value problem (1.7.40), (1.7.42) admits solutions with $\lambda = \beta^2/\sigma^2 > 0$. The corresponding values of σ are therefore real and come in pairs, one positive and one negative. Thus, we can conclude that in this case the flow is unstable.

If however the inner cylinder is motionless, and the outer cylinder rotates with angular velocity Ω_2, then

$$\Phi = 4\Omega_2^2 \frac{R_2^4}{(R_2^2 - R_1^2)^2} \left(\frac{\hat{r}}{R_1} - \frac{R_1}{\hat{r}} \right) \frac{R_1}{\hat{r}}.$$

Bearing in mind that everywhere between the cylinders, $\hat{r}/R_1 > 1$, we can see that Φ now stays positive throughout the flow field. In this case the eigen-value problem (1.7.40), (1.7.42) only admits solutions with $\lambda = \beta^2/\sigma^2 < 0$. Consequently, σ is imaginary, which (in the framework of inviscid stability theory) corresponds to a stable flow.

Rayleigh's physical arguments

To determine the conditions under which Taylor vortices can form, Rayleigh (1916) used the principle of minimum energy. According to this principle, a physical system always tends to minimize its energy. For the inviscid fluid flow considered here, the principle may be expressed in the form of the following question. What are the conditions under which the kinetic energy of the fluid undergoing laminar circumferential motion between two cylinders can be released and converted into the energy of Taylor vortices? To answer this question, Rayleigh considered two fluid rings lying in the planes perpendicular to the axis of rotation (see Figure 1.36), and considered what would happen if the rings exchanged their positions. The task was to see whether this would lead to a reduction in the energy of the laminar flow.

Let $\hat{r}_1$ be the radius of the first ring, $\widehat{V}_{\phi 1}$ the azimuthal velocity of the fluid in the ring, and τ_1 its volume. Then the kinetic energy of the ring is calculated as

$$E_1 = \tfrac{1}{2}\rho\tau_1 \widehat{V}_{\phi 1}^2.$$

Similarly, for the second ring we have,

$$E_2 = \tfrac{1}{2}\rho\tau_2 \widehat{V}_{\phi 2}^2.$$

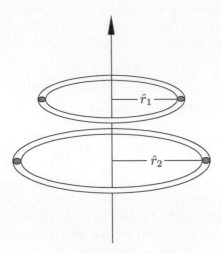

Fig. 1.36: Two fluid rings.

Thus, the total kinetic energy of the fluid in the two rings is given by

$$E = \tfrac{1}{2}\rho\tau_1\widehat{V}_{\phi 1}^2 + \tfrac{1}{2}\rho\tau_2\widehat{V}_{\phi 2}^2. \qquad (1.7.44)$$

Let us now see what happens if the radius of a ring changes with time. Assuming the flow to be inviscid, we set $\nu = 0$ in the circumferential momentum equation (1.7.1a). We have

$$\frac{\partial\widehat{V}_\phi}{\partial\hat{t}} + \widehat{V}_r\frac{\partial\widehat{V}_\phi}{\partial\hat{r}} + \widehat{V}_z\frac{\partial\widehat{V}_\phi}{\partial\hat{z}} + \frac{\widehat{V}_r\widehat{V}_\phi}{\hat{r}} = 0.$$

Multiplying all the terms in this equation by $\hat{r}$, we have

$$\frac{\partial}{\partial\hat{t}}\left(\hat{r}\widehat{V}_\phi\right) + \widehat{V}_r\frac{\partial}{\partial\hat{r}}\left(\hat{r}\widehat{V}_\phi\right) + \widehat{V}_z\frac{\partial}{\partial\hat{z}}\left(\hat{r}\widehat{V}_\phi\right) = 0. \qquad (1.7.45)$$

The expression on the left-hand side of (1.7.45) represents the total derivative of $\hat{r}\widehat{V}_\phi$ for an axisymmetric flow. As it is zero, we can conclude that $\hat{r}\widehat{V}_\phi$ remains constant for each ring. Consequently, if we denote by $\widehat{V}'_{\phi 1}$ the fluid velocity in the first ring after the exchange of the rings is completed, then we will have

$$\hat{r}_1\widehat{V}_{\phi 1} = \hat{r}_2\widehat{V}'_{\phi 1}.$$

The kinetic energy of the first ring becomes

$$E'_1 = \tfrac{1}{2}\rho\tau_1\widehat{V}'^2_{\phi 1} = \tfrac{1}{2}\rho\tau_1\frac{\hat{r}_1^2}{\hat{r}_2^2}\widehat{V}_{\phi 1}^2.$$

Similarly, the new velocity of the fluid in the second ring, $\widehat{V}'_{\phi 2}$, is calculated from the equation

$$\hat{r}_2\widehat{V}_{\phi 2} = \hat{r}_1\widehat{V}'_{\phi 2},$$

and its kinetic energy will be

$$E_2' = \tfrac{1}{2}\rho\tau_2\widehat{V}_{\phi2}'^2 = \tfrac{1}{2}\rho\tau_2\frac{\hat{r}_2^2}{\hat{r}_1^2}\widehat{V}_{\phi2}^2.$$

Consequently, for the total energy of the two rings we have

$$E' = E_1' + E_2' = \tfrac{1}{2}\rho\tau_1\frac{\hat{r}_1^2}{\hat{r}_2^2}\widehat{V}_{\phi1}^2 + \tfrac{1}{2}\rho\tau_2\frac{\hat{r}_2^2}{\hat{r}_1^2}\widehat{V}_{\phi2}^2. \qquad (1.7.46)$$

Let us now note that, due to the mass conservation law, one ring can displace the other only if they have the same volume, $\tau_1 = \tau_2$. Keeping this in mind, and using (1.7.44) and (1.7.46), we find that the energy of the two rings changes by the amount

$$\Delta E = E' - E = \tfrac{1}{2}\rho\tau_1\left(\frac{\hat{r}_1^2}{\hat{r}_2^2}\widehat{V}_{\phi1}^2 - \widehat{V}_{\phi1}^2 + \frac{\hat{r}_2^2}{\hat{r}_1^2}\widehat{V}_{\phi2}^2 - \widehat{V}_{\phi2}^2\right)$$

$$= \tfrac{1}{2}\rho\left[(\hat{r}_2\widehat{V}_{\phi2})^2 - (\hat{r}_1\widehat{V}_{\phi1})^2\right]\left(\frac{1}{\hat{r}_1^2} - \frac{1}{\hat{r}_2^2}\right)\tau_1. \qquad (1.7.47)$$

Without loss of generality we can assume that $\hat{r}_1 < \hat{r}_2$. Then it follows from (1.7.47) that the rings' exchange results in a release of energy if $(\hat{r}_2\widehat{V}_{\phi2})^2 < (\hat{r}_1\widehat{V}_{\phi1})^2$. In differential form this requirement may be written as

$$\frac{d}{d\hat{r}}\left[(\hat{r}\widehat{V}_\phi)^2\right] < 0,$$

in accordance with Theorem 1.6.

1.7.2 Görtler vortices

Another important example where the flow is subject to centrifugal instability is the boundary layer on a concave wall, including the lower side of an aircraft wing and the flow near the wing flaps. Here we shall consider a cylindrical surface S, the cross-section of which is shown in Figure 1.37 with the generatrix being perpendicular to the sketch plane. For simplicity, we shall assume that the oncoming flow is oriented in such a way that the laminar basic flow in the boundary layer on surface S is two-dimensional.

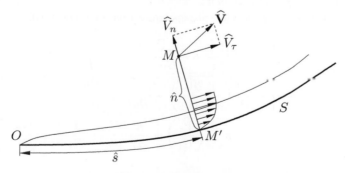

Fig. 1.37: Flow layout; side view.

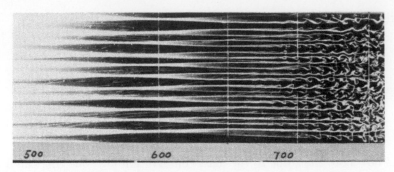

Fig. 1.38: Visualization of Görtler vortices; top view (Nakayama, 1988).

If we think of the outer edge of the boundary layer as an inner cylinder, and the body surface S as an outer cylinder, then by analogy with the flow in Figure 1.34, we should expect a centrifugal instability to develop. Indeed, experimental observations confirm this expectation, as Figure 1.38 shows. We see that the boundary layer loses stability some distance from the leading edge, which results in the formation of steady longitudinal vortices in the boundary layer, known as *Görtler vortices*. These grow downstream ultimately leading to laminar-turbulent transition. Here, we shall focus on the initial linear stage of this process.

To describe the flow it is convenient to use body-fitted coordinates. For a detailed description of these coordinates, the reader is referred to Section 1.8 in Part 1 of this book series. When dealing with such coordinates, one starts by choosing the coordinate origin O on the body contour, say, at the leading edge. Then the position of any point M in the flow field is defined by the distance $\hat{s}$ measured along the body contour from O to point M' which is obtained by dropping the perpendicular from M onto the body surface. The second coordinate $\hat{n}$ is the distance between M and the body surface. The spanwise coordinate $\hat{z}$ is measured along the surface generatrix, i.e. perpendicular to the plane of the sketch in Figure 1.37.

The unit vector triad $(\mathbf{e}_1, \mathbf{e}_2, \mathbf{e}_3)$ is oriented at point M such that $\mathbf{e}_1$ is parallel to the tangent to the body contour drawn through point M'; $\mathbf{e}_2$ is directed along the $\hat{n}$-axis, and $\mathbf{e}_3$ along the body generatrix. The velocity vector at point M is decomposed as $\widehat{\mathbf{V}} = \widehat{V}_\tau \mathbf{e}_1 + \widehat{V}_n \mathbf{e}_2 + \widehat{V}_z \mathbf{e}_3$, with $\widehat{V}_\tau$ termed the tangential velocity, $\widehat{V}_n$ the normal velocity, and $\widehat{V}_z$ the spanwise velocity.

In these coordinates the Navier–Stokes equations have the form[17]

$$
\frac{\partial \widehat{V}_\tau}{\partial \hat{t}} + \frac{\widehat{V}_\tau}{H_1}\frac{\partial \widehat{V}_\tau}{\partial \hat{s}} + \widehat{V}_n \frac{\partial \widehat{V}_\tau}{\partial \hat{n}} + \widehat{V}_z \frac{\partial \widehat{V}_\tau}{\partial \hat{z}} + \frac{\kappa \widehat{V}_\tau \widehat{V}_n}{H_1} = f_\tau - \frac{1}{\rho H_1}\frac{\partial \hat{p}}{\partial \hat{s}} +
$$
$$
+ \nu \left[\frac{1}{H_1}\frac{\partial}{\partial \hat{s}}\left(\frac{1}{H_1}\frac{\partial \widehat{V}_\tau}{\partial \hat{s}} \right) + \frac{\partial^2 \widehat{V}_\tau}{\partial \hat{n}^2} + \frac{\partial^2 \widehat{V}_\tau}{\partial \hat{z}^2} + \right.
$$
$$
\left. + \kappa \frac{\partial}{\partial \hat{n}}\left(\frac{\widehat{V}_\tau}{H_1} \right) + \frac{\kappa}{H_1^2}\frac{\partial \widehat{V}_n}{\partial \hat{s}} + \frac{1}{H_1}\frac{\partial}{\partial \hat{s}}\left(\frac{\kappa \widehat{V}_n}{H_1} \right) \right], \quad (1.7.48a)
$$

---

[17]See equations (1.8.49) in Part 1 of this book series.

$$\frac{\partial \widehat{V}_n}{\partial \hat{t}} + \frac{\widehat{V}_\tau}{H_1} \frac{\partial \widehat{V}_n}{\partial \hat{s}} + \widehat{V}_n \frac{\partial \widehat{V}_n}{\partial \hat{n}} + \widehat{V}_z \frac{\partial \widehat{V}_n}{\partial \hat{z}} - \frac{\kappa \widehat{V}_\tau^2}{H_1} = f_n - \frac{1}{\rho} \frac{\partial \hat{p}}{\partial \hat{n}} +$$

$$+ \nu \left[\frac{1}{H_1} \frac{\partial}{\partial \hat{s}} \left(\frac{1}{H_1} \frac{\partial \widehat{V}_n}{\partial \hat{s}} \right) + \frac{\partial^2 \widehat{V}_n}{\partial \hat{n}^2} + \frac{\partial^2 \widehat{V}_n}{\partial \hat{z}^2} + \right.$$

$$\left. + \kappa \frac{\partial}{\partial \hat{n}} \left(\frac{\widehat{V}_n}{H_1} \right) - \frac{\kappa}{H_1^2} \frac{\partial \widehat{V}_\tau}{\partial \hat{s}} - \frac{1}{H_1} \frac{\partial}{\partial \hat{s}} \left(\frac{\kappa \widehat{V}_\tau}{H_1} \right) \right], \qquad (1.7.48b)$$

$$\frac{\partial \widehat{V}_z}{\partial \hat{t}} + \frac{\widehat{V}_\tau}{H_1} \frac{\partial \widehat{V}_z}{\partial \hat{s}} + \widehat{V}_n \frac{\partial \widehat{V}_z}{\partial \hat{n}} + \widehat{V}_z \frac{\partial \widehat{V}_z}{\partial \hat{z}} = f_z - \frac{1}{\rho} \frac{\partial \hat{p}}{\partial \hat{z}} +$$

$$+ \nu \left[\frac{1}{H_1} \frac{\partial}{\partial \hat{s}} \left(\frac{1}{H_1} \frac{\partial \widehat{V}_z}{\partial \hat{s}} \right) + \frac{\partial^2 \widehat{V}_z}{\partial \hat{n}^2} + \frac{\partial^2 \widehat{V}_z}{\partial \hat{z}^2} + \frac{\kappa}{H_1} \frac{\partial \widehat{V}_z}{\partial \hat{n}} \right], \qquad (1.7.48c)$$

$$\frac{1}{H_1} \frac{\partial \widehat{V}_\tau}{\partial \hat{s}} + \frac{\partial \widehat{V}_n}{\partial \hat{n}} + \frac{\partial \widehat{V}_z}{\partial \hat{z}} + \frac{\kappa \widehat{V}_n}{H_1} = 0. \qquad (1.7.48d)$$

Here H_1 is the Lamé coefficient,

$$H_1 = 1 + \kappa(\hat{s})\hat{n},$$

with $\kappa(\hat{s})$ being the curvature of the body surface; it is positive for a convex wall, and negative for a concave wall, such as the one in Figure 1.37. With R being the local radius of the body contour, the curvature is calculated as

$$\kappa = \mathrm{sgn}(\kappa) \frac{1}{R}. \qquad (1.7.49)$$

Experimentally observed Görtler vortices are usually steady. Keeping this in mind we shall disregard the time derivatives in the Navier–Stokes equations (1.7.48). We shall also neglect the body force **f**, and restrict our attention to the inviscid stability of the flow by setting $\nu = 0$. Under these assumptions, (1.7.48) reduce to

$$\frac{\widehat{V}_\tau}{H_1} \frac{\partial \widehat{V}_\tau}{\partial \hat{s}} + \widehat{V}_n \frac{\partial \widehat{V}_\tau}{\partial \hat{n}} + \widehat{V}_z \frac{\partial \widehat{V}_\tau}{\partial \hat{z}} + \frac{\kappa \widehat{V}_\tau \widehat{V}_n}{H_1} = -\frac{1}{\rho H_1} \frac{\partial \hat{p}}{\partial \hat{s}}, \qquad (1.7.50a)$$

$$\frac{\widehat{V}_\tau}{H_1} \frac{\partial \widehat{V}_n}{\partial \hat{s}} + \widehat{V}_n \frac{\partial \widehat{V}_n}{\partial \hat{n}} + \widehat{V}_z \frac{\partial \widehat{V}_n}{\partial \hat{z}} - \frac{\kappa \widehat{V}_\tau^2}{H_1} = -\frac{1}{\rho} \frac{\partial \hat{p}}{\partial \hat{n}}, \qquad (1.7.50b)$$

$$\frac{\widehat{V}_\tau}{H_1} \frac{\partial \widehat{V}_z}{\partial \hat{s}} + \widehat{V}_n \frac{\partial \widehat{V}_z}{\partial \hat{n}} + \widehat{V}_z \frac{\partial \widehat{V}_z}{\partial \hat{z}} = -\frac{1}{\rho} \frac{\partial \hat{p}}{\partial \hat{z}}, \qquad (1.7.50c)$$

$$\frac{1}{H_1} \frac{\partial \widehat{V}_\tau}{\partial \hat{s}} + \frac{\partial \widehat{V}_n}{\partial \hat{n}} + \frac{\partial \widehat{V}_z}{\partial \hat{z}} + \frac{\kappa \widehat{V}_n}{H_1} = 0. \qquad (1.7.50d)$$

Inspection analysis

Let us consider a position on the body surface at a distance L from the leading edge O. The thickness of the boundary layer at this position is estimated as

$$\delta \sim L Re^{-1/2}.$$

Here the Reynolds number

$$Re = \frac{U_e L}{\nu}$$

is calculated using the local value of the velocity, U_e, at the outer edge of the bound-ary layer. Since the Görtler vortices are confined to the boundary layer, the normal coordinate $\hat{n} \sim \delta$. Hence, in all physically realistic situations, where $\delta \ll R$, the Lamé coefficient

$$H_1 = 1 + \text{sgn}(\kappa)\frac{\hat{n}}{R} = 1 + \cdots.$$

The Görtler vortices are formed as a result of the penetration of high energy fluid particles (initially situated near the outer edge of the boundary layer) into the region closer to the wall, where the kinetic energy of the fluid particles is low. This process is described by the second term on the left-hand side of the longitudinal momentum equation (1.7.50a). The first term in (1.7.50a) is 'responsible' for the spatial growth/decay of the vortices. Therefore, instead of (1.7.11), we shall now write

$$\frac{\widehat{V}_\tau}{H_1}\frac{\partial \widehat{V}_\tau}{\partial \hat{s}} \sim \widehat{V}_n \frac{\partial \widehat{V}_\tau}{\partial \hat{n}}. \tag{1.7.51}$$

Keeping in mind that in the boundary layer

$$\widehat{V}_\tau \sim U_e, \tag{1.7.52}$$

and approximating the derivatives in (1.7.51) by finite differences, we have

$$\frac{U_e}{\Delta \hat{s}} \sim \frac{\widehat{V}_n}{\delta}. \tag{1.7.53}$$

As for the case of Taylor vortices, the spanwise velocity $\widehat{V}_z$ can only be generated by a corresponding pressure gradient, which means that the principal balance in (1.7.50c) is

$$\frac{\widehat{V}_\tau}{H_1}\frac{\partial \widehat{V}_z}{\partial \hat{s}} \sim \frac{1}{\rho}\frac{\partial \hat{p}}{\partial \hat{z}},$$

which is written in finite-difference form as

$$U_e \frac{\widehat{V}_z}{\Delta \hat{s}} \sim \frac{1}{\rho}\frac{\Delta \hat{p}}{\Delta \hat{z}}. \tag{1.7.54}$$

The pressure perturbations associated with the Görtler vortices are caused by the centrifugal effect, which means that the last term on the left-hand side of equation (1.7.50b) has to be in balance with the normal pressure gradient on the right-hand side:

$$\frac{\kappa \widehat{V}_\tau^2}{H_1} \sim \frac{1}{\rho}\frac{\partial \hat{p}}{\partial \hat{n}}. \tag{1.7.55}$$

Using (1.7.52) and (1.7.49), we can write (1.7.55) as

$$\frac{U_e^2}{R} \sim \frac{1}{\rho}\frac{\Delta \hat{p}}{\delta}. \tag{1.7.56}$$

Another principle balance in the normal momentum equation (1.7.50b) is between the first term, describing growth/decay of the perturbations with distance $\hat{s}$ along the body contour, and the centrifugal term:

$$\frac{\widehat{V}_\tau}{H_1}\frac{\partial \widehat{V}_n}{\partial \hat{s}} \sim \frac{\kappa \widehat{V}_\tau^2}{H_1}. \tag{1.7.57}$$

Using again the fact that $\widehat{V}_\tau \sim U_e$, we can deduce from (1.7.57) that

$$\frac{\widehat{V}_n}{\Delta \hat{s}} \sim \frac{U_e}{R}. \tag{1.7.58}$$

It remains to analyse the continuity equation (1.7.50d). We shall require that

$$\frac{1}{H_1}\frac{\partial \widehat{V}_\tau}{\partial \hat{s}} \sim \frac{\partial \widehat{V}_z}{\partial \hat{z}}.$$

Representing the derivatives in this equation by finite differences, we have

$$\frac{U_e}{\Delta \hat{s}} \sim \frac{\widehat{V}_z}{\Delta \hat{z}}. \tag{1.7.59}$$

Equations (1.7.53), (1.7.54), (1.7.56), (1.7.58), and (1.7.59), considered together, form a closed set of order-of-magnitude equations for five unknown quantities: $\Delta \hat{s}$, $\Delta \hat{z}$, $\widehat{V}_n$, $\widehat{V}_z$, and $\Delta \hat{p}$. Solving these equations we find that

$$\Delta \hat{s} \sim \sqrt{R\delta}, \quad \Delta \hat{z} \sim \delta, \quad \widehat{V}_n \sim U_e\sqrt{\frac{\delta}{R}}, \quad \widehat{V}_z \sim U_e\sqrt{\frac{\delta}{R}}, \quad \Delta \hat{p} \sim \rho U_e^2 \frac{\delta}{R}.$$

Correspondingly, the non-dimensional variables are introduced as

$$\left. \begin{aligned} \widehat{V}_\tau = U_e u, \quad \widehat{V}_n = U_e\sqrt{\frac{\delta^*}{R}}\,v, \quad \widehat{V}_z = U_e\sqrt{\frac{\delta^*}{R}}\,w, \quad \hat{p} = \hat{p}_0 + \rho U_e^2 \frac{\delta^*}{R}\,p, \\ \hat{s} = \hat{s}_0 + \sqrt{R\delta^*}\,x, \quad \hat{n} = \delta^* y, \quad \hat{z} = \delta^* z. \end{aligned} \right\} \tag{1.7.60}$$

Here U_e and $\hat{p}_0$ are the inviscid flow velocity and pressure at the outer edge of the boundary layer, δ^* is the displacement thickness and R is the local radius of curvature of the body contour S. In what follows, we shall make use of the quasi-parallel approximation, which holds provided that the characteristic length scale $\Delta \hat{s} \sim \sqrt{R\delta}$ associated with the Görtler vortices' growth is small compared to the distance L from the leading edge O to the point of observation M'; see Figure 1.37. This requirement is written as

$$\sqrt{R\delta} \ll L.$$

Under this condition, U_e, δ^*, and R may be treated as independent of x. Substitution of (1.7.60) into the Euler equations (1.7.50) results in

$$\frac{u}{H_1}\frac{\partial u}{\partial x} + v\frac{\partial u}{\partial y} + w\frac{\partial u}{\partial z} + \frac{\delta^*}{R}\mathrm{sgn}(\kappa)\frac{uv}{H_1} = -\frac{\delta^*}{R}\frac{1}{H_1}\frac{\partial p}{\partial x}, \tag{1.7.61a}$$

$$\frac{u}{H_1}\frac{\partial v}{\partial x} + v\frac{\partial v}{\partial y} + w\frac{\partial v}{\partial z} - \mathrm{sgn}(\kappa)\frac{u^2}{H_1} = -\frac{\partial p}{\partial y}, \tag{1.7.61b}$$

$$\frac{u}{H_1}\frac{\partial w}{\partial x} + v\frac{\partial w}{\partial y} + w\frac{\partial w}{\partial z} = -\frac{\partial p}{\partial z}, \tag{1.7.61c}$$

$$\frac{1}{H_1}\frac{\partial u}{\partial x} + \frac{\partial v}{\partial y} + \frac{\partial w}{\partial z} + \frac{\delta^*}{R}\mathrm{sgn}(\kappa)\frac{v}{H_1} = 0, \tag{1.7.61d}$$

with the Lamé coefficient

$$H_1 = 1 + \frac{\delta^*}{R}\mathrm{sgn}(\kappa)y.$$

Setting $\delta^*/R \to 0$ allows us to reduce (1.7.61) to

$$u\frac{\partial u}{\partial x} + v\frac{\partial u}{\partial y} + w\frac{\partial u}{\partial z} = 0, \tag{1.7.62a}$$

$$u\frac{\partial v}{\partial x} + v\frac{\partial v}{\partial y} + w\frac{\partial v}{\partial z} - \mathrm{sgn}(\kappa)u^2 = -\frac{\partial p}{\partial y}, \tag{1.7.62b}$$

$$u\frac{\partial w}{\partial x} + v\frac{\partial w}{\partial y} + w\frac{\partial w}{\partial z} = -\frac{\partial p}{\partial z}, \tag{1.7.62c}$$

$$\frac{\partial u}{\partial x} + \frac{\partial v}{\partial y} + \frac{\partial w}{\partial z} = 0. \tag{1.7.62d}$$

Linear stability analysis

Let $U(y)$ be the longitudinal velocity profile in the boundary layer at the chosen position on the body surface. To study the stability of the flow, we shall seek the solution of (1.7.62) in the form

$$u = U(y) + \varepsilon u', \quad v = \varepsilon v', \quad w = \varepsilon w', \quad p = P(y) + \varepsilon p', \tag{1.7.63}$$

where the amplitude of the perturbations ε is presumed small. Substituting (1.7.63) into (1.7.62) and working with the $O(\varepsilon)$ terms, we arrive at the following set of linear equations

$$U\frac{\partial u'}{\partial x} + \frac{dU}{dy}v' = 0, \tag{1.7.64a}$$

$$U\frac{\partial v'}{\partial x} - 2\,\mathrm{sgn}(\kappa)Uu' = -\frac{\partial p'}{\partial y}, \tag{1.7.64b}$$

$$U\frac{\partial w'}{\partial x} = -\frac{\partial p'}{\partial z}, \tag{1.7.64c}$$

$$\frac{\partial u'}{\partial x} + \frac{\partial v'}{\partial y} + \frac{\partial w'}{\partial z} = 0. \tag{1.7.64d}$$

We further introduce the normal-mode representation of the perturbations

$$u' = e^{\sigma x + i\beta z}\breve{u}(y), \qquad v' = e^{\sigma x + i\beta z}\breve{v}(y), \Bigg\}$$
$$w' = e^{\sigma x + i\beta z}\breve{w}(y), \qquad p' = e^{\sigma x + i\beta z}\breve{p}(y), \Bigg\} \qquad (1.7.65)$$

with real σ and β. Here, it is taken into account that the Görtler vortices are steady, periodic in the spanwise direction, and may grow/decay in x. Substitution of (1.7.65) into (1.7.64) results in

$$\sigma U \breve{u} + \frac{dU}{dy}\breve{v} = 0, \qquad (1.7.66a)$$

$$\sigma U \breve{v} - 2\,\mathrm{sgn}(\kappa) U \breve{u} = -\frac{d\breve{p}}{dy}, \qquad (1.7.66b)$$

$$\sigma U \breve{w} = -i\beta\breve{p}, \qquad (1.7.66c)$$

$$\sigma \breve{u} + \frac{d\breve{v}}{dy} + i\beta\breve{w} = 0. \qquad (1.7.66d)$$

We shall now show that the set of equations (1.7.66) may be reduced to a single equation for $\breve{v}$. We start by eliminating the function $\breve{w}$. It follows from (1.7.66d) that

$$\breve{w} = \frac{i\sigma}{\beta}\breve{u} + \frac{i}{\beta}\frac{d\breve{v}}{dy},$$

which, after being substituted into (1.7.66c), yields

$$\breve{p} = -\frac{\sigma^2}{\beta^2}U\breve{u} - \frac{\sigma}{\beta^2}U\frac{d\breve{v}}{dy}. \qquad (1.7.67)$$

It follows from (1.7.66a) that

$$U\breve{u} = -\frac{1}{\sigma}\frac{dU}{dy}\breve{v}. \qquad (1.7.68)$$

Substitution of (1.7.68) into (1.7.67) allows us to express $\breve{p}$ in terms of $\breve{v}$ and its derivative $d\breve{v}/dy$:

$$\breve{p} = \frac{\sigma}{\beta^2}\left(\frac{dU}{dy}\breve{v} - U\frac{d\breve{v}}{dy}\right). \qquad (1.7.69)$$

It remains to substitute (1.7.69) and (1.7.68) into (1.7.66b). We find that the sought equation for $\breve{v}$ has the form

$$U\frac{d^2\breve{v}}{dy^2} - \left(\frac{d^2U}{dy^2} + 2\frac{\beta^2}{\sigma^2}\mathrm{sgn}(\kappa)\frac{dU}{dy} + \beta^2 U\right)\breve{v} = 0. \qquad (1.7.70a)$$

It should be solved subject to the impermeability condition on the body surface

$$\breve{v} = 0 \quad \text{at} \quad y = 0, \qquad (1.7.70b)$$

and the disturbance attenuation condition at the outer edge of the boundary layer

$$\breve{v} \to 0 \quad \text{as} \quad y \to \infty. \qquad (1.7.70c)$$

Remarkably, the solution of the eigen-value problem (1.7.70) may be constructed analytically for an arbitrary basic velocity $U(y)$; see Denier *et al.* (1991). We seek the solution in the form

$$\breve{v} = U(y)\Phi(y). \tag{1.7.71}$$

Substitution of (1.7.71) into (1.7.70a) results in

$$U^2\frac{d^2\Phi}{dy^2} + 2U\frac{dU}{dy}\frac{d\Phi}{dy} = \beta^2 U^2\Phi + 2\frac{\beta^2}{\sigma^2}\mathrm{sign}(\kappa)U\frac{dU}{dy}\Phi. \tag{1.7.72}$$

Clearly, equation (1.7.72) is satisfied if

$$\frac{d^2\Phi}{dy^2} = \beta^2\Phi \tag{1.7.73}$$

and

$$\frac{d\Phi}{dy} = \frac{\beta^2}{\sigma^2}\mathrm{sgn}(\kappa)\Phi. \tag{1.7.74}$$

The general solution of equation (1.7.73) is written as

$$\Phi = C_1 e^{\beta y} + C_2 e^{-\beta y}.$$

Without loss of generality we can assume that the spanwise wave number β is positive. Then to satisfy the attenuation condition (1.7.70c) we have to set $C_1 = 0$, and we can write

$$\Phi = C_2 e^{-\beta y}. \tag{1.7.75}$$

Substituting (1.7.75) back into (1.7.71), we have

$$\breve{v} = C_2 U(y)e^{-\beta y}. \tag{1.7.76}$$

Since the basic flow velocity $U(y)$ satisfies the no-slip condition on the body surface, it is clear that for (1.7.76) the boundary condition (1.7.70b) is satisfied automatically.

It remains to substitute (1.7.75) into (1.7.74), which yields

$$\sigma^2 = -\beta\,\mathrm{sgn}(\kappa). \tag{1.7.77}$$

Thus, if $\mathrm{sgn}(\kappa)$ is negative (which corresponds to a concave wall), then the two solutions of (1.7.77) are real,

$$\sigma_{1,2} = \pm\sqrt{\beta},$$

with the positive solution ($\sigma = \sqrt{\beta}$) representing growing perturbations. In particular, the velocity component normal to the wall is written as

$$v' = e^{\sigma x + i\beta z}\breve{v}(y) = C_2 e^{\sqrt{\beta}x + i\beta z}e^{-\beta y}.$$

If, on the other hand, $\mathrm{sgn}(\kappa)$ is positive (the case of a convex wall), then the two solutions of (1.7.77) are imaginary,

$$\sigma_{1,2} = \pm i\sqrt{\beta},$$

which, in the framework of inviscid stability theory, corresponds to a stable flow.

Exercises 6

1. Consider the motion of a fluid contained between two coaxial cylinders of radii R_1 and R_2; see Figure 1.34. Assume that the outer cylinder is motionless, while the inner cylinder rotates with angular velocity Ω_1. Assume further that the gap between the cylinders, $h = R_2 - R_1$, is small compared with the radius of the inner cylinder, namely, $h/R_1 \ll 1$.

   You task is to analyse the basic laminar flow between the cylinders, and to show that in the narrow gap case, the circumferential velocity $\widehat{U}$ and the pressure $\widehat{P}$ can be written as

   $$\widehat{U} = \Omega_1 R_1 (1 - y), \qquad \widehat{P} = \widehat{P}_0 - \tfrac{1}{3}\rho\Omega_1^2 h R_1 (1 - y)^3, \qquad (1.7.78)$$

   where y is the scaled radial coordinate defined by

   $$\hat{r} = R_1 + hy. \qquad (1.7.79)$$

   Suggestion: You may use without derivation the fact that in the general case, $\widehat{V}_\phi$ is given by (1.7.3a), (1.7.5). Perform your analysis according to the following steps:

   (a) Express the radius of the outer cylinder in the form $R_2 = R_1 + h$, and show that

   $$C_1 = -\frac{\Omega_1 R_1}{2h\left(1 + \frac{h}{2R_1}\right)}.$$

   Then use the fact that h/R_1 is small, and deduce that

   $$C_1 = -\frac{\Omega_1 R_1}{2h}\left(1 - \frac{h}{2R_1} + \cdots\right). \qquad (1.7.80)$$

   You may use without proof the fact that for small ε

   $$\frac{1}{1 + \varepsilon} = 1 - \varepsilon + O(\varepsilon^2).$$

   (b) Similarly, show that

   $$C_2 = \frac{\Omega_1 R_1^3}{2h}\left(1 + \frac{3h}{2R_1} + \cdots\right). \qquad (1.7.81)$$

   (c) Substitute (1.7.80), (1.7.81) together with (1.7.79) into (1.7.3a) and simplify the resulting expression for $\widehat{U}$ by disregarding the $O(h^2/R_1^2)$ and smaller terms. You should find that the circumferential velocity is indeed given by

   $$\widehat{U} = \Omega_1 R_1 (1 - y). \qquad (1.7.82)$$

   (d) To obtain the corresponding solution for the pressure $\widehat{P}$, use the radial momentum equation (1.7.1b). Show that for the basic flow it reduces to

   $$\hat{r}\frac{d\widehat{P}}{d\hat{r}} = \rho\widehat{U}^2. \qquad (1.7.83)$$

   Perform the substitution of variables (1.7.79) on the left-hand side of (1.7.83) and use (1.7.82) for the right-hand side.

2. Here, your task is to prove Theorem 1.6, namely, you need to show that the eigen-values, λ, of the boundary-value problem

$$\frac{d^2\breve{v}}{d\hat{r}^2} + \frac{1}{\hat{r}}\frac{d\breve{v}}{d\hat{r}} - \left(\beta^2 + \frac{1}{\hat{r}^2}\right)\breve{v} = \lambda\Phi\breve{v}, \qquad (1.7.84a)$$

$$\breve{v}\Big|_{\hat{r}=R_1} = \breve{v}\Big|_{\hat{r}=R_2} = 0, \qquad (1.7.84b)$$

are all negative if $\Phi > 0$ in the entire interval $\hat{r} \in [R_1, R_2]$, and they are all positive if $\Phi < 0$ throughout the interval $[R_1, R_2]$.

Suggestion: You may perform this task in the following steps:

(a) Multiply all the terms of equation (1.7.84a) by $\hat{r}$, and show that

$$\frac{d}{d\hat{r}}\left(\hat{r}\frac{d\breve{v}}{d\hat{r}}\right) - \left(\beta^2\hat{r} + \frac{1}{\hat{r}}\right)\breve{v} = \lambda\hat{r}\Phi\breve{v}. \qquad (1.7.85)$$

(b) Multiply all the terms in (1.7.85) by the function $\bar{\breve{v}}$ which is the complex conjugate of $\breve{v}$, and integrate the resulting equation between the two cylinders. Show that

$$\lambda\int_{R_1}^{R_2} \hat{r}\Phi|\breve{v}|^2\,d\hat{r} = -\int_{R_1}^{R_2} \hat{r}\left|\frac{d\breve{v}}{d\hat{r}}\right|^2 d\hat{r} - \int_{R_1}^{R_2} \left(\beta^2\hat{r} + \frac{1}{\hat{r}}\right)|\breve{v}|^2\,d\hat{r}.$$

(c) Hence, argue that Theorem 1.6 holds.

2
High-Reynolds-Number Analysis of Parallel and Shear Flow Instabilities

We have seen in Chapter 1 that, for basic non-inflexional shear flows such as Blasius boundary-layer flow and plane Poiseuille flow in a channel, we can represent their susceptibility to instability by computing a neutral curve inside of which the flow is unstable to infinitesimal disturbances. There is a characteristic shape to such a marginal curve, as can be seen in Figures 1.8 and 1.12, with a critical Reynolds number below which the flow is linearly stable and upper and lower branches along which the disturbance wave number and phase speed approach zero as $Re \to \infty$, thereby limiting instability to an ever narrower region in this limit. In this chapter we will investigate the asymptotic structure of the neutral modes on both branches of the neutral curve. We will begin in Section 2.1 by studying Blasius flow, and then apply the same techniques to plane Poiseuille flow (Section 2.5). The mathematical approach adopted involves analysing the Orr–Sommerfeld equation at high Reynolds number and we shall use for this purpose the theory developed by Lin (1946). It should be noted that, in his study, Lin (1946) did not follow a strict asymptotic procedure, and we shall retain, to a certain degree, his style of analysis here. The approach will lead, on both branches, to an eigenrelation expressing the disturbance phase speed in terms of its wave number. In fact this relation allows us to study not only the neutral modes themselves but also spatially or temporally growing modes in the vicinity of these branches. Indeed, the lower branch analysis will prove to cover almost the entire unstable region between the two branches.

It transpires that the modes possess a complicated multi-layer structure with different physical balances dominating in particular regions of the flow field. As expected, viscous effects are important near solid boundaries but we will also discover that on the upper branch, viscosity plays a crucial role in smoothing out an inviscid singularity that arises at the location where the unperturbed flow velocity is equal to the disturbance phase speed. The region in which this regularization occurs is referred to as a *critical layer* and in Section 2.6 we investigate in some detail the physical balances within this layer and show how the inclusion of viscosity allows us to continue the predominantly inviscid solution across the layer. We bring the chapter to a close by discussing more recent research which explores how the dynamics of the critical layer changes as the disturbance size is increased. In particular we focus on the way in which these changes are communicated to the bulk of the flow outside of the critical layer via an increase in the distortion of the basic flow.

2.1 Problem Formulation

We shall consider as an example the flow past a flat plate; see Figure 1.10. Remember that for this flow, the basic velocity profile is given by the Blasius solution (1.3.2), (1.3.3):

$$\hat{u} = V_\infty \varphi'(\eta), \qquad \eta = \sqrt{\frac{V_\infty}{\nu}} \, \frac{\hat{y}}{\sqrt{\hat{x}}}. \tag{2.1.1}$$

Here the function $\varphi(\eta)$ satisfies the Blasius equation

$$\varphi''' + \frac{1}{2} \varphi \varphi'' = 0,$$

which is solved subject to the boundary conditions

$$\varphi(0) = \varphi'(0) = 0, \qquad \varphi'(\infty) = 1.$$

For future reference we recall that near the plate surface[1]

$$\varphi(\eta) = \frac{1}{2} \lambda \eta^2 - \frac{\lambda^2}{240} \eta^5 + \cdots \quad \text{as} \quad \eta \to 0, \tag{2.1.2}$$

with constant $\lambda \simeq 0.3321$. Near the outer edge of the boundary layer[2]

$$\varphi(\eta) = \eta - A' + \frac{C}{(\eta - A')^2} e^{-(\eta - A')^2/4} + \cdots \quad \text{as} \quad \eta \to \infty, \tag{2.1.3}$$

where $A' \simeq 1.7208$. Using (2.1.3) it may be found that the displacement thickness δ^* of the boundary layer is given by[3]

$$\delta^* = A' \sqrt{\frac{\nu \hat{x}}{V_\infty}}. \tag{2.1.4}$$

When analysing the stability of this flow we shall rely on the parallel flow approximation. We choose a position $\hat{x} = \hat{x}_0$ on the plate surface, calculate the corresponding value of the displacement thickness (2.1.4), and assuming δ^* constant, introduce non-dimensional variables as follows:

$$\hat{x} = \hat{x}_0 + \delta^* x, \qquad \hat{y} = \delta^* y, \qquad \hat{t} = \frac{\delta^*}{V_\infty} t,$$

$$\hat{u} = V_\infty u, \qquad \hat{v} = V_\infty v, \qquad \hat{p} = p_\infty + \rho V_\infty^2 p.$$

Expressed in these variables, the basic velocity profile (2.1.1) is given by

$$U(y) = \frac{d\varphi}{d\eta}\bigg|_{\eta = A'y}. \tag{2.1.5}$$

---

[1]See Problem 2 in Exercises 1 in Part 3 of this book series.
[2]See equation (1.1.53) in Part 3 of this book series.
[3]See equations (1.3.7), (1.3.8) in this volume.

In Section 1.3.3 we showed that if we superimpose small disturbances on the basic flow by writing

$$u = U(y) + \varepsilon e^{i\alpha(x-ct)}\breve{u}(y), \qquad v = \varepsilon e^{i\alpha(x-ct)}\breve{v}(y), \qquad p = \varepsilon e^{i\alpha(x-ct)}\breve{p}(y),$$

then the lateral velocity perturbation $\breve{v}$ satisfies the Orr–Sommerfeld equation

$$\frac{1}{i\alpha Re_*}\left(\frac{d^4\breve{v}}{dy^4} - 2\alpha^2\frac{d^2\breve{v}}{dy^2} + \alpha^4\breve{v}\right) = (U-c)\left(\frac{d^2\breve{v}}{dy^2} - \alpha^2\breve{v}\right) - \frac{d^2U}{dy^2}\breve{v}, \qquad (2.1.6)$$

where Re_* is the Reynolds number based on displacement thickness δ^*:

$$Re_* = \frac{V_\infty \delta^*}{\nu}.$$

The boundary conditions for (2.1.6) are

$$\breve{v} = \frac{d\breve{v}}{dy} = 0 \quad \text{at} \quad y = 0, \qquad\qquad (2.1.7)$$

$$\breve{v} \to 0 \qquad \text{as} \quad y \to \infty. \qquad\qquad (2.1.8)$$

In what follows we will take the wave number α to be real, but for later convenience allow it to either be positive or negative. It transpires that the equations we derive are still valid for complex values of α, allowing us to investigate fixed frequency modes that exhibit spatial growth or decay. The phase speed c is considered to be a complex quantity in general, although on the neutral curve itself it is of course real, and in its proximity the imaginary part remains small.

2.2 Asymptotic Analysis of the Orr–Sommerfeld Equation for Blasius Flow

Since the Blasius velocity profile does not have an inflexion point, the instability we observe must be due to viscous effects. We can see from Figure 1.15 that the wave number α becomes progressively smaller as the Reynolds number Re_* increases. Hence, in what follows we shall use the limit

$$Re_* \to \infty, \quad \alpha \to 0.$$

2.2.1   Outer region

We start with the region that lies outside the boundary layer (we shall refer to it as the 'outer region'). We expect the flow in this region to be inviscid. Therefore, we shall disregard the left-hand side of the Orr–Sommerfeld equation (2.1.6). We further take into account that in the Blasius solution, d^2U/dy^2 tends to zero exponentially as the outer edge of the boundary layer is approached, as is apparent from (2.1.3). The leading-order governing equation is therefore

$$\frac{d^2\breve{v}}{dy^2} - \alpha^2\breve{v} = 0,$$

with solution

$$\breve{v} = C_1 e^{-\alpha y} + C_2 e^{\alpha y}. \tag{2.2.1}$$

To satisfy boundary condition (2.1.8), we have to set $C_2 = 0$ if $\alpha > 0$ or $C_1 = 0$ if $\alpha < 0$. This reduces (2.2.1) to

$$\breve{v} = Ce^{-|\alpha|y}.$$

Finally, keeping in mind that the boundary-value problem (2.1.6)–(2.1.8) is homogeneous and only defines $\breve{v}$ to within an arbitrary multiplicative factor, we normalize the solution by choosing $C = 1$. Hence

$$\breve{v} = e^{-|\alpha|y}. \tag{2.2.2}$$

It can be easily seen from (2.2.2) that the characteristic thickness of the outer region is $y = O(1/|\alpha|) \gg 1$.

2.2.2   Main part of the boundary layer

Let us now consider the main part of the boundary layer. In this region $y = O(1)$. Since the small parameter α appears in the Orr–Sommerfeld equation (2.1.6) through α^2 and α^4, it is natural to seek the solution in the main part of the boundary layer in the form

$$\breve{v} = \breve{v}_0(y) + \alpha^2 \breve{v}_1(y) + \cdots. \tag{2.2.3}$$

Substituting (2.2.3) into (2.1.6) and neglecting as before the viscous terms, we find that $\breve{v}_0$ and $\breve{v}_1$ satisfy the following equations:

$$(U - c)\frac{d^2\breve{v}_0}{dy^2} - \frac{d^2U}{dy^2}\breve{v}_0 = 0, \tag{2.2.4}$$

$$(U - c)\frac{d^2\breve{v}_1}{dy^2} - \frac{d^2U}{dy^2}\breve{v}_1 = (U - c)\breve{v}_0. \tag{2.2.5}$$

Strictly speaking, the asymptotic expansion (2.2.3) should also include the $O(\alpha)$ term, as will become evident when performing the matching with the solution (2.2.2) in the outer region. However, the equation for this term coincides with (2.2.4), and therefore, it may be 'absorbed' into $\breve{v}_0$. Similarly, the $O(\alpha^3)$ term may be incorporated into (2.2.3) as a correction to $\breve{v}_1$.

Equation (2.2.4) is solved in the following two steps. First, it can be directly integrated with respect to y to yield

$$(U - c)\frac{d\breve{v}_0}{dy} - \frac{d(U - c)}{dy}\breve{v}_0 = b_0, \tag{2.2.6}$$

where b_0 is the integration constant. Secondly, dividing both sides of (2.2.6) by $(U-c)^2$, we have

$$\frac{1}{(U - c)}\frac{d\breve{v}_0}{dy} - \frac{1}{(U - c)^2}\frac{d(U - c)}{dy}\breve{v}_0 = \frac{b_0}{(U - c)^2},$$

or, equivalently,

$$\frac{d}{dy}\left(\frac{\check{v}_0}{U-c}\right) = \frac{b_0}{(U-c)^2}. \tag{2.2.7}$$

The result of the integration of (2.2.7) can be expressed in the form

$$\frac{\check{v}_0}{U-c} = a_0 + \frac{b_0}{(1-c)^2}\,y + b_0\int\limits_\infty^y\left[\frac{1}{(U-c)^2} - \frac{1}{(1-c)^2}\right]dy_1. \tag{2.2.8}$$

Here we have added and subtracted the quantity $b_0/(1-c)^2$ to the right-hand side of (2.2.7) to make the integral in (2.2.8) convergent. We can conclude that the general solution of the leading-order equation (2.2.4) may be written as

$$\check{v}_0 = a_0\Phi + b_0\Psi, \tag{2.2.9}$$

where $\Phi(y)$ and $\Psi(y)$ are the two complementary solutions:

$$\Phi = U - c, \tag{2.2.10a}$$

$$\Psi = (U-c)\int\limits_\infty^y\left[\frac{1}{(U-c)^2} - \frac{1}{(1-c)^2}\right]dy_1 + \frac{U-c}{(1-c)^2}\,y. \tag{2.2.10b}$$

The solution of the second-order equation (2.2.5) is found in the same way, and proves to be[4]

$$\check{v}_1 = a_1\Phi + b_1\Psi + a_0\Phi g_1 + b_0\Phi h_1, \tag{2.2.11}$$

where

$$g_1 = \int\limits_\infty^y\frac{1}{(U-c)^2}\left\{\int\limits_\infty^{y_1}\left[\Phi^2 - (1-c)^2\right]dy_2\right\}dy_1$$

$$+ (1-c)^2\int\limits_\infty^y\left[\frac{1}{(U-c)^2} - \frac{1}{(1-c)^2}\right]y_1\,dy_1 + \frac{1}{2}y^2. \tag{2.2.12}$$

and

$$h_1 = \int\limits_\infty^y\frac{1}{(U-c)^2}\left[\int\limits_\infty^{y_1}(\Phi\Psi - y_2)\,dy_2\right]dy_1$$

$$+ \int\limits_\infty^y\left[\frac{1}{(U-c)^2} - \frac{1}{(1-c)^2}\right]\frac{y_1^2}{2}\,dy_1 + \frac{y^3}{6(1-c)^2}. \tag{2.2.13}$$

---

[4]See Problem 1 in Exercises 7.

Matching with the solution in the outer region

To find the constants a_0, b_0, a_1, and b_1 in (2.2.11) we need to match the solution in the main part of the boundary layer with the solution in the outer region.[5] For small y the outer solution (2.2.2) may be written in the form of the Taylor expansion

$$\check{v} = 1 - |\alpha|y + \frac{1}{2}\alpha^2 y^2 - \frac{1}{6}\alpha^2|\alpha|y^3 + \cdots , \qquad (2.2.14)$$

which represents the 'inner expansion of the outer solution'.

Now we need to see what happens to the solution in the main part of the boundary layer as $y \to \infty$. Using the procedure outlined in Problem 1(a,b) in Exercises 7, it may be shown that all integrals in (2.2.10b), (2.2.12), and (2.2.13) are exponentially small, and therefore, to the leading-order approximation:

$$\Phi = 1 - c, \qquad \Psi = \frac{y}{1-c}, \qquad g_1 = \frac{y^2}{2}, \qquad h_1 = \frac{y^3}{6(1-c)^2}. \qquad (2.2.15)$$

Substituting (2.2.15) into (2.2.9) and (2.2.11), and then into (2.2.3), we have

$$\check{v} = a_0(1-c) + \frac{b_0}{1-c}y$$
$$+ \alpha^2\left[a_1(1-c) + \frac{b_1}{1-c}y + a_0\frac{1-c}{2}y^2 + b_0\frac{1}{6(1-c)}y^3\right] + \cdots , \qquad (2.2.16)$$

which is the 'outer expansion of the inner solution'. It should coincide with (2.2.14). Comparing the first two terms in (2.2.14) and (2.2.16) we see that

$$a_0 = \frac{1}{1-c}, \qquad b_0 = -|\alpha|(1-c). \qquad (2.2.17a)$$

With (2.2.17a), the last two terms in (2.2.16) prove to be identical to the third and fourth terms in (2.2.14). The remaining two terms in (2.2.16) should vanish as they do not have counterparts in (2.2.14). This is achieved by setting

$$a_1 = b_1 = 0. \qquad (2.2.17b)$$

Using (2.2.17a) and (2.2.17b) in (2.2.9) and (2.2.11), we can conclude that the solution (2.2.3) in the main part of the boundary layer is written as

$$\check{v} = \frac{\Phi}{1-c} - |\alpha|(1-c)\Psi + \alpha^2\frac{\Phi g_1}{1-c} + O(\alpha^3). \qquad (2.2.18)$$

Singularity at the critical point

It is evident from (2.2.10b), (2.2.12), and (2.2.13) that if c is real to leading order, then the solution (2.2.18) develops a singularity on approach to the point $y = y_*$ where

---

[5]A detailed discussion of the matching procedure may be found in Section 1.4 in Part 2 of this book series.

$U - c = 0$. This point is termed the *critical point*. We shall see that a small vicinity of this point is occupied by the *critical layer* where the viscous forces can no longer be disregarded. Before analysing the flow in the critical layer, we need to determine the form of the singularity in the solution (2.2.18) just above the critical layer.

When performing this task we shall assume, subject to subsequent confirmation, that the phase speed c is small. In this case, when solving the equation

$$U(y_*) = c, \tag{2.2.19}$$

we can use the near-wall representation (2.1.2) of the Blasius function. Substitution of (2.1.2) into (2.1.5) results in

$$U = \lambda_1 y + \frac{\lambda_4}{24}y^4 + \cdots, \tag{2.2.20}$$

where

$$\lambda_1 = \lambda A', \qquad \lambda_4 = -\frac{1}{2}\lambda^2 A'^4. \tag{2.2.21}$$

Using (2.2.20) in (2.2.19) we arrive at the following equation for y_*:

$$\lambda_1 y_* + \frac{\lambda_4}{24}y_*^4 + \cdots = c. \tag{2.2.22}$$

It shows that y_* is of the same order of magnitude as c. Consequently, in the leading-order of approximation, the second term on the left-hand side of (2.2.22) may be disregarded, and it follows that

$$y_* = \frac{c}{\lambda_1} + \cdots. \tag{2.2.23}$$

In order to determine the behaviour of the three terms in (2.2.18) on approach to the critical point $y = y_*$, we use the fact that the basic velocity profile $U(y)$ is a smooth function and can be represented near $y = y_*$ by the Taylor expansion

$$U = c + \tilde{\lambda}(y - y_*) + \tilde{\mu}(y - y_*)^2 + \cdots. \tag{2.2.24}$$

Here the coefficients $\tilde{\lambda}$ and $\tilde{\mu}$ are calculated, using (2.2.20), as

$$\tilde{\lambda} = U'(y_*) = \lambda_1 + \frac{\lambda_4}{6}y_*^3 + \cdots, \qquad \tilde{\mu} = \frac{1}{2}U''(y_*) = \frac{\lambda_4}{4}y_*^2 + \cdots. \tag{2.2.25}$$

Using (2.2.24) in (2.2.10a), we see that the first term in (2.2.18) is written as

$$\frac{\Phi}{1-c} = \frac{\tilde{\lambda}}{1-c}(y - y_*) + \frac{\mu}{1-c}(y - y_*)^2 + \cdots. \tag{2.2.26}$$

When dealing with the second term in (2.2.18), we need to evaluate the integral

$$\mathcal{I}_\Psi = \int\limits_\infty^y \left[\frac{1}{(U-c)^2} - \frac{1}{(1-c)^2}\right] dy_1. \tag{2.2.27}$$

It follows from (2.2.24) that

$$\frac{1}{(U-c)^2} = \frac{1}{\tilde{\lambda}^2(y_1-y_*)^2} - \frac{2\tilde{\mu}}{\tilde{\lambda}^3}\frac{1}{y_1-y_*} + O(1) \quad \text{as} \quad y_1 \to y_* + 0. \tag{2.2.28}$$

To make the corresponding integral convergent at $y_1 = \infty$ we modify the second term on the right-hand side of (2.2.28) as follows:

$$\frac{1}{(U-c)^2} = \frac{1}{\tilde{\lambda}^2(y_1-y_*)^2} - \frac{2\tilde{\mu}}{\tilde{\lambda}^3}\frac{1}{(y_1-y_*)(1+y_1-y_*)} + O(1). \tag{2.2.29}$$

Now we add and subtract the two terms on the right-hand side of (2.2.29) to the integrand in (2.2.27), and perform the integration. We find

$$\mathcal{I}_\Psi = -\frac{1}{\tilde{\lambda}^2(y-y_*)} - \frac{2\tilde{\mu}}{\tilde{\lambda}^3}\ln\frac{y-y_*}{1+y-y_*} + B(y), \tag{2.2.30}$$

where

$$B(y) = \int\limits_\infty^y \left[\frac{1}{(U-c)^2} - \frac{1}{(1-c)^2} - \frac{1}{\tilde{\lambda}^2(y_1-y_*)^2} + \frac{2\tilde{\mu}}{\tilde{\lambda}^3}\frac{1}{(y_1-y_*)(1+y_1-y_*)}\right]dy_1$$

stays finite as $y \to y_*$. It remains to substitute (2.2.30) together with (2.2.24) into (2.2.10b), and we arrive at the conclusion that

$$\Psi = -\frac{1}{\tilde{\lambda}} - \frac{2\tilde{\mu}}{\tilde{\lambda}^2}(y-y_*)\ln(y-y_*)$$

$$+ \left[\tilde{\lambda}B(y_*) - \frac{\tilde{\mu}}{\tilde{\lambda}^2} + \frac{\tilde{\lambda}y_*}{(1-c)^2}\right](y-y_*) + O\left[(y-y_*)^2\ln(y-y_*)\right]. \tag{2.2.31}$$

Finally, we need to consider the third term in the solution (2.2.18) for the main part of the boundary layer. Remember that the function $g_1(y)$ is given by (2.2.12). The two integrals in (2.2.12) are analysed in the same manner as the integral (2.2.27). We find that[6]

$$g_1 = -\frac{D(y_*)+(1-c)^2 y_*}{\tilde{\lambda}^2}\left[\frac{1}{y-y_*} + \frac{2\tilde{\mu}}{\tilde{\lambda}}\ln(y-y_*)\right] + O(1) \quad \text{as} \quad y \to y_*, \tag{2.2.32}$$

where

$$D(y_*) = \int\limits_\infty^{y_*}\left[(U-c)^2 - (1-c)^2\right]dy.$$

It remains to substitute (2.2.26), (2.2.31), and (2.2.32) into (2.2.18). When performing this task we use the Taylor expansion (2.2.24) for $\Phi = U - c$. As a result the following representation of $\check{v}$ on approach to the critical point is obtained:

---

[6]For a detailed analysis of the first integral in (2.2.12) the reader is referred to Problem 2 in Exercises 7.

$$\check{v} = S_1 + S_2(y - y_*)\ln(y - y_*) + S_3(y - y_*) + O\big[(y - y_*)^2 \ln(y - y_*)\big]. \qquad (2.2.33)$$

Here

$$S_1 = |\alpha|\frac{1-c}{\tilde{\lambda}} - \alpha^2 \frac{\overbrace{D(y_*) + (1-c)^2 y_*}}{(1-c)\tilde{\lambda}}, \qquad (2.2.34\text{a})$$

$$S_2 = |\alpha|\frac{2(1-c)\tilde{\mu}}{\tilde{\lambda}^2} - \alpha^2 \frac{2\tilde{\mu}}{1-c}\frac{\overbrace{D(y_*) + (1-c)^2 y_*}}{\tilde{\lambda}^2}, \qquad (2.2.34\text{b})$$

$$S_3 = \frac{\tilde{\lambda}}{1-c} - |\alpha|(1-c)\left[\tilde{\lambda}B(y_*) - \frac{\overbrace{\tilde{\mu}}}{\tilde{\lambda}^2} + \frac{\overbrace{\tilde{\lambda}y_*}}{(1-c)^2}\right]$$

$$\qquad - \alpha^2 \frac{\tilde{\mu}}{1-c}\frac{\overbrace{D(y_*) + (1-c)^2 y_*}}{\tilde{\lambda}^2}. \qquad (2.2.34\text{c})$$

Equations (2.2.34) allow for the following simplifications. First, it follows from (2.2.23) that y_* is small for small values of the phase speed c. Second, using (2.2.25) and (2.2.23), we can see that

$$\frac{2\tilde{\mu}}{\tilde{\lambda}^2} = O(c^2),$$

which is even smaller. On this basis, we disregard the 'over-braced' terms in (2.2.34), and we are left with

$$S_1 = |\alpha|\underbrace{\frac{1-c}{\lambda_1}}_{} - \alpha^2 \frac{D(0)}{(1-c)\lambda_1}, \qquad (2.2.35\text{a})$$

$$S_2 = \left[|\alpha|(1-c) - \alpha^2 \frac{D(0)}{1-c}\right]\frac{\lambda_4 c^2}{2\lambda_1^4}, \qquad (2.2.35\text{b})$$

$$S_3 = \underbrace{\frac{\lambda_1}{1-c}}_{} - |\alpha|(1-c)\lambda_1 B(0). \qquad (2.2.35\text{c})$$

The significance of the 'under-braced' terms in (2.2.35a) and (2.2.35c) will be made clear in Section 2.3.

Viscous effects

The logarithmic singularity in (2.2.33) is smoothed out in the critical layer where the viscous forces become important. The characteristic thickness δ of the critical layer may be found, as usual, by comparing the viscous and inviscid terms in the Orr–Sommerfeld equation (2.1.6). When performing this task, one needs to keep in mind that each time $\check{v}$ is differentiated with respect to y, its order of magnitude increases by a factor of δ^{-1}. This means that the leading-order viscous term on the left-hand side of (2.1.6) is

$$\frac{1}{i\alpha Re_*}\frac{d^4\check{v}}{dy^4}. \qquad (2.2.36)$$

In the critical layer it should be in balance with the leading-order inertial term

$$(U - c)\frac{d^2 \breve{v}}{dy^2}, \tag{2.2.37}$$

Approximating the derivatives by finite differences, we can estimate the viscous term (2.2.36) as

$$\frac{1}{i\alpha Re_*}\frac{d^4 \breve{v}}{dy^4} \sim \frac{1}{\alpha Re_*}\frac{\breve{v}}{\delta^4}. \tag{2.2.38}$$

To deduce the corresponding estimate for the inertial term (2.2.37), we recall that near the critical point, the basic velocity U is given by the equation (2.2.24), from which it follows that in the critical layer $U - c \sim \delta$. Hence,

$$(U - c)\frac{d^2 \breve{v}}{dy^2} \sim \delta\frac{\breve{v}}{\delta^2}. \tag{2.2.39}$$

It is now easily seen that the viscous term (2.2.38) is in balance with the inertial term (2.2.39) provided that

$$\delta \sim (\alpha Re_*)^{-1/3}. \tag{2.2.40}$$

Given the phase speed c, one can find the position of the critical layer above the wall using equation (2.2.23) which allows us to estimate y_* as

$$y_* \sim c. \tag{2.2.41}$$

If $c \gg (\alpha Re_*)^{-1/3}$, then the thickness of the region lying between the critical layer and the wall is much larger than the thickness of the critical layer itself. In these conditions, the viscous terms in the Orr–Sommerfeld equation (2.1.6) may be disregarded and the flow below the critical layer may be treated as inviscid. Still, to satisfy the no-slip conditions, viscous effects must be reinstated in the immediate proximity of the wall where the *Stokes layer* forms. The characteristic thickness δ_{St} of the Stokes layer may be found by balancing again the viscous and inertial terms in the Orr–Sommerfeld equation (2.1.6):

$$\frac{1}{i\alpha Re_*}\frac{d^4 \breve{v}}{dy^4} \sim (U - c)\frac{d^2 \breve{v}}{dy^2}. \tag{2.2.42}$$

However, now it should be taken into account that, close to the wall, the basic velocity U is small compared to the phase speed c, which allows us to express the order-of-magnitude equation (2.2.42) in the form

$$\frac{1}{\alpha Re_*}\frac{\breve{v}}{\delta_{St}^4} \sim c\frac{\breve{v}}{\delta_{St}^2}.$$

Solving this equation for δ_{St} we find that

$$\delta_{St} \sim (\alpha c Re_*)^{-1/2}. \tag{2.2.43}$$

Since the neutral curve has two branches (see Figure 1.12) one should expect two distinguished asymptotic models for describing the upper and lower branches. We shall see that on the upper branch the critical layer and the Stokes layer are separated as described above, while on the lower branch

$$c \sim (\alpha Re_*)^{-1/3}, \qquad (2.2.44)$$

in which case (2.2.41) and (2.2.43) become the same order

$$y_* \sim \delta_{\mathrm{St}} \sim (\alpha Re_*)^{-1/3},$$

and coincide with (2.2.40), which means that in this case the critical layer is absorbed by the Stokes layer.

We shall now study in detail these two situations starting with the one that corresponds to the lower branch of the stability curve. In this case, since the singularity has effectively been removed by the smoothing effects of viscosity, we can relax the assumption that c is real to leading order and allow for disturbances that exhibit significant growth or decay, in addition to those that remain neutral.

2.3   Critical Layer and Stokes Layer Coincident: Lower Branch

In this flow regime, in addition to the outer flow region, and the main part of the boundary layer, we need to consider the viscous layer that forms on the plate surface. In the outer flow (here we shall refer to it as region 1) the asymptotic solution of the Orr–Sommerfeld equation is given by (2.2.2). In the main part of the boundary layer (denoted region 2), the solution was represented in the form of the asymptotic expansion (2.2.3), and our main focus was on the behaviour of the solution on approach to the viscous layer. We found that immediately above the viscous layer, $\breve{v}$ is represented by (2.2.33), (2.2.35). It should be noted that when dealing with the lower branch of the neutral curve, only the first term in (2.2.3) is needed, which produces the under-braced terms in (2.2.35). Substituting these into (2.2.33) and disregarding the rest of the terms, we have

$$\breve{v} = |\alpha| \frac{1-c}{\lambda_1} + \frac{\lambda_1}{1-c}(y - y_*) + \cdots . \qquad (2.3.1)$$

2.3.1   The viscous sublayer

Let us now analyse the flow in the viscous sublayer, which we shall call region 3. Guided by (2.2.40) we scale the independent variable y as

$$y = (\alpha Re_*)^{-1/3} Y_*. \qquad (2.3.2)$$

We also scale the phase speed according to (2.2.44):

$$c = (\alpha Re_*)^{-1/3} c', \qquad (2.3.3)$$

with c' being an order one parameter. The basic velocity profile in region 3 is found by substituting (2.3.2) into (2.2.20). Restricting our attention to the leading-order approximation we have

$$U = (\alpha Re_*)^{-1/3} \lambda_1 Y_* + \cdots . \qquad (2.3.4)$$

Substituting (2.3.2), (2.3.3), and (2.3.4) into the Orr–Sommerfeld equation (2.1.6) we arrive at the following equation for region 3:

$$\frac{d^4\breve{v}}{dY_*^4} = i(\lambda_1 Y_* - c')\frac{d^2\breve{v}}{dY_*^2}.$$

(2.3.5)

It should be solved subject to the no-slip conditions (2.1.7). In the variables used here they are written as

$$\breve{v} = \frac{d\breve{v}}{dY_*} = 0 \quad \text{at} \quad Y_* = 0.$$

(2.3.6)

When dealing with equation (2.3.5) it is convenient to introduce a new independent variable

$$z = z_0 + \theta Y_*,$$

(2.3.7)

where

$$\theta = (i\lambda_1)^{1/3}, \qquad z_0 = -\frac{c'}{\lambda_1}\theta.$$

This transforms (2.3.5) into the Airy equation for the function $d^2\breve{v}/dz^2$:

$$\frac{d^4\breve{v}}{dz^4} = z\frac{d^2\breve{v}}{dz^2},$$

(2.3.8)

with the boundary conditions (2.3.6) now written as

$$\breve{v} = \frac{d\breve{v}}{dz} = 0 \quad \text{at} \quad z = z_0.$$

(2.3.9)

It is well known (see, for example, Abramowitz and Stegun, 1965) that the general solution of the Airy equation

$$\frac{d^2w}{dz^2} - zw = 0$$

(2.3.10)

may be expressed in the form

$$w = C_1^* Ai(z) + C_2^* Bi(z).$$

(2.3.11)

Here $Ai(z)$ and $Bi(z)$ are two complementary solutions of (2.3.10). They exhibit the following behaviour at large values of z:

$$Ai(z) = \frac{z^{-1/4}}{2\sqrt{\pi}}e^{-\zeta} + \cdots, \qquad Bi(z) = \frac{z^{-1/4}}{\sqrt{\pi}}e^{\zeta} + \cdots,$$

(2.3.12)

where $\zeta = \frac{2}{3}z^{3/2}$. Keeping in mind that, with an exponentially growing function such as $Bi(z)$, the matching with the solution in the main part of the boundary layer (region 2) is not possible,[7] we shall set $C_2^* = 0$, which reduces (2.3.11) to

$$\frac{d^2\breve{v}}{dz^2} = C_1^* Ai(z).$$

(2.3.13)

---

[7]See Section 1.4.2 in Part 2 of this book series.

Integrating (2.3.13), with the boundary condition $d\breve{v}/dz = 0$ at $z = z_0$, we have

$$\frac{d\breve{v}}{dz} = C_1^* \int_{z_0}^{z} Ai(\zeta)\, d\zeta. \tag{2.3.14}$$

We need to integrate (2.3.14) once again, now with the boundary condition $\breve{v} = 0$ at $z = z_0$. Taking into account that $Ai(z)$ satisfies the Airy equation (2.3.10), we can express the result of this second integration in the form

$$\breve{v} = C_1^* \left[z \int_{z_0}^{z} Ai(\zeta)\, d\zeta - Ai'(z) + Ai'(z_0) \right]. \tag{2.3.15}$$

To complete the analysis, we need to perform the matching of the solutions in regions 2 and 3. We start with the solution (2.3.15) in region 3. Since $Ai(z)$ decays exponentially, the two-term asymptotic expansion of (2.3.15) at large values of z is written as

$$\breve{v} = C_1^* \left[z \int_{z_0}^{\infty} Ai(z)\, dz + Ai'(z_0) + \cdots \right]. \tag{2.3.16}$$

It remains to recall that

$$z = z_0 + \theta Y_* = \theta Y_* - \frac{c'}{\lambda_1}\theta,$$

and we find that the 'outer expansion of the inner solution' has the form

$$\breve{v} = C_1^* Y_* \theta \int_{z_0}^{\infty} Ai(z)\, dz + C_1^* \left[-\frac{c'}{\lambda_1}\theta \int_{z_0}^{\infty} Ai(z)\, dz + Ai'(z_0) \right] + \cdots . \tag{2.3.17}$$

Let us now return to the solution (2.3.1) for region 2. Taking into account that the phase speed c is small, we write it as

$$\breve{v} = \frac{|\alpha|}{\lambda_1} + \lambda_1(y - y_*) + \cdots . \tag{2.3.18}$$

Here y_* is given by the equation (2.2.23), which enables us to express (2.3.18) in the form

$$\breve{v} = \lambda_1 y + \left(\frac{|\alpha|}{\lambda_1} - c \right) + \cdots . \tag{2.3.19}$$

As the next step of the matching procedure, we need to represent (2.3.19) in terms of the variables of region 3. We substitute (2.3.2) and (2.3.3) into (2.3.19), which yields the 'inner expansion of the outer solution':

$$\breve{v} = (\alpha Re_*)^{-1/3}\lambda_1 Y_* + \left[\frac{|\alpha|}{\lambda_1} - (\alpha Re_*)^{-1/3}c' \right] + \cdots . \tag{2.3.20}$$

The requirement that (2.3.20) should coincide with the 'outer expansion of the inner solution' (2.3.17) leads to the following two equations:

$$(\alpha Re_*)^{-1/3}\lambda_1 = C_1^* \theta \int\limits_{z_0}^{\infty} Ai(z)\,dz, \tag{2.3.21}$$

$$\frac{|\alpha|}{\lambda_1} - (\alpha Re_*)^{-1/3}c' = C_1^*\left[-\frac{c'}{\lambda_1}\theta\int\limits_{z_0}^{\infty} Ai(z)\,dz + Ai'(z_0)\right]. \tag{2.3.22}$$

Elimination of C_1^* from (2.3.21) and (2.3.22) yields the following *dispersion equation*

$$\frac{\lambda_1^2}{|\alpha|}(\alpha Re_*)^{-1/3}Ai'(z_0) = \theta\int\limits_{z_0}^{\infty} Ai(z)\,dz, \tag{2.3.23}$$

where

$$z_0 = -\frac{c'}{\lambda_1}\theta. \tag{2.3.24}$$

For future applications it is convenient to express (2.3.24) in terms of the frequency ω. When doing this, we shall change the sign of ω in the normal-mode representation (1.3.17), so that we now seek perturbations proportional to $e^{i(\alpha x + \omega t)}$. In this case the phase speed is calculated as

$$c = -\frac{\omega}{\alpha}. \tag{2.3.25}$$

It then follows from (2.3.3) and (2.3.25) that

$$c' = -(\alpha Re_*)^{1/3}\frac{\omega}{\alpha}. \tag{2.3.26}$$

Using (2.3.26) and the fact that $\theta = (i\lambda_1)^{1/3}$, we can cast (2.3.23) and (2.3.24) in the form

$$\alpha^{1/3}|\alpha|Re_*^{1/3}\int\limits_{z_0}^{\infty} Ai(z)\,dz + (i\lambda_1)^{5/3}Ai'(z_0) = 0, \tag{2.3.27}$$

$$z_0 = \frac{i\omega}{(i\alpha\lambda_1)^{2/3}}Re_*^{1/3}. \tag{2.3.28}$$

2.3.2   Canonical form of the dispersion equation

To preserve the full family of solutions of the dispersion equation, we need to avoid degeneration in (2.3.27), (2.3.28) as $Re_* \to \infty$. First, we need to ensure that z_0 remains finite, which is only possible if

$$\frac{\omega Re_*^{1/3}}{\alpha^{2/3}} = O(1). \tag{2.3.29}$$

Second, we need the two terms in (2.3.27) to remain comparable to one another, which leads to the requirement

$$\alpha^{4/3} Re_*^{1/3} = O(1). \tag{2.3.30}$$

It immediately follows from (2.3.30) that

$$\alpha = O(Re_*^{-1/4}), \tag{2.3.31}$$

and then, using (2.3.31) in (2.3.29), we find that

$$\omega = O(Re_*^{-1/2}). \tag{2.3.32}$$

Guided by (2.3.31) and (2.3.32) we shall represent the wave number α and the frequency of oscillations ω in the form

$$\alpha = \lambda_1^{5/4} Re_*^{-1/4} \alpha_*, \qquad \omega = \lambda_1^{3/2} Re_*^{-1/2} \omega_*. \tag{2.3.33}$$

The coefficients $\lambda_1^{5/4}$ and $\lambda_1^{3/2}$ are introduced in (2.3.33) to eliminate the skin friction parameter λ_1 from (2.3.27) and (2.3.28), which makes the dispersion equation applicable not only to the Blasius boundary layer, but also to any other boundary layer with a different value of λ_1 in the representation (2.2.20) of the basic velocity profile near the body surface. Upon substitution of (2.3.33) into (2.3.27) and (2.3.28) we have the dispersion equation in its canonical form

$$(i\alpha_*)^{1/3} |\alpha_*| \int_{z_0}^{\infty} Ai(z) \, dz - Ai'(z_0) = 0, \qquad z_0 = \frac{i\omega_*}{(i\alpha_*)^{2/3}}. \tag{2.3.34}$$

Although the dispersion relation (2.3.34) was derived on the basis that the wave number α_* is a real quantity, it will be seen below that we can employ analytic continuation into the complex α_*-plane, enabling us to study disturbances that grow or decay in the downstream direction.

Clarifying remarks

When solving equation (2.3.34) we shall assume that the frequency ω_* is real and positive. Our task is to find (for each ω_*) the wave number α_* which is expected to be a complex quantity in general, but of course will be real for the mode which lies on the neutral curve itself. When working in the complex α_*-plane, we, first, need to define the analytic continuation of $|\alpha_*|$ from the real axis into the complex plane. Second, we need to clarify how $(i\alpha_*)^{1/3}$ is calculated in the complex plane. Starting with $|\alpha_*|$, we note that we represented the perturbations in the form

$$v' = e^{i(\alpha x + \omega t)} \breve{v}(y) = e^{-\alpha_i x} e^{i(\alpha_r x + \omega t)} \breve{v}(y). \tag{2.3.35}$$

The Tollmien–Schlichting waves propagate downstream in the boundary layer, which means that for positive ω_*, the real part of α_* is expected to be negative. An analytic continuation of $|\alpha_*|$ from the real negative semi-axis is written as

$$|\alpha_*| = -\alpha_*. \tag{2.3.36}$$

Turning to the cubic root $(i\alpha_*)^{1/3}$, one should remember that it was first introduced in the definition (2.3.7) of z. Then, when analysing the solution (2.3.11) of the

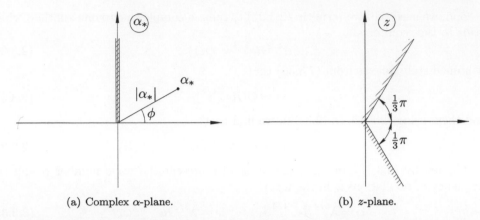

(a) Complex α-plane.   (b) z-plane.

Fig. 2.1: How to choose an analytical branch of $(i\alpha)^{1/3}$.

Airy equation, we noted that $Ai(z)$ tends to zero as $z \to \infty$, while $Bi(z)$ becomes exponentially large. For this reason the function $Bi(z)$ was discarded, leading to (2.3.13). Now, we shall look at these arguments more closely. The fact is that the behaviour of $Ai(z)$ and $Bi(z)$ at large z depends on the direction along which z tends to infinity in the complex plane. Using (2.3.2) and (2.3.33), we can express (2.3.7) in the form

$$z = z_0 + \varkappa (i\alpha_*)^{1/3} y, \qquad (2.3.37)$$

where $\varkappa = (\lambda_1^3 Re_*)^{1/4}$ is real and positive. For our purposes it is convenient to choose an analytic branch of $(i\alpha_*)^{1/3}$ in the following way. We make a branch cut in the complex α_*-plane along the positive imaginary semi-axis (see Figure 2.1a) and write α_* in the exponential form, $\alpha_* = |\alpha_*| e^{i\phi}$. Then we will have

$$(i\alpha_*)^{1/3} = \left(e^{i\pi/2} |\alpha_*| e^{i\phi}\right)^{1/3} = |\alpha_*|^{1/3} e^{i(\pi/6 + \phi/3)}. \qquad (2.3.38)$$

Here ϕ belongs to the interval

$$\phi \in \left(-\frac{3}{2}\pi, \ \frac{1}{2}\pi\right).$$

Correspondingly, for large values of y, when z_0 can be disregarded in (2.3.37), the argument of z will lie in the interval

$$\arg z \in \left(-\frac{1}{3}\pi, \ \frac{1}{3}\pi\right),$$

as shown on Figure 2.1(b). Under this restriction, the real part of $\zeta = \frac{2}{3} z^{3/2}$ remains positive, and it follows from (2.3.12) that $Ai(z)$ does indeed tend to zero, while $Bi(z)$ is exponentially growing for any point in the complex α_*-plane of Figure 2.1(a).

2.3.3 Numerical solution of the dispersion equation

From the second equation in (2.3.34) we have

$$(i\alpha_*)^{2/3} = \frac{i\omega_*}{z_0}. \tag{2.3.39}$$

Substitution of (2.3.36) and (2.3.39) into the first equation in (2.3.34) allows us to write it in the form

$$\omega_*^2 \int_{z_0}^{\infty} Ai(z)\,dz - iz_0^2 Ai'(z_0) = 0. \tag{2.3.40}$$

This equation defines z_0 as a function of ω_*. Once it is solved, that is, z_0 is found for a given value of ω_*, we can find the corresponding value of α_* using equation (2.3.39).

It is convenient to start the calculations at small values of ω_*. If we assume that z_0 remains finite as $\omega_* \to 0$, then the dispersion equation (2.3.40) reduces to

$$Ai'(z_0) = 0. \tag{2.3.41}$$

It is known (see, for example, Abramowitz and Stegun, 1965) that there are an infinite (countable) number of points in the complex z_0-plane, where the derivative of the Airy function vanishes; all lie on the negative real semi-axis. The first five are

$$z_0^{(1)} \simeq -1.01879,$$

$$z_0^{(2)} \simeq -3.24819,$$

$$z_0^{(3)} \simeq -4.82009,$$

$$z_0^{(4)} \simeq -6.16330,$$

$$z_0^{(5)} \simeq -7.37217.$$

Each of these may be used as an 'initial guess' in the iterative solution of the dispersion equation (2.3.40) for a small value of ω_*. The results presented in Figure 2.2 were obtained using Newton iteration. During the iteration process, the frequency ω_* was kept fixed until convergence is achieved to give the corresponding value of z_0. Then ω_* was increased by a small step, and the iteration process was repeated. When performing the Newton iterations, the values of the Airy function and its derivative at the point z_0 in the complex z-plane were found by solving the initial-value problem for the Airy equation along a straight line connecting the origin $z = 0$ and $z = z_0$, with the Airy function and its derivative assigned the well known values at z = 0:

$$Ai(0) \simeq 0.355028, \qquad Ai'(0) \simeq -0.258819.$$

The integral in (2.3.40) was calculated as[8]

---

[8]Here we take into account a well known fact that

$$\int_0^{\infty} Ai(z)\,dz = \frac{1}{3}.$$

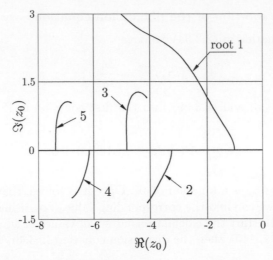

Fig. 2.2: The trajectories of the first five roots of the dispersion equation (2.3.40) in the z_0-plane as ω_* is increased.

$$\int\limits_{z_0}^{\infty} Ai(z)\,dz = \frac{1}{3} - \int\limits_{0}^{z_0} Ai(z)\,dz.$$

As ω_* becomes large, all the roots, except the first one, approach the points in the complex z_0-plane, where

$$\int\limits_{z_0}^{\infty} Ai(z)\,dz = 0. \tag{2.3.42}$$

Indeed, assuming z_0 finite and setting $\omega_* \to \infty$, reduces the dispersion equation (2.3.40) to (2.3.42). It is known that the integral of the Airy function vanishes at an infinite (countable) number of points; these come in complex conjugate pairs and all lie in the left half of the complex plane. The first four of these are

$$z_0^{(2)} \simeq -4.106992 - i \cdot 1.144158,$$

$$z_0^{(3)} \simeq -4.106992 + i \cdot 1.144158,$$

$$z_0^{(4)} \simeq -6.798114 - i \cdot 1.035140,$$

$$z_0^{(5)} \simeq -6.798114 + i \cdot 1.035140.$$

These are the points where the roots 2 to 5 terminate as $\omega_* \to \infty$. As far as root 1 is concerned, its behaviour is different. With increasing ω_* it moves towards infinity (see Figure 2.2). In order to study the behaviour of this root we shall assume that z_0 in (2.3.40) is large, in which case the derivative of the Airy function and its integral may be represented by their asymptotic expansions

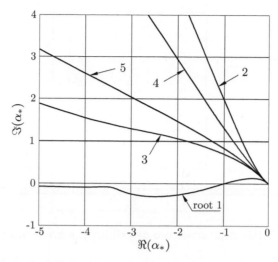

Fig. 2.3: The trajectories of the first five roots of (2.3.34) in the complex α_*-plane.

$$\left.\begin{array}{l} Ai'(z) = -\dfrac{1}{2\sqrt{\pi}}\, z^{1/4} e^{-\frac{2}{3}z^{3/2}} + \cdots, \\[3mm] \displaystyle\int\limits_{z}^{\infty} Ai(\zeta)d\zeta = \dfrac{1}{2\sqrt{\pi}}\, z^{-3/4} e^{-\frac{2}{3}z^{3/2}} + \cdots \end{array}\right\} \quad \text{as} \quad z \to \infty. \qquad (2.3.43)$$

Substitution of (2.3.43) into (2.3.40) leads to the equation

$$z_0^3 = i\omega_*^2,$$

which confirms that $z_0 \to \infty$ as $\omega_* \to \infty$.

The trajectories of the first five roots of the dispersion equation in the complex α_*-plane are shown in Figure 2.3. All the roots start (at $\omega_* = 0$) from the coordinate origin, and all of them, except the first one, remain in the second quadrant for all $\omega_* \in [0, \infty)$, which means that the corresponding perturbations decay with x; see equation (2.3.35). The behaviour of the first root is different. It stays in the second quadrant until the frequency reaches its critical (neutral) value,

$$\omega_{*n} \simeq 2.29797, \qquad (2.3.44)$$

when it crosses the real axis at the point

$$\alpha_{*n} \simeq -1.00049, \qquad (2.3.45)$$

and after that remains in the third quadrant for $\omega_* \in (\omega_{*n}, \infty)$. Clearly, this must be the root that represents the Tollmien–Schlichting waves. For subcritical values of the frequency, it lies to the left of the neutral curve (see Figure 1.14a). Then at (2.3.44) it reaches the lower branch of the neutral curve and, as ω_* increases further, it crosses into the region between the two branches of the neutral curve. Figure 2.3 shows that

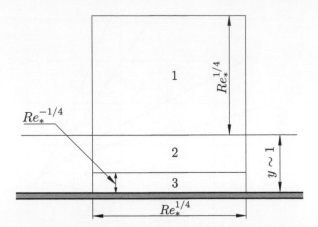

Fig. 2.4: Triple-deck structure describing the lower branch modes.

this root captures the entire spectrum of perturbations including those with maximum amplification rate. Thus the theory presented here describes not only a vicinity of the lower branch of the neutral curve, but is also applicable to almost the entire region between the two branches, except the immediate vicinity of the upper branch, which will be discussed in detail in the next section.

In conclusion, let us return to equations (2.3.33). If we consider the second of these equations, and combine it with (2.3.44), then we will have

$$\omega = \lambda_1^{3/2} Re_*^{-1/2} \omega_{*n} \quad \text{as} \quad Re_* \to \infty, \tag{2.3.46}$$

which gives the asymptotic behaviour of the lower branch of the neutral curve in the (ω, Re_*)-plane. To obtain the corresponding formula for the (α, Re_*)-plane, we need to substitute (2.3.45) into the first equation in (2.3.33). We have

$$\alpha = \lambda_1^{5/4} Re_*^{-1/4} \alpha_{*n} \quad \text{as} \quad Re_* \to \infty. \tag{2.3.47}$$

It should be noted that the sign of α depends on the choice of the sign of ω; see Problem 5 in Exercises 7. For negative ω, equation (2.3.47) assumes the form

$$\alpha = \lambda_1^{5/4} Re_*^{-1/4} |\alpha_{*n}| \quad \text{as} \quad Re_* \to \infty. \tag{2.3.48}$$

Another important observation is that the Tollmien–Schlichting waves have a *triple-deck structure*; see Figure 2.4. It is composed of the inviscid region 1 situated outside the boundary layer, the main part of the boundary layer (region 2), and the near-wall viscous sublayer (region 3). It follows from (2.2.2) and (2.3.48) that the characteristic thickness of region 1 can be estimated as

$$y_{(1)} = O(Re_*^{1/4}). \tag{2.3.49}$$

The characteristic thickness of region 2 is

$$y_{(2)} = O(1). \tag{2.3.50}$$

Finally, using (2.3.48) in (2.2.40) allows us to see that the thickness of region 3 is estimated as

$$y_{(3)} = O(Re_*^{-1/4}). \tag{2.3.51}$$

As far as the longitudinal extent of the triple-deck is concerned, it may be measured in terms of the wavelength of the Tollmien–Schlichting wave:

$$\Delta x \sim \frac{1}{\alpha} = O(Re_*^{1/4}). \tag{2.3.52}$$

Exercises 7

1. With the help of (2.2.9) and (2.2.10a), equation (2.2.5) may be cast in the form

$$(U - c)\frac{d^2 \breve{v}_1}{dy^2} - \frac{d^2 U}{dy^2}\breve{v}_1 = a_0 \Phi^2 + b_0 \Phi \Psi. \tag{2.3.53}$$

Your task is to find the general solution of (2.3.53). Perform this task in the following steps:

(a) Substitute (2.1.3) into (2.1.5) and show that

$$U(y) = 1 - \frac{C}{2A'(y-1)}e^{-A'^2(y-1)^2/4} + \cdots \quad \text{as} \quad y \to \infty.$$

Show further that

$$\frac{1}{(U-c)^2} = \frac{1}{(1-c)^2} + \frac{C}{A'(1-c)^3(y-1)}e^{-A'^2(y-1)^2/4} + \cdots .$$

(b) Using integration by parts,[9] show that

$$\int\limits_\infty^y \left[\frac{1}{(U-c)^2} - \frac{1}{(1-c)^2}\right]dy_1 = -\frac{2C}{A'^3(1-c)^3(y-1)^2}e^{-A'^2(y-1)^2/4} + \cdots .$$

Hence, conclude that the complementary solutions (2.2.10) of the homogeneous equation display the following behaviour at the outer edge of the boundary layer

$$\left.\begin{aligned} \Phi &= 1 - c + \cdots, \\ \Psi &= \frac{1}{1-c}y + \cdots \end{aligned}\right\} \quad \text{as} \quad y \to \infty.$$

(c) Keeping this in mind, integrate (2.3.53), and show that

$$(U - c)\frac{d\breve{v}_1}{dy} - \frac{dU}{dy}v_1 = b_1 + a_0 \int\limits_\infty^y \left[\Phi^2 - (1-c)^2\right]dy_1 + a_0(1-c)^2 y$$

$$+ b_0 \int\limits_\infty^y (\Phi\Psi - y_1)\, dy_1 + \frac{1}{2}b_0 y^2. \tag{2.3.54}$$

[9]See Section 1.1.2 in Part 2 of this book series.

(d) Divide all terms in (2.3.54) by $(U - c)^2$ and integrate the resulting equation with respect to y. You should find that

$$\frac{\check{v}_1}{U - c} = a_1 + b_1 \left\{ \int_\infty^y \left[\frac{1}{(U - c)^2} - \frac{1}{(1 - c)^2} \right] dy_1 + \frac{y}{(1 - c)^2} \right\}$$

$$+ a_0 g_1 + b_0 h_1,$$

where g_1 and h_1 are given by (2.2.12) and (2.2.13) respectively.

2. Consider the integral

$$I_g = \int_\infty^y \frac{1}{(U - c)^2} \left\{ \int_\infty^{y_1} \left[(U - c)^2 - (1 - c)^2 \right] dy_2 \right\} dy_1.$$

Your task is to find the asymptotic behaviour of this integral on approach to the critical point $y = y_*$.

You may perform this task in the following steps:

(a) Introduce the function

$$D(y_1) = \int_\infty^{y_1} \left[(U - c)^2 - (1 - c)^2 \right] dy_2,$$

and show that its Taylor expansion near the critical point is written as

$$D(y_1) = D(y_*) - (1 - c)^2 (y_1 - y_*) + \cdots.$$

(b) Using the fact that

$$U(y_1) - c = \tilde{\lambda}(y_1 - y_*) + \tilde{\mu}(y_1 - y_*)^2 + \cdots \quad \text{as} \quad y_1 \to y_*,$$

deduce that

$$\frac{D(y_1)}{(U - c)^2} = \frac{D(y_*)}{\tilde{\lambda}^2} \frac{1}{(y_1 - y_*)^2}$$

$$- \left[\frac{2\tilde{\mu} D(y_*)}{\tilde{\lambda}^3} + \frac{(1 - c)^2}{\tilde{\lambda}^2} \right] \frac{1}{y_1 - y_*} + \cdots. \tag{2.3.55}$$

(c) By integrating (2.3.55), show that

$$I_g = -\frac{D(y_*)}{\tilde{\lambda}^2} \frac{1}{y - y_*}$$

$$- \left[\frac{2\tilde{\mu} D(y_*)}{\tilde{\lambda}^3} + \frac{(1 - c)^2}{\tilde{\lambda}^2} \right] \ln(y - y_*) + O(1).$$

3. Show that in dimensional variables $(\hat{x}, \hat{y})$, the estimates (2.3.49)–(2.3.51) for the widths of the three layers in the triple-deck structure (see Figure 2.4), its length (2.3.52), and the frequency (2.3.46) may be written as

$$\hat{y}_{(1)}/\hat{x}_0 \sim Re^{-3/8}, \quad \hat{y}_{(2)}/\hat{x}_0 \sim Re^{-1/2}, \quad \hat{y}_{(3)}/\hat{x}_0 \sim Re^{-5/8}, \\ \Delta\hat{x}/\hat{x}_0 \sim Re^{-3/8}, \quad \hat{\omega}/\Omega \sim Re^{1/4}, \tag{2.3.56}$$

where $\hat{x}_0$ is the distance between the leading edge of the plate and the interaction region, $\Omega = V_\infty/\hat{x}_0$ and $Re = V_\infty \hat{x}_0/\nu$.

When performing this task, you may use the fact that the dimensionless coordinates (x, y) are introduced through the scalings

$$\hat{x} = \delta^* x, \qquad \hat{y} = \delta^* y,$$

where δ^* is the displacement thickness of the boundary layer given by (2.1.4), that is

$$\delta^* = A'\sqrt{\frac{\nu \hat{x}_0}{V_\infty}}.$$

You may also use without proof the fact that $Re_* = A' Re^{1/2}$.

4. Refine the asymptotic analysis of root 1 of the dispersion equation

$$w_*^2 \int_{z_0}^{\infty} Ai(z)\, dz - i z_0^2 Ai'(z_0) = 0 \tag{2.3.57}$$

by including the next-order terms in the asymptotic expansions (2.3.43):

$$Ai'(z) = -\frac{1}{2\sqrt{\pi}} z^{1/4} e^{-\frac{2}{3}z^{3/2}} \left(1 + \frac{7}{48} z^{-3/2} + \cdots\right), \\ \int_z^{\infty} Ai(\zeta)\, d\zeta = \frac{1}{2\sqrt{\pi}} z^{-3/4} e^{-\frac{2}{3}z^{3/2}} \left(1 - \frac{41}{48} z^{-3/2} + \cdots\right). \tag{2.3.58}$$

Perform this task in the following steps:

(a) By substituting (2.3.58) into (2.3.57) deduce that

$$w_*^2 + i z_0^3 \left(1 + z_0^{-3/2} + \cdots\right) = 0 \quad \text{as} \quad z_0 \to \infty. \tag{2.3.59}$$

(b) Notice that, in the leading-order approximation, (2.3.59) reduces to the equation

$$z_0^3 = i w_*^2,$$

which has three roots:

$$z_{01} = e^{i\frac{\pi}{6}} w_*^{2/3}, \qquad z_{02} = e^{i\frac{5}{6}\pi} w_*^{2/3}, \qquad z_{03} = -i w_*^{2/3}.$$

Based on the results of the numerical analysis of the dispersion equation (Figure 2.2) choose the second of these, and seek the solution to equation (2.3.59) in the form

$$z_0 = e^{i\frac{5}{6}\pi} w_*^{2/3} + z_0', \tag{2.3.60}$$

where z_0' is a small correction. Substitute (2.3.60) into (2.3.59), and disregarding the squares of perturbations, deduce that

$$z_0' = -\frac{1}{3} e^{-i\frac{5}{12}\pi} \omega_*^{-1/3}. \tag{2.3.61}$$

(c) Finally, consider equation (2.3.39):

$$(i\alpha_*)^{2/3} = \frac{i\omega_*}{z_0}. \tag{2.3.62}$$

Substitute (2.3.61) into (2.3.60) and then into (2.3.62). Solve the resulting equation for α_*. You should have

$$\alpha_* = -\omega_*^{1/2} + \frac{1-i}{2\sqrt{2}} \omega_*^{-1/2} + \cdots \quad \text{as} \quad \omega_* \to \infty.$$

(d) What is the amplification rate of the perturbations represented by root 1 at large values of the frequency ω_*?

5. Consider the dispersion equation (2.3.34):

$$(i\alpha_*)^{1/3}|\alpha_*| \int\limits_{z_0}^{\infty} Ai(z)\, dz - Ai'(z_0) = 0, \qquad z_0 = \frac{i\omega_*}{(i\alpha_*)^{2/3}}. \tag{2.3.63}$$

Suppose that (2.3.63) admits a real negative solution, α_*, at a real positive value of ω_*. Prove that for $\omega_*' = -\omega_*$, there is a solution α_*' of (2.3.63) such that $\alpha_*' = -\alpha_*$.

Suggestion: First, prove that changing the sign of α_* turns z_0 into its complex conjugate. To perform this task use equation (2.3.38), where ϕ should be chosen to be zero for positive α_*, and $\phi = -\pi$ for negative α_*. Second, take complex conjugates on both sides of the first equation in (2.3.63). When performing this task, use (without proof) the *Schwarz reflection principle*. The latter is formulated as follows. Suppose that $F(z)$ is a function of a complex variable z, analytic in a domain that lies in the upper half plane ($\Im(z) > 0$). Suppose further that $F(z)$ assumes real values on the real axis. Then an analytic continuation of $F(z)$ into the lower half plane is given by

$$F(\bar{z}) = \overline{F(z)},$$

where the 'overline' denotes the complex conjugate.

2.4   Critical Layer and Stokes Layer Distinct: Upper Branch

In Section 2.3, the asymptotic analysis of the Orr–Sommerfeld equation (2.1.6) was performed under the assumption that the phase speed c satisfied the condition (2.2.44). This assumption allowed us to find the solution corresponding to the lower branch of the neutral curve. When deriving this solution, the triple-deck model (shown in Figure 2.4) had to be used.

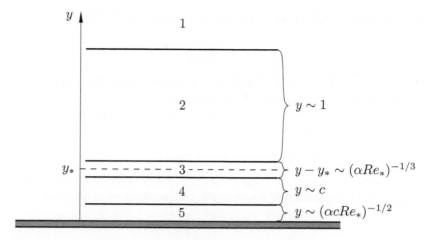

Fig. 2.5: The five-tier structure describing the upper branch modes.

Here, to construct the second solution of the Orr–Sommerfeld equation we shall assume that

$$|c| \gg (\alpha Re_*)^{-1/3}. \tag{2.4.1}$$

In this case the thickness of the critical layer (2.2.40) is much less than the distance (2.2.41) between this layer and the body surface, which gives rise to two more regions: the Stokes layer (region 5) adjacent to the wall and the inviscid region 4 that lies between the critical layer and the Stokes layer. As a result, the five-tier structure shown in Figure 2.5 forms. Here region 1 and region 2 play the same role as their counterparts in the triple-deck theory. Moreover, the solution (2.2.2) for region 1 remains valid, as does the solution (2.2.18) for region 2 together with its asymptotic form (2.2.33), (2.2.35) which can be used for matching with the critical layer (region 3).

We shall precede the formal mathematical analysis of the critical layer with the following 'intuitive' arguments. The existence of the critical layer is revealed by the second term in (2.2.33) with logarithmic singularity. However, as was mentioned before, S_2 is small. Hence, in the first instance we can disregard the logarithmic term in (2.2.33) and adopt the solution in the form of (2.3.1), that is

$$\check{v} = \frac{\alpha(1-c)}{\lambda_1} + \frac{\lambda_1}{1-c}(y - y_*) + O(\alpha^2). \tag{2.4.2}$$

Here and in the analysis that follows we will take $\alpha > 0$ without loss of generality and assume the phase speed c to be real at leading order but allow for the possibility of a small imaginary component in order to describe the slow growth or decay of disturbances in the vicinity of the upper branch of the neutral curve.

The solution (2.4.2) is unaffected by the existence of the critical layer, and may be used not only in region 2 but also in region 4. Since the flow in region 4 is inviscid, (2.4.2) should satisfy the impermeability condition on the body surface

$$\check{v}\Big|_{y=0} = 0. \tag{2.4.3}$$

Substituting (2.2.23) into (2.4.2) and then into (2.4.3) yields

$$c = \frac{\alpha}{\lambda_1} + O(\alpha^2). \qquad (2.4.4)$$

This confirms that c is small and real to leading order. To find the growth rate of the perturbations which is given by the imaginary part c_i of the phase velocity c, we need to consider the higher order terms and include the contributions of the critical layer and the Stokes layer.

2.4.1   Contribution of the critical layer

The characteristic thickness of the critical layer (region 3 in Figure 2.5) is given by (2.2.40). Correspondingly, we scale the independent variable y as

$$y = y_* + (\alpha Re_*)^{-1/3} Y, \qquad (2.4.5)$$

with Y being an order one quantity in region 3. We seek the solution of the Orr–Sommerfeld equation (2.1.6) in region 3 in the form

$$\breve{v} = V_0(Y) + \chi V_1(Y) + \cdots, \qquad (2.4.6)$$

where χ is a small parameter to be determined as part of the analysis of the Orr–Sommerfeld equation.[10] Using (2.4.5), (2.4.6) together with (2.2.24), (2.2.25) the three terms in this equation are calculated as

$$\frac{1}{i\alpha Re_*}\left(\frac{d^4\breve{v}}{dy^4} - 2\alpha^2\frac{d^2\breve{v}}{dy^2} + \alpha^4\breve{v}\right) = \frac{(\alpha Re_*)^{4/3}}{i\alpha Re_*}\left(\frac{d^4 V_0}{dY^4} + \chi\frac{d^4 V_1}{dY^4} + \cdots\right),$$

$$(U - c)\left(\frac{d^2\breve{v}}{dy^2} - \alpha^2\breve{v}\right) = \left[\left(\lambda_1 + \frac{\lambda_4}{6}y_*^3 + \cdots\right)(\alpha Re_*)^{-1/3}Y\right.$$

$$\left. + \frac{\lambda_4}{4}y_*^2(\alpha Re_*)^{-2/3}Y^2 + \cdots\right]$$

$$\times (\alpha Re_*)^{2/3}\left(\frac{d^2 V_0}{dY^2} + \chi\frac{d^2 V_1}{dY^2} + \cdots\right), \qquad (2.4.7)$$

$$\frac{d^2 U}{dy^2}\breve{v} = \frac{\lambda_4}{2}y_*^2 V_0 + \cdots.$$

We see that in the leading-order approximation the Orr–Sommerfeld equation reduces to

$$\frac{d^4 V_0}{dY^4} - i\lambda_1 Y\frac{d^2 V_0}{dY^2} = 0. \qquad (2.4.8)$$

The substitution of the independent variable

$$Z = (i\lambda_1)^{1/3}Y$$

---

[10]We shall see that the equation for V_0 is linear and homogeneous, which allows us to leave the task of defining the appropriate scaling for V_0 until the matching with the solution in region 2 is performed.

turns (2.4.8) into the Airy equation

$$\frac{d^4 V_0}{dZ^4} - Z\frac{d^2 V_0}{dZ^2} = 0$$

for $d^2 V_0/dZ^2$. The general solution to this equation is written as

$$\frac{d^2 V_0}{dZ^2} = C_1 Ai(Z) + C_2 Bi(Z). \tag{2.4.9}$$

It should allow for matching with the solutions in region 2 and region 4. Hence, we need to ensure that (2.4.9) does not grow exponentially as $Y \to \pm\infty$. At the upper edge of region 3, where Y is real and positive, Z lies on the ray $Z = e^{i\pi/6}\lambda_1^{1/3}Y$ in the complex Z-plane. It is easily seen from (2.3.12) that along this ray, $Ai(Z)$ decays while $Bi(Z)$ is exponentially growing. This means that we have to set $C_2 = 0$. Now, we consider the lower edge of region 3, where Y is real and negative, and Z lies on the ray $Z = e^{-i5\pi/6}\lambda_1^{1/3}|Y|$. Using again (2.3.12) we see that $Ai(Z)$ is exponentially growing as $|Y| \to \infty$ unless $C_1 = 0$. Thus, we can conclude that

$$\frac{d^2 V_0}{dY^2} = (i\lambda_1)^{2/3}\frac{d^2 V_0}{dZ^2} = 0, \tag{2.4.10}$$

and therefore

$$V_0 = b_0 + b_1 Y. \tag{2.4.11}$$

The constants b_0 and b_1 are found from matching with the solution (2.2.33), (2.2.35) in region 2. To perform the matching we need to obtain the 'inner expansion of the outer solution'. This is done by substituting (2.4.5) into (2.2.33), which leads to

$$\breve{v} = S_1 + S_2(\alpha Re_*)^{-1/3}Y\ln\left[(\alpha Re_*)^{-1/3}Y\right] + S_3(\alpha Re_*)^{-1/3}Y + \cdots, \tag{2.4.12}$$

and assuming $Y = O(1)$. It follows from (2.2.35a) and (2.4.4) that the first term in (2.4.12) is estimated as

$$S_1 = O(c). \tag{2.4.13}$$

According to (2.2.35c),

$$S_3 = O(1),$$

and therefore the third term in (2.4.12) is an order $(\alpha Re_*)^{-1/3}$ quantity, which, in view of (2.4.1), is much smaller than the first term (2.4.13). Using similar arguments, it is easily shown that the second term in (2.4.12) is even smaller. Hence, in the leading-order approximation the 'inner expansion of the outer solution' is written as

$$\breve{v} = S_1 + \cdots. \tag{2.4.14}$$

Now we need to consider the solution in region 3. Substituting (2.4.11) into (2.4.6) and restricting our attention to the leading-order approximation, we have the solution in region 3 in the form

$$\breve{v} = b_0 + b_1 Y.$$

It should coincide with (2.4.14). Hence

$$b_0 = \mathcal{S}_1, \qquad b_1 = 0,$$

which confirms our prediction that $\breve{v}$ does not change across the critical layer in the leading-order approximation.

Let us now turn to the next order term in (2.4.6). In order to avoid a trivial solution for V_1 we choose parameter χ to be $\chi = c^2 (\alpha Re_*)^{-1/3}$. Then, using (2.4.10) in (2.4.7) we find that V_1 satisfies the equation

$$\frac{d^4 V_1}{dY^4} - i\lambda_1 Y \frac{d^2 V_1}{dY^2} = -\frac{i\lambda_4}{2\lambda_1^2} b_0. \tag{2.4.15}$$

As with (2.4.8), we are interested in the solution of the equation (2.4.15) that does not grow exponentially as $|Y| \to \infty$. The behaviour of $V_1(Y)$ at the upper and lower edges of the critical layer may be easily ascertained with the help of the following arguments. Since the flow in regions 2 and 4 is inviscid, we expect the viscous term in (2.4.15) to become negligible as $Y \to \pm\infty$. We have

$$\frac{d^2 V_1}{dY^2} = \frac{b_0 \lambda_4}{2\lambda_1^3} Y^{-1} + \cdots \quad \text{as} \quad Y \to \pm\infty.$$

Integrating the above equation twice yields

$$V_1 = \frac{b_0 \lambda_4}{2\lambda_1^3} Y \left[\ln |Y| + h^\pm \right] + \cdots \quad \text{as} \quad Y \to \pm\infty,$$

where the plus/minus signs in the constant $h^\pm$ refer to the upper and lower edges of the critical layer. The solution of equation (2.4.15) will be given in Section 2.6. The key result is that $h^\pm$ experiences a jump (phase shift) across the critical layer, namely,

$$h^+ - h^- = i\pi. \tag{2.4.16}$$

This means that in order to continue the solution (2.2.33), (2.2.35) for region 2 through the critical layer into region 4 we simply need to substitute $\ln(y - y_*)$ in (2.2.33) by $\ln(y_* - y) - i\pi$. We find that in region 4

$$\breve{v} = \mathcal{S}_1 + \mathcal{S}_2 (y - y_*) \left[\ln(y_* - y) - i\pi \right] + \mathcal{S}_3 (y - y_*) + \cdots, \tag{2.4.17}$$

where $\mathcal{S}_1$, $\mathcal{S}_2$, and $\mathcal{S}_3$ are still given by (2.2.35).

2.4.2   Analysis of the Stokes layer

To complete the analysis of the five-tier structure, it remains to consider the Stokes layer (region 5 in Figure 2.5). The characteristic thickness of the Stokes layer is given by (2.2.43). Keeping in mind that, according to (2.4.4), the phase speed $c = O(\alpha)$, we scale the coordinate y in region 5 as

$$y = (\alpha^2 Re_*)^{-1/2} \bar{Y}, \tag{2.4.18}$$

with $\bar{Y}$ being an order one quantity.

We seek the solution for $\breve{v}$ in the form

$$\breve{v} = \bar{V}_0(\bar{Y}) + \cdots .\tag{2.4.19}$$

Substitution of (2.4.19) together with (2.4.18) into the Orr–Sommerfeld equation (2.1.6) yields

$$\frac{d^4\bar{V}_0}{d\bar{Y}^4} + i\frac{c}{\alpha}\frac{d^2\bar{V}_0}{d\bar{Y}^2} = 0,\tag{2.4.20}$$

where c/α can be taken to be real at leading order. The general solution of the equation (2.4.20) is written as

$$\bar{V}_0 = C_1 + C_2\bar{Y} + C_3 e^{-(1-i)\sqrt{c/2\alpha}\,\bar{Y}} + C_4 e^{(1-i)\sqrt{c/2\alpha}\,\bar{Y}}.\tag{2.4.21}$$

It is easily seen that the last term in (2.4.21) grows exponentially as $\bar{Y} \to \infty$, and therefore, we have to set $C_4 = 0$. It further follows from the no-slip conditions (2.1.7) that

$$\bar{V}_0 = \frac{d\bar{V}_0}{d\bar{Y}} = 0 \quad \text{at} \quad \bar{Y} = 0.\tag{2.4.22}$$

Substituting (2.4.21) into (2.4.22) we find that

$$C_1 = -C_3, \qquad C_2 = (1-i)\sqrt{\frac{c}{2\alpha}}\, C_3,$$

which allows us to write (2.4.21) in the form

$$\bar{V}_0 = C_3\left[(1-i)\sqrt{\frac{c}{2\alpha}}\,\bar{Y} - 1 + e^{-(1-i)\sqrt{c/2\alpha}\,\bar{Y}}\right].\tag{2.4.23}$$

2.4.3 The dispersion relations for the upper branch

To complete the flow analysis, we need to perform the matching of the solution in region 4 with the solution in the Stokes layer (region 5). The former is given by (2.4.17). The two-term Taylor expansion of (2.4.17) for small y is written as

$$\breve{v} = \left[\mathcal{S}_1 - \mathcal{S}_2 y_*(\ln y_* - i\pi) - \mathcal{S}_3 y_*\right] + \left[\mathcal{S}_2(\ln y_* - i\pi + 1) + \mathcal{S}_3\right]y + \cdots .\tag{2.4.24}$$

Substituting (2.4.18) into (2.4.24), we have the 'inner expansion of the outer solution' in the form

$$\breve{v} = \mathcal{F} + \mathcal{G}\bar{Y} + \cdots ,\tag{2.4.25}$$

where

$$\mathcal{F} = \mathcal{S}_1 - \mathcal{S}_2 y_*(\ln y_* - i\pi) - \mathcal{S}_3 y_*,\tag{2.4.26}$$

$$\mathcal{G} = \left[\mathcal{S}_2(\ln y_* - i\pi + 1) + \mathcal{S}_3\right](\alpha^2 Re_*)^{-1/2}.\tag{2.4.27}$$

To obtain the 'outer expansion of the inner solution' we need to examine the behaviour of (2.4.23) as $\bar{Y} \to \infty$. We discard the exponentially small term in (2.4.23) and substitute the remainder into (2.4.19). We have

$$\check{v} = -C_3 + C_3(1-i)\sqrt{\frac{c}{2\alpha}}\,\bar{Y} + \cdots,$$

which should coincide with (2.4.25). This requirement leads to the following two equations

$$\mathcal{F} = -C_3, \qquad \mathcal{G} = C_3(1-i)\sqrt{\frac{c}{2\alpha}}.$$

Elimination of the amplitude factor C_3 between these equations yields the *dispersion equation for the upper branch* of the neutral curve:

$$\alpha^{1/2}(1+i)\mathcal{G} = -\sqrt{2c}\,\mathcal{F} \qquad (2.4.28)$$

For any point that lies on the neutral curve or very close to it, we can substitute the phase speed c in (2.4.28) by its real part c_r. We then separate the real and imaginary parts in (2.4.28), which leads to

$$(2c_r)^{1/2}\mathcal{F}_r = \alpha^{1/2}(\mathcal{G}_i - \mathcal{G}_r), \qquad (2.4.29)$$

$$(2c_r)^{1/2}\mathcal{F}_i = -\alpha^{1/2}(\mathcal{G}_r + \mathcal{G}_i). \qquad (2.4.30)$$

It follows from (2.4.26) and (2.4.27) that

$$\mathcal{F}_r = \mathcal{S}_{1r} - y_*\big(\mathcal{S}_{2r}\ln y_* + \pi\mathcal{S}_{2i} + \mathcal{S}_{3r}\big), \qquad (2.4.31a)$$

$$\mathcal{F}_i = \mathcal{S}_{1i} - y_*\big(\mathcal{S}_{2i}\ln y_* - \pi\mathcal{S}_{2r} + \mathcal{S}_{3i}\big), \qquad (2.4.31b)$$

$$\mathcal{G}_r = \big[\mathcal{S}_{2r}(\ln y_* + 1) + \pi\mathcal{S}_{2i} + \underbrace{\mathcal{S}_{3r}}\big](\alpha^2 Re_*)^{-1/2}, \qquad (2.4.31c)$$

$$\mathcal{G}_i = \big[\mathcal{S}_{2i}(\ln y_* + 1) - \pi\mathcal{S}_{2r} + \mathcal{S}_{3i}\big](\alpha^2 Re_*)^{-1/2}. \qquad (2.4.31d)$$

It further follows from (2.2.35) that, upon neglecting some higher-order terms:

$$\left.\begin{aligned}
&\mathcal{S}_{1r} = \frac{\alpha(1-c_r)}{\lambda_1} - \frac{\alpha^2 D(0)}{\lambda_1}, &&\mathcal{S}_{1i} = -\frac{\alpha}{\lambda_1}c_i, \\[2mm]
&\mathcal{S}_{2r} = \frac{\lambda_4 \alpha c_r^2}{2\lambda_1^4}, &&\mathcal{S}_{2i} = -\frac{\lambda_4 \alpha c_r^2}{2\lambda_1^4}c_i, \\[2mm]
&\mathcal{S}_{3r} = \frac{\lambda_1}{1-c_r} - \alpha\lambda_1 B(0), &&\mathcal{S}_{3i} = \lambda_1 c_i.
\end{aligned}\right\} \qquad (2.4.32)$$

Examining the three terms in the equation (2.4.29) with the help of (2.4.31) and (2.4.32) we see that $\mathcal{G}_r$ and $\mathcal{G}_i$ are small compared to $\mathcal{F}_r$ which allows us to disregard the right-hand side of (2.4.29) leading to

$$\mathcal{S}_{1r} - y_*\big(\underbrace{\mathcal{S}_{2r}\ln y_*} + \underbrace{\pi\mathcal{S}_{2i}} + \mathcal{S}_{3r}\big) = 0.$$

Further simplifications arise from the fact that the under-braced terms in the above equation are small compared to $\mathcal{S}_{1r}$ and $\mathcal{S}_{3r}$. Hence, we can write

$$\mathcal{S}_{1r} - y_*\mathcal{S}_{3r} = 0. \qquad (2.4.33)$$

It remains to substitute $\mathcal{S}_{1r}$ and $\mathcal{S}_{3r}$ from (2.4.32) into (2.4.33) and make use of the equation (2.2.23) for y_* where, in the leading order, the phase speed c may be approximated by its real part c_r. We find that

$$\frac{\alpha(1-c_r)}{\lambda_1} - \frac{\alpha^2 D(0)}{\lambda_1} - \frac{c_r}{1-c_r} + \alpha c_r B(0) = 0.$$

To solve the above equation for c_r we multiply it through by $1 - c_r$ and write

$$c_r = \frac{\alpha(1-c_r)^2}{\lambda_1} - \frac{\alpha^2 D(0)(1-c_r)}{\lambda_1} + \alpha c_r (1-c_r) B(0). \qquad (2.4.34)$$

Obviously, the leading-order solution of (2.4.34) for small α is

$$c_r = \frac{\alpha}{\lambda_1} + O(\alpha^2), \qquad (2.4.35)$$

which confirms the validity of (2.4.4). To improve the accuracy of this solution, we use (2.4.35) on the right-hand side of (2.4.34). Disregarding the $O(\alpha^3)$ and higher order terms, we have

$$c_r = \frac{\alpha}{\lambda_1}\left\{1 - \alpha\left[\frac{2}{\lambda_1} + D(0) - B(0)\right]\right\} + O(\alpha^3). \qquad (2.4.36)$$

Equation (2.4.30) is dealt with in a similar way. We start with the function $\mathcal{F}_i$ on the left-hand side of (2.4.30). It follows from the equations for $\mathcal{S}_{1i}$ and $\mathcal{S}_{2i}$ in (2.4.32) that $y_* \mathcal{S}_{2i} \ln y_* \ll \mathcal{S}_{1i}$, which allows us to write (2.4.31b) as

$$\mathcal{F}_i = \mathcal{S}_{1i} + \pi y_* \mathcal{S}_{2r} - y_* \mathcal{S}_{3i}. \qquad (2.4.37)$$

The first term on the right-hand side of (2.4.37) can be taken directly from (2.4.32):

$$\mathcal{S}_{1i} = -\frac{\alpha}{\lambda_1} c_i. \qquad (2.4.38)$$

When dealing with the second term, we use again the fact that to leading order $y_* = c_r/\lambda_1$. We have

$$y_* \mathcal{S}_{2r} = \frac{\lambda_4 \alpha c_r^3}{2\lambda_1^5}. \qquad (2.4.39)$$

Finally, with the help of (2.4.35) the third term in (2.4.37) may be expressed in the form

$$y_* \mathcal{S}_{3i} = \frac{c_r}{\lambda_1}\lambda_1 c_i = c_r c_i = \frac{\alpha}{\lambda_1} c_i. \qquad (2.4.40)$$

Substitution of (2.4.38), (2.4.39), and (2.4.40) into (2.4.37) yields

$$\mathcal{F}_i = -2\frac{\alpha}{\lambda_1} c_i + \frac{\pi\lambda_4 \alpha c_r^3}{2\lambda_1^5}. \qquad (2.4.41)$$

Turning now to the right-hand side of (2.4.30) we need to examine the six terms in the square brackets in (2.4.31c) and (2.4.31d). Using (2.4.32) it is easily seen that the under-braced term in (2.4.31c) is dominant, while the remaining five terms may be disregarded in the leading-order approximation. Retaining the leading-order part of $\mathcal{S}_{3r}$ we therefore can conclude that

$$\mathcal{G}_r + \mathcal{G}_i = \lambda_1 (\alpha^2 Re_*)^{-1/2}. \tag{2.4.42}$$

Substituting (2.4.41) and (2.4.42) into (2.4.30) and solving the resulting equation for c_i we find that near the upper branch of the neutral curve the disturbances' amplification rate is given by

$$c_i = \frac{\pi \lambda_4 c_r^3}{4 \lambda_1^4} + \frac{\lambda_1^2}{2\sqrt{2}} (\alpha^3 c_r Re_*)^{-1/2} + \cdots . \tag{2.4.43}$$

It follows from the second equation in (2.2.21) that λ_4 is negative and so the first term on the right-hand side of (2.4.43) represents a stabilizing effect. The contribution from the second term is positive, leading to instability. To obtain the neutral curve we set c_i to zero and use the leading-order approximation for c_r from (2.4.36). We see that on the upper branch of the neutral curve:

$$\left. \begin{aligned} \alpha &= \frac{2^{1/10} \lambda_1^{19/10}}{\pi^{1/5} (-\lambda_4)^{1/5}} Re_*^{-1/10} + \cdots, \\ c_r &= \frac{2^{1/10} \lambda_1^{9/10}}{\pi^{1/5} (-\lambda_4)^{1/5}} Re_*^{-1/10} + \cdots \end{aligned} \right\} \quad \text{as} \quad Re_* \to \infty.$$

We see that if $\alpha \ll Re_*^{-1/10}$ then $c_i > 0$ and so the flow is unstable, while if $\alpha \gg Re_*^{-1/10}$ then $c_i < 0$, leading to stability. In Section 2.3 we saw that $\alpha \sim Re_*^{-1/4}$ on the lower branch of the neutral curve. We therefore can conclude that for $Re_* \gg 1$, the unstable region between the two branches of the neutral curve for Blasius flow lies in the wave number interval

$$O(Re_*^{-1/4}) < \alpha < O(Re_*^{-1/10}).$$

2.5   Asymptotic Analysis of the Orr–Sommerfeld Equation for Plane Poiseuille Flow

In the previous sections in this chapter we have seen how we can use perturbation methods to analyse the linear stability of Blasius flow at high Reynolds number. We have derived eigenrelations expressing the behaviour of the instability modes near the lower and upper branches of the neutral curve. In this section we will outline how these methods can be adapted to derive the corresponding neutral stability criteria for plane Poiseuille flow with

$$U = 1 - y^2 \quad \text{for} \quad y \in [-1, 1]. \tag{2.5.1}$$

As in the case of the Blasius boundary layer we will assume $Re \to \infty$ and start by considering the flow away from the boundaries, and then analyse separately the situations where the critical layer and near-wall Stokes layer are coincident (the lower branch) and distinct (the upper branch). We will concentrate here on simply describing

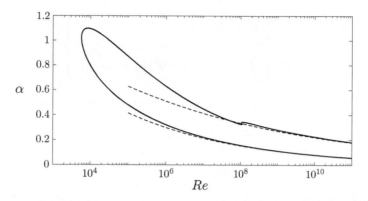

Fig. 2.6: The linear neutral curve for plane Poiseuille flow. The dashed curves show the large Reynolds number asymptotic predictions for the lower and upper branches.

the neutral modes, and therefore assume α and c real, taking $\alpha > 0$ without loss of generality.

As in the previous sections, our starting point is the Orr–Sommerfeld equation

$$(U - c)\left(\frac{d^2\breve{v}}{dy^2} - \alpha^2\breve{v}\right) - \frac{d^2U}{dy^2}\breve{v} = \frac{1}{i\alpha Re}\left(\frac{d^4\breve{v}}{dy^4} - 2\alpha^2\frac{d^2\breve{v}}{dy^2} + \alpha^4\breve{v}\right), \qquad (2.5.2)$$

where the Reynolds number Re is calculated based on the centreline velocity and channel half-width. We will consider modes where $\breve{v}$ is even about the channel centreline $y = 0$, so that

$$\frac{d\breve{v}}{dy} = \frac{d^3\breve{v}}{dy^3} = 0 \quad \text{on} \quad y = 0. \qquad (2.5.3)$$

These so-called *sinuous modes* are the ones that become unstable, as was established in Section 1.2.2. Our attention can therefore be restricted to the upper half of the channel $0 \le y \le 1$, where we need to solve equation (2.5.2) subject to (2.5.3) and the no-slip conditions on the upper plate:

$$\breve{v} = \frac{d\breve{v}}{dy} = 0 \quad \text{on} \quad y = 1. \qquad (2.5.4)$$

We anticipate, from observation of the neutral curve (Figure 2.6), and the earlier analysis for Blasius flow, that the wave number α and phase speed c will be small on both branches of the neutral curve: in view of this, we pose the following expansions[11]

$$\breve{v} = \breve{v}_0(y) + \alpha^2\breve{v}_1(y) + \cdots, \qquad c = \alpha^2 c_0 + \cdots, \qquad (2.5.5)$$

valid in the core of the flow where $y = O(1)$. Substitution of (2.5.5) into (2.5.2) yields

---

[11]Note that in Section 2.2.2 we performed our study of the Blasius flow by initially assuming that the phase speed $c = O(1)$. We then took the advantage of the fact that c was small in the later stages of the analysis.

$$\left. \begin{aligned} (1 - y^2)\frac{d^2 \breve{v}_0}{dy^2} + 2\breve{v}_0 &= 0, \\ (1 - y^2)\frac{d^2 \breve{v}_1}{dy^2} + 2\breve{v}_1 &= (1 - y^2)\breve{v}_0 + c_0 \frac{d^2 \breve{v}_0}{dy^2}, \end{aligned} \right\} \tag{2.5.6}$$

where we have assumed $\alpha^2 \gg (\alpha Re)^{-1}$, so that viscous effects can be neglected in this region to the order we work.

Equations (2.5.6) are dealt with in the same way as equations (2.2.4), (2.2.5). We find that the solutions of (2.5.6) satisfying the conditions (2.5.3) may be expressed in the form

$$\breve{v}_0 = a_0(1 - y^2), \tag{2.5.7}$$

$$\breve{v}_1 = a_1(1 - y^2) - \frac{a_0}{30}\left[30c_0 - 8 + 3(1 - y^2)^2 + 4(1 - y^2)\ln(1 - y^2)\right], \tag{2.5.8}$$

where a_0, a_1 are, generally speaking, arbitrary complex constants. However, keeping in mind that the Orr–Sommerfeld equation (2.5.2) is linear and homogeneous, as are the boundary conditions (2.5.3) and (2.5.4), we can 'normalize' the function $\breve{v}$ in (2.5.5) with a factor $\kappa = \kappa_0 + \alpha^2 \kappa_1 + \cdots$. Choosing $\kappa_0 = 1/a_0$ and $\kappa_1 = -a_1/a_0^2$ makes[12]

$$a_0 = 1, \qquad a_1 = 0.$$

The appearance of the logarithmic term in (2.5.8) signifies the formation of a critical layer in the vicinity of the upper wall. Examining (2.5.7), (2.5.8) it is easily found that the asymptotic expansions of the functions $v_0(y)$ and $v_1(y)$ on approach to the upper wall are

$$\left. \begin{aligned} \breve{v}_0 &= 2(1 - y) - (1 - y)^2 + \cdots, \\ \breve{v}_1 &= \left(\frac{4}{15} - c_0\right) - \frac{4}{15}(1 - y)\ln(1 - y) + \cdots \end{aligned} \right\} \quad \text{as} \quad y \to 1, \tag{2.5.9}$$

thereby demonstrating that the presumed hierarchy of the terms in (2.5.5) is violated when $(1 - y)$ becomes small enough. Indeed, if $(1 - y) = O(\alpha^2)$, then $\alpha^2 \breve{v}_1$ is no longer small compared to $\breve{v}_0$, and a new near-wall region should be considered where the independent variable Y_* is defined as

$$y = 1 - \alpha^2 Y_*. \tag{2.5.10}$$

Substitution of (2.5.10) into (2.5.9) and then into (2.5.5) yields

$$\breve{v} = \alpha^2\left(2Y_* + \frac{4}{15} - c_0\right) + \alpha^4 \ln\alpha\left(-\frac{8}{15}Y_*\right)$$
$$+ \alpha^4\left(-Y_*^2 - \frac{4}{15}Y_* \ln Y_*\right) + \cdots. \tag{2.5.11}$$

This suggests that the solution in the near-wall region should be sought in the form

---

[12]See Problem 1 in Exercises 8.

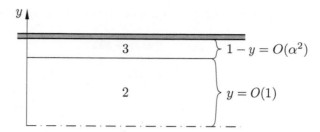

Fig. 2.7: Two-tier flow structure for the lower branch of the neutral curve for plane Poiseuille flow.

$$\breve{v} = \alpha^2 V_0(Y_*) + \alpha^4 \ln \alpha V_1(Y_*) + \alpha^4 V_2(Y_*) + \cdots . \qquad (2.5.12)$$

It further follows from (2.5.11) that the matching conditions with the solution in the core region are

$$\left.\begin{aligned}
V_0 &= 2Y_* + \frac{4}{15} - c_0 + \cdots , \\
V_1 &= -\frac{8}{15} Y_* + \cdots , \\
V_2 &= -Y_*^2 - \frac{4}{15} Y_* \ln Y_* + \cdots
\end{aligned}\right\} \quad \text{as} \quad Y_* \to \infty. \qquad (2.5.13)$$

Before deriving the equations for V_0, V_1, and V_2 we note that as in the Blasius flow case, there are two possibilities. In the first one, the entire near-wall layer is viscous, and the behaviour of the flow can be described in the framework of the two-tier structure shown in Figure 2.7, where region 2 is the core flow region, while region 3 is the near-wall region. The second possibility corresponds to the situation where region 3 is predominantly inviscid, and the viscous effects are confined to the critical layer and the Stokes layer forming inside region 3. We shall start with the former case which describes the lower branch of the neutral curve.

2.5.1   Derivation of the lower branch eigenrelation

If the wave number scales with Reynolds number in the following way:

$$\alpha^6 = O(\alpha Re)^{-1}, \qquad (2.5.14)$$

then viscous effects become significant at leading order within the sublayer. Indeed, substitution of (2.5.12) into (2.5.2) shows that V_0 satisfies

$$\alpha^6 (2Y_* - c_0) \frac{d^2 V_0}{dY_*^2} = \frac{1}{i\alpha Re} \frac{d^4 V_0}{dY_*^4}. \qquad (2.5.15)$$

Here it is taken into account that the basic velocity $U = 1 - y^2$ is written in terms of (2.5.10) as

$$U = \alpha^2 2Y_* + \cdots . \qquad (2.5.16)$$

We can then proceed along the lines of Section 2.3.1 by introducing the transformation

$$z = z_0 + \theta Y_*, \tag{2.5.17}$$

with

$$\theta = (2i\alpha^7 Re)^{1/3}, \qquad z_0 = -\frac{c_0}{2}\theta, \tag{2.5.18}$$

which reduces equation (2.5.15) to Airy's equation for $d^2 V_0 / dz^2$:

$$\frac{d^4 V_0}{dz^4} - z \frac{d^2 V_0}{dz^2} = 0. \tag{2.5.19}$$

The no-slip conditions (2.5.4) are written in the new variables as

$$V_0 = \frac{dV_0}{dz} = 0 \quad \text{at} \quad z = z_0. \tag{2.5.20}$$

The solution of (2.5.19) that does not grow exponentially as $z \to \infty$ and satisfies the boundary conditions (2.5.20) is given by (2.3.15):

$$V_0 = C_1^* \left[z \int_{z_0}^{z} Ai(\zeta)\, d\zeta - Ai'(z) + Ai'(z_0) \right]. \tag{2.5.21}$$

To deduce the desired dispersion equation, it remains to consider the matching condition for V_0 in (2.5.13). For z large, (2.5.21) implies

$$V_0 = C_1^* \left[z \int_{z_0}^{\infty} Ai(z)\, dz + Ai'(z_0) \right] + \cdots. \tag{2.5.22}$$

Now we substitute (2.5.17) into (2.5.22), which leads to

$$V_0 = C_1^* Y_* \theta \int_{z_0}^{\infty} Ai(z)\, dz + C_1^* \left[z_0 \int_{z_0}^{\infty} Ai(z)\, dz + Ai'(z_0) \right] + \cdots. \tag{2.5.23}$$

We see that the matching condition (2.5.13) for V_0 is satisfied provided that

$$C_1^* \theta \int_{z_0}^{\infty} Ai(z)\, dz = 2, \tag{2.5.24}$$

$$C_1^* \left[z_0 \int_{z_0}^{\infty} Ai(z)\, dz + Ai'(z_0) \right] = \frac{4}{15} - c_0. \tag{2.5.25}$$

Eliminating C_1^* from (2.5.24), (2.5.25) and using (2.5.18) we arrive at the conclusion that the desired *dispersion equation* is

$$(2i\alpha^7 Re)^{1/3} \int_{z_0}^{\infty} Ai(z)\, dz = \frac{15}{2} Ai'(z_0), \qquad z_0 = -\frac{c_0}{2}(2i\alpha^7 Re)^{1/3}. \tag{2.5.26}$$

This is analogous to the Blasius lower branch eigenrelation (2.3.27), (2.3.28) derived in Section 2.3 and can be solved by a similar computational method in order to investigate spatial or temporal growth in the vicinity of the lower branch. Our focus here however, is on neutral solutions, which may be easily deduced from the corresponding solution for the Blasius flow. Using (2.3.44), (2.3.45) in (2.3.34) we have

$$\frac{Ai'(z_0)}{\int_{z_0}^{\infty} Ai(z)\, dz} = e^{-i\pi/6}d_1, \qquad z_0 = e^{i5\pi/6}d_2, \tag{2.5.27}$$

where

$$d_1 = |\alpha_{*n}|^{4/3} \simeq 1.001, \qquad d_2 = \frac{\omega_{*n}}{|\alpha_{*n}|^{2/3}} \simeq 2.297. \tag{2.5.28}$$

Remember that when dealing with the stability of Blasius flow in Section 2.3, we assumed the wave number α negative. Here, α is assumed positive. Hence, to make (2.5.27) applicable to the dispersion equation (2.5.26) for plane Poiseuille flow we have to take the complex conjugates of the two equations in (2.5.27); see Problem 1 in Exercises 1. This turns (2.5.27) into

$$\frac{Ai'(z_0)}{\int_{z_0}^{\infty} Ai(z)\, dz} = e^{i\pi/6}d_1, \qquad z_0 = e^{-i5\pi/6}d_2, \tag{2.5.29}$$

Comparing (2.5.26) with (2.5.29) we can now see that

$$(2\alpha^7 Re)^{1/3} = \frac{15}{2}d_1, \qquad c_0 = \frac{4d_2}{15d_1}. \tag{2.5.30}$$

It remains to solve the first equation in (2.5.30) for α and combine the second equation with the asymptotic expansion for c in (2.5.5). We come to the conclusion that the asymptotic behaviour of the wave number α and the phase speed c for the neutral oscillations on the lower branch for plane Poiseuille flow are given by

$$\left.\begin{array}{l} \alpha = 2.149\, Re^{-1/7} + \cdots, \\ c = 2.826\, Re^{-2/7} + \cdots \end{array}\right\} \qquad \text{as} \quad Re \to \infty. \tag{2.5.31}$$

The prediction for α is plotted as the lower dashed curve on Figure 2.6, and demonstrates good agreement with the lower branch of the finite-Reynolds-number neutral curve.

2.5.2 Derivation of the upper branch eigenrelation

We revisit the sublayer expansion (2.5.12), but now assume that

$$\alpha^6 \gg \alpha Re^{-1},$$

in contrast to the lower branch scaling (2.5.14). Under these conditions, the flow in the sublayer appears to be predominantly inviscid, i.e. the viscous term on the right-hand side of equation (2.5.15) can be disregarded reducing this equation to

$$\frac{d^2 V_0}{dY_*^2} = 0. \tag{2.5.32}$$

The general solution of (2.5.32) is written as

$$V_0 = A_0 Y_* + B_0.$$

The constants A_0 and B_0 are easily found from the condition for V_0 in (2.5.13). We see that

$$A_0 = 2, \qquad B_0 = \frac{4}{15} - c_0.$$

Hence, we can conclude that

$$V_0 = 2Y_* + \frac{4}{15} - c_0. \tag{2.5.33}$$

While we expect the critical layer to form inside the region we are dealing with, the solution for $V_0(Y_*)$ does not involve the logarithmic singularity characteristic of the critical layer. This means that (2.5.33) remains unaffected by the critical layer and may be used both above and below it. Furthermore, since the equation (2.5.32) for $V_0(Y_*)$ is 'inviscid', its solution should satisfy the impermeability condition on the channel wall:

$$V_0 = 0 \quad \text{on} \quad Y_* = 0,$$

which produces the following leading-order dispersion equation:

$$c_0 = \frac{4}{15}. \tag{2.5.34}$$

Now we turn our attention to the second and third terms in (2.5.12). Supplementing the asymptotic expansion of the phase speed c in (2.5.5) with the higher order terms

$$c = \alpha^2 c_0 + \alpha^4 \ln \alpha\, c_1 + \alpha^4 c_2 + \cdots, \tag{2.5.35}$$

and substituting (2.5.12), (2.5.16), and (2.5.35) together with (2.5.10) into the Orr–Sommerfeld equation (2.5.2) we find that

$$\frac{d^2 V_1}{dY_*^2} = 0, \qquad (2Y_* - c_0)\frac{d^2 V_2}{dY_*^2} = -2V_0. \tag{2.5.36}$$

The solution for V_1 is given as

$$V_1 = A_1 Y_* + B_1. \tag{2.5.37}$$

The matching condition for V_1 in (2.5.13) allows us to determine the first constant $A_1 = -8/15$. The second constant, B_1, is found using the impermeability condition

$$V_1 = 0 \quad \text{at} \quad Y_* = 0,$$

and we see that $B_1 = 0$.

Finally, we consider the third term in (2.5.12). With (2.5.33) and (2.5.34), the equation for V_2 in (2.5.36) assumes the form

$$\frac{d^2V_2}{dY_*^2} = -\frac{4Y_*}{2Y_* - c_0}.$$

Integrating it twice we find that

$$V_2 = -c_0\left(Y_* - \frac{1}{2}c_0\right)\ln\left(Y_* - \frac{1}{2}c_0\right) - Y_*^2 + A_2Y_* + B_2. \tag{2.5.38}$$

It is easy to see that (2.5.38) satisfies the matching condition (2.5.13) automatically without imposing restriction on constants A_2 and B_2. To find A_2, B_2 one needs to refine the conditions (2.5.13) which can be done by analysing the next order terms in the solution (2.5.5) for the core region. Fortunately, there is no need to perform such an analysis. For future purposes we only need to know that A_2 and B_2 are real, which can be shown by extending the analysis in Problem 1, Exercises 8 to higher order terms.

The solution (2.5.38) is applicable to the region $Y_* > c_0/2$. To continue this solution across the critical line to the region $Y_* < c_0/2$, the jump condition (2.4.16) should be used, which turns (2.5.38) into[13]

$$V_2 = -c_0\left(Y_* - \frac{1}{2}c_0\right)\left[\ln\left(\frac{1}{2}c_0 - Y_*\right) - i\pi\right] - Y_*^2 + A_2Y_* + B_2.$$

To complete the analysis we need to consider the Stokes layer adjacent to the wall. We notice that on approach to this layer, both V_0 and V_1 decay linearly with Y_*:

$$V_0 = 2Y_* + \cdots, \qquad V_1 = -\frac{8}{15}Y_* + \cdots, \tag{2.5.39}$$

while V_2 remains finite:

$$V_2 = \frac{1}{2}c_0^2\left[\ln\left(\frac{c_0}{2}\right) - i\pi\right] + B_2 + \cdots \quad \text{as} \quad Y_* \to 0. \tag{2.5.40}$$

This makes the expansion (2.5.12) disordered in the near-wall region where Y_* becomes as small as $O(\alpha^2)$. To reexamine the Orr–Sommerfeld equation (2.5.2) in this region we scale the independent variable as

$$y = 1 - \alpha^2 Y_* = 1 - \alpha^4 \bar{Y}. \tag{2.5.41}$$

We then express (2.5.39), (2.5.40) in terms of $\bar{Y}$ and substitute into (2.5.12). As a result, we obtain the 'inner expansion of the outer solution':

$$\check{v} = \alpha^4\left\{2\bar{Y} + \frac{1}{2}c_0^2\left[\ln\left(\frac{c_0}{2}\right) - i\pi\right] + B_2\right\} + \cdots. \tag{2.5.42}$$

---

[13] As has been already mentioned the singularity at $Y = c_0/2$ necessitates the inclusion of a linear critical layer where viscous effects smooth out the irregular behaviour of V_2. A detailed analysis of the critical layer is given in Section 2.6.

Guided by (2.5.42) we represent the solution in the Stokes layer in the form

$$\check{v} = \alpha^4 \bar{V}(\bar{Y}) + \cdots, \qquad (2.5.43)$$

where the function $\bar{V}(\bar{Y})$ is such that

$$\bar{V}(\bar{Y}) = 2\bar{Y} + \frac{1}{2}c_0^2 \left[\ln\left(\frac{c_0}{2}\right) - i\pi \right] + B_2 + \cdots \quad \text{as} \quad \bar{Y} \to \infty. \qquad (2.5.44)$$

Substitution of (2.5.43), (2.5.41) together with (2.5.1) and (2.5.35) into the Orr–Sommerfeld equation (2.5.2) yields at leading order

$$\frac{d^4\bar{V}}{d\bar{Y}^4} + i c_0 \alpha^{11} Re \frac{d^2\bar{V}}{d\bar{Y}^2} = 0. \qquad (2.5.45)$$

The general solution of equation (2.5.45) is written as

$$\bar{V} = C_1 + C_2\bar{Y} + C_3 e^{-(1-i)\sqrt{\varkappa/2}\,\bar{Y}} + C_4 e^{(1-i)\sqrt{\varkappa/2}\,\bar{Y}}, \qquad (2.5.46)$$

where

$$\varkappa = c_0 \alpha^{11} Re. \qquad (2.5.47)$$

We start the analysis of this solution by noticing that the matching condition (2.5.44) cannot be satisfied if $\bar{V}$ grows exponentially at the outer edge of the Stokes layer. We therefore have to set $C_4 = 0$. We then use the no-slip conditions on the channel wall

$$\bar{V} = \frac{d\bar{V}}{d\bar{Y}} = 0 \quad \text{on} \quad \bar{Y} = 0. \qquad (2.5.48)$$

Substituting (2.5.46) into (2.5.48) and solving the resulting equations for C_2 and C_3 we find that

$$C_2 = -(1-i)\sqrt{\frac{\varkappa}{2}}\, C_1, \qquad C_3 = -C_1.$$

This allows us to write the solution (2.5.46) in the Stokes layer in the form

$$\bar{V} = C_1 \left[1 - (1-i)\sqrt{\frac{\varkappa}{2}}\,\bar{Y} - e^{-(1-i)\sqrt{\varkappa/2}\,\bar{Y}} \right]. \qquad (2.5.49)$$

It remains to apply the matching condition (2.5.44). Setting $\bar{Y} \to \infty$ in (2.5.49), and comparing the result with (2.5.44), we see that

$$-C_1(1-i)\sqrt{\frac{\varkappa}{2}} = 2, \qquad C_1 = \frac{1}{2}c_0^2 \left[\ln\left(\frac{c_0}{2}\right) - i\pi \right] + B_2.$$

Elimination of C_1 in the above equations yields

$$-(1+i)\sqrt{\frac{2}{\varkappa}} = \frac{1}{2}c_0^2 \left[\ln\left(\frac{c_0}{2}\right) - i\pi \right] + B_2. \qquad (2.5.50)$$

Taking the imaginary parts on both sides of the equation (2.5.50), we have

$$\sqrt{\frac{2}{\varkappa}} = \frac{\pi}{2}c_0^2. \qquad (2.5.51)$$

Finally, we substitute $\varkappa$ and c_0 from (2.5.47) and (2.5.34), which then leads us to the *upper branch asymptote*

$$\alpha = 1.789\, Re^{-1/11} + \cdots \quad \text{as} \quad Re \to \infty. \qquad (2.5.52)$$

We plot this as the upper dashed curve in Figure 2.6. Remember that the solid curve represents the results of numerical solution of the Orr–Sommerfeld equation performed under the assumption of a finite Reynolds number. We observe excellent agreement between the two provided Re is sufficiently large.

To obtain the corresponding asymptote for the phase speed c, we consider the leading-order term in (2.5.35) and then, using (2.5.34) and (2.5.52), we find that

$$c = 0.854\, Re^{-2/11} + \cdots \quad \text{as} \quad Re \to \infty.$$

The small growth/decay in the vicinity of the upper branch can also be modelled by this approach by retaining more terms in the asymptotic expansions.

Exercises 8

1. *Normalization of the solution for the core region*

   Express the solution (2.5.5), (2.5.7), (2.5.8) for the core region describing the linear stability of plane Poiseuille flow in the form

   $$\breve{v} = a_0 F_0(y) + \alpha^2 \Big[a_1 F_0(y) + a_0 G_1(y) \Big] + \cdots, \qquad (2.5.53)$$

   where $F_0(y)$ and $G_1(y)$ are real-valued functions and a_0, a_1 are complex constants.
      Show that the multiplicative factor $\kappa = \kappa_0 + \alpha^2 \kappa_1 + \cdots$ with $\kappa_0 = 1/a_0$ and $\kappa_1 = -a_1/a_0^2$ turns (2.5.53) into

   $$\breve{v} = F_0(y) + \alpha^2 G_1(y) + \cdots,$$

   thereby making $\breve{v}$ real.

2. *The upper branch eigenrelation for a pressure-gradient-driven boundary layer*

   Adjust the stability analysis of Sections 2.2 and 2.4 for the boundary layer exposed to a favourable pressure gradient. You may use without proof the fact that in this case the Taylor expansion of the basic velocity near the body surface (2.2.20) assumes the form

   $$U = \lambda_1 y - \frac{1}{2}\beta y^2 + \cdots \quad \text{as} \quad y \to 0,$$

   where both the skin friction parameter λ_1 and the pressure gradient parameter β are positive and may be assumed constant in a vicinity of the observation point $\hat{x}_0$.

Your task is to derive the large Re_* approximation for the wave number α on the upper branch of the neutral stability curve by solving the Orr–Sommerfeld equation (2.1.6) subject to the no-slip condition (2.1.7) on the body surface and attenuation in the far-field (2.1.8). You may accomplish this in the following steps:

(a) Guided by the corresponding analysis for the Blasius boundary layer (Section 2.2), consider $\alpha \ll 1$ and pose a phase speed expansion of the form

$$c = \alpha c_0 + \cdots .$$

Show that in an outer region, where $Y_1 = \alpha y$ is an $O(1)$ quantity, the leading-order solution of (2.1.6) which decays in the far-field is

$$\breve{v} = a_0 e^{-Y_1}. \tag{2.5.54}$$

Explain why we can take the constant $a_0 = 1$ without loss of generality. What inequality needs to hold between Re_* and α to ensure that viscosity can be neglected?

(b) Now consider the main part of the flow field where $y = O(1)$. Pose the expansion

$$\breve{v} = \breve{v}_0 + \alpha \breve{v}_1 + \cdots , \tag{2.5.55}$$

and, again under the assumption that viscosity is negligible, derive the following equations for $\breve{v}_0$ and $\breve{v}_1$:

$$U(y)\frac{d^2 \breve{v}_0}{dy^2} - \frac{d^2 U}{dy^2}\breve{v}_0 = 0, \qquad U(y)\frac{d^2 \breve{v}_1}{dy^2} - \frac{d^2 U}{dy^2}\breve{v}_1 = c_0 \frac{d^2 \breve{v}_0}{dy^2}.$$

Show that the solutions which match appropriately to (2.5.54) are

$$\breve{v}_0 = U(y), \qquad \breve{v}_1 = c_0\big[U(y) - 1\big] - U(y)\big[I(y) + y\big],$$

where

$$I(y) = \int_{\infty}^{y} \left[\frac{1}{U(y_1)^2} - 1\right] dy_1.$$

(c) By using a similar approach to that outlined in Section 2.2.2, show that $I(y)$ can be written in the form

$$I(y) = -\frac{1}{\lambda_1^2 y} + \frac{\beta}{\lambda_1^3}\ln\left(\frac{y}{1+y}\right) + B(y),$$

where $B(y)$, which is finite at $y = 0$, is an integral to be identified. Deduce that

$$\breve{v} = \lambda_1 y + \alpha\left(-c_0 + \frac{1}{\lambda_1} - \frac{\beta}{\lambda_1^2}y\ln y\right) + \cdots \quad \text{as} \quad y \to 0,$$

and hence that the expansion (2.5.55) becomes disordered when $y = O(\alpha)$.

(d) In the new region set $y = \alpha Y_*$, with $Y_* = O(1)$, and deduce that $\breve{v}$ expands in this region in the form

$$\breve{v} = \alpha V_0(Y_*) + \alpha^2 \ln \alpha V_1(Y_*) + \alpha^2 V_2(Y_*) + \cdots . \qquad (2.5.56)$$

Again, neglecting viscous effects, show that

$$\frac{d^2 V_0}{dY_*^2} = 0, \qquad \frac{d^2 V_1}{dY_*^2} = 0, \qquad (\lambda_1 Y_* - c_0)\frac{d^2 V_2}{dY_*^2} = -\beta V_0.$$

Write down the matching conditions to be satisfied by V_0, V_1, and V_2 as $Y_* \to \infty$.

Show that solutions for V_0, V_1 which satisfy the impermeability condition on the body surface only exist provided

$$c_0 = \lambda_1^{-1}.$$

Establish that the corresponding solution for V_2 is

$$V_2 = -\frac{\beta Y_*^2}{2} - \frac{\beta c_0}{\lambda_1}\left(Y_* - \frac{c_0}{\lambda_1}\right)\ln\left(Y_* - \frac{c_0}{\lambda_1}\right) + A_2 Y_* + B_2,$$

valid for $Y_* > c_0/\lambda_1$, where A_2, B_2 are real constants. Deduce the existence of a critical layer at $Y_* = c_0/\lambda_1$.

(e) By using the linear critical layer jump property (2.4.16), write down the corresponding solution for V_2 below the critical layer.

Analyse the behaviour of V_0, V_1, and V_2 on approach to the body surface, and observe that the asymptotic expansion (2.5.56) becomes disordered when Y_* decreases to $O(\alpha)$.

(f) Hence, introduce the Stokes layer, and argue that in this layer the solution should be sought in the form

$$\breve{v} = \alpha^2 \bar{V}(\bar{Y}) + \cdots , \qquad \bar{Y} = \alpha^{-2} y.$$

Show that if

$$\alpha^5 = O(\alpha Re_*)^{-1},$$

then viscous effects are present at leading order in the Stokes layer, and the function $\bar{V}(\bar{Y})$ satisfies the equation

$$-\alpha^5 c_0 \frac{d^2 \bar{V}}{d\bar{Y}^2} = \frac{1}{i\alpha Re_*}\frac{d^4 \bar{V}}{d\bar{Y}^4}. \qquad (2.5.57)$$

Formulate the no-slip conditions on the body surface and the conditions of matching with the solution outside the Stokes layer. Show that the solution of (2.5.57) subject to these conditions is only possible if

$$\frac{\pi\beta c_0^2}{\lambda_1^2} = \frac{\lambda_1}{(2c_0)^{1/2}}(\alpha^6 Re_*)^{-1/2}.$$

Hence deduce that the upper branch of the neutral stability curve at large Re_* is given by

$$\alpha = 2^{-1/6}(\pi\beta)^{-1/3}\lambda_1^{11/6} Re_*^{-1/6}.$$

2.6 Critical Layer Theory

We have seen in the preceding sections, when discussing the asymptotic structure of the upper branch mode, how it is necessary to introduce a *critical layer* at the location where the basic flow is equal to the phase speed of the disturbance in order to smooth out the logarithmic singularity that occurs in the absence of viscosity. In this section we will investigate in detail the internal structure of the critical layer and discuss the role it plays in the evolution of unsteady disturbances. As part of this process we will derive the jump condition (2.4.16) that we made use of in Sections 2.4 and 2.5 in order to study the upper branch instability for the Blasius boundary layer and plane Poiseuille flow, respectively. We will also see how this jump condition is modified as the disturbance size is increased. The critical layers considered in Sections 2.4 and 2.5 were situated close to a boundary due to the assumed smallness of the phase speed c, but we will now consider the more general case where c and the wave number α are $O(1)$ quantities and the critical layer itself may be situated well away from any boundaries. The Reynolds number will still be assumed large: $Re \gg 1$.

We will take as our starting point the Orr-Sommerfeld equation (1.2.35)

$$(U - c)\left(\frac{d^2\phi}{dy^2} - \alpha^2\phi\right) - \frac{d^2U}{dy^2}\phi = \frac{1}{i\alpha Re}\left(\frac{d^4\phi}{dy^4} - 2\alpha^2\frac{d^2\phi}{dy^2} + \alpha^4\phi\right), \qquad (2.6.1)$$

expressed here in terms of the perturbations of the stream function

$$\psi' = e^{i\alpha(x-ct)}\phi(y) + (c.c.). \qquad (2.6.2)$$

For the moment, we will neglect the viscous terms in (2.6.1), leaving us with the Rayleigh equation

$$(U - c)\left(\frac{d^2\phi}{dy^2} - \alpha^2\phi\right) - \frac{d^2U}{dy^2}\phi = 0, \qquad (2.6.3)$$

which has to be solved subject to the impermeability conditions on the boundaries, namely

$$\phi = 0 \quad \text{at} \quad y = y_1 \text{ and } y = y_2. \qquad (2.6.4)$$

Given a basic flow $U(y)$ and wave number α, the equation (2.6.3) considered together with the boundary conditions (2.6.4) comprises an eigen-value problem to determine the complex phase speed $c = c(\alpha)$. If we write $c = c_r + ic_i$ and assume α real positive, then for instability we require $c_i > 0$. From Theorem 1.2 in Section 1.4.1 we know that a necessary condition for instability is that $U(y)$ must contain a point of inflexion. We also have Howard's semicircle theorem (see Problem 3 in Exercises 3) which tells us that the eigen-values must lie inside the semi-circle in the complex c-plane given by

$$\left[c_r - \frac{1}{2}(U_{\max} + U_{\min})\right]^2 + c_i^2 = \left[\frac{1}{2}(U_{\max} - U_{\min})\right]^2,$$

with $U_{\min}$ and $U_{\max}$ being the minimum and maximum values of $U(y)$ over the interval $y_1 \le y \le y_2$. It therefore follows that, for a neutral mode with $c_i = 0$, there is a point

$y = y_*$ where $U = c_r$. In the vicinity of this point, the basic flow velocity $U(y)$ may be represented by the Taylor expansion

$$U = c + \tilde{\lambda}(y - y_*) + \tilde{\mu}(y - y_*)^2 + \cdots, \qquad (2.6.5)$$

with

$$\tilde{\lambda} = U'(y_*), \qquad \tilde{\mu} = \frac{1}{2}U''(y_*).$$

This means that $y = y_*$ is a regular singular point of Rayleigh's equation (2.6.3) provided $U''(y_*) \neq 0$. In fact it can be shown that the local *Frobenius solution* of (2.6.3) near $y = y_*$ takes the form (see Problem 1, Exercises 9)

$$\phi = A^{\pm}\phi_1(y) + B^{\pm}\phi_2(y), \qquad (2.6.6)$$

where

$$\phi_1(y) = \sum_{n=0}^{\infty} a_n(y - y_*)^{n+1}, \qquad (2.6.7)$$

with $a_0 = 1$, and

$$\phi_2(y) = \sum_{n=0}^{\infty} e_n(y - y_*)^n + \sigma\phi_1(y)\ln|y - y_*|, \qquad (2.6.8)$$

with $e_0 = 1$, $e_1 = 0$, and $\sigma = U''(y_*)/U'(y_*) = 2\tilde{\mu}/\tilde{\lambda}$. The $\pm$ signs on the coefficients $A^{\pm}$, $B^{\pm}$ are used in anticipation of a discontinuity in the solution across $y = y_*$.

We note that the derivatives of ϕ_2 and hence ϕ become singular as $y \to y_*$. This means that there is a small vicinity of $y = y_*$ where the viscous terms in (2.6.1) cannot be neglected. Indeed, keeping in mind that $U - c \sim y - y_*$, we can estimate the inertial term in (2.6.1) as

$$(U - c)\frac{d^2\phi}{dy^2} = O\left[|y - y_*|^{-1}\right],$$

while the viscous term

$$\frac{1}{i\alpha Re}\frac{d^4\phi}{dy^4} = O\left[(\alpha Re)^{-1}|y - y_*|^{-4}\right].$$

These become same order quantities in the *critical layer* whose thickness is estimated as

$$|y - y_*| = O(\alpha Re)^{-1/3}. \qquad (2.6.9)$$

Before proceeding further with the critical layer analysis, it is useful to revert from using the linear Orr–Sommerfeld equation as our starting point to the nonlinear Navier–Stokes equations. This will facilitate in Section 2.6.2 the inclusion of nonlinearity into the critical layer equations. For two-dimensional flow of an incompressible fluid, the Navier–Stokes equations are written as

$$\frac{\partial u}{\partial t} + u\frac{\partial u}{\partial x} + v\frac{\partial u}{\partial y} = -\frac{\partial p}{\partial x} + \frac{1}{Re}\left(\frac{\partial^2 u}{\partial x^2} + \frac{\partial^2 u}{\partial y^2}\right), \tag{2.6.10a}$$

$$\frac{\partial v}{\partial t} + u\frac{\partial v}{\partial x} + v\frac{\partial v}{\partial y} = -\frac{\partial p}{\partial y} + \frac{1}{Re}\left(\frac{\partial^2 v}{\partial x^2} + \frac{\partial^2 v}{\partial y^2}\right), \tag{2.6.10b}$$

$$\frac{\partial u}{\partial x} + \frac{\partial v}{\partial y} = 0. \tag{2.6.10c}$$

For our purposes it is convenient to express these equations in terms of the vorticity

$$\omega = \frac{\partial v}{\partial x} - \frac{\partial u}{\partial y}, \tag{2.6.11}$$

and the stream function ψ. Remember that ψ is related to the velocity components via

$$u = \frac{\partial \psi}{\partial y}, \qquad v = -\frac{\partial \psi}{\partial x}. \tag{2.6.12}$$

Differentiation of the x-momentum equation (2.6.10a) with respect to y yields

$$\frac{\partial}{\partial t}\left(\frac{\partial u}{\partial y}\right) + u\frac{\partial}{\partial x}\left(\frac{\partial u}{\partial y}\right) + \underbrace{\frac{\partial u}{\partial y}\frac{\partial u}{\partial x}} + v\frac{\partial}{\partial y}\left(\frac{\partial u}{\partial y}\right) + \underbrace{\frac{\partial v}{\partial y}\frac{\partial u}{\partial y}}$$

$$= -\frac{\partial^2 p}{\partial x \partial y} + \frac{1}{Re}\left[\frac{\partial^2}{\partial x^2}\left(\frac{\partial u}{\partial y}\right) + \frac{\partial^2}{\partial y^2}\left(\frac{\partial u}{\partial y}\right)\right], \tag{2.6.13}$$

where the under-braced terms can be discarded in view of the continuity equation (2.6.10c). Similarly, differentiating the y-momentum equation (2.6.10b) with respect to x, we have

$$\frac{\partial}{\partial t}\left(\frac{\partial v}{\partial x}\right) + u\frac{\partial}{\partial x}\left(\frac{\partial v}{\partial x}\right) + v\frac{\partial}{\partial y}\left(\frac{\partial v}{\partial x}\right)$$

$$= -\frac{\partial^2 p}{\partial x \partial y} + \frac{1}{Re}\left[\frac{\partial^2}{\partial x^2}\left(\frac{\partial v}{\partial x}\right) + \frac{\partial^2}{\partial y^2}\left(\frac{\partial v}{\partial x}\right)\right]. \tag{2.6.14}$$

We can now eliminate the pressure p by subtracting (2.6.13) from (2.6.14). This yields the following equation for the vorticity:

$$\frac{\partial \omega}{\partial t} + u\frac{\partial \omega}{\partial x} + v\frac{\partial \omega}{\partial y} = \frac{1}{Re}\left(\frac{\partial^2 \omega}{\partial x^2} + \frac{\partial^2 \omega}{\partial y^2}\right). \tag{2.6.15}$$

The equation

$$\omega = -\left(\frac{\partial^2 \psi}{\partial x^2} + \frac{\partial^2 \psi}{\partial y^2}\right) \tag{2.6.16}$$

relating the vorticity ω to the stream function ψ is obtained by substituting (2.6.12) into (2.6.11).

As usual, to perform the stability analysis, we represent the fluid-dynamic functions in the form

$$\left.\begin{array}{ll} \psi = \Psi(y) + \varepsilon\psi'(t,x,y), & \omega = \Omega(y) + \varepsilon\omega'(t,x,y), \\ u = U(y) + \varepsilon u'(t,x,y), & v = \varepsilon v'(t,x,y). \end{array}\right\} \tag{2.6.17}$$

We shall work under the parallel flow assumption, in which case U, Ω, and Ψ are functions of y only, and

$$U = \frac{d\Psi}{dy}, \qquad \Omega = -\frac{dU}{dy}.$$

Substitution of (2.6.17) into (2.6.15), (2.6.16), and (2.6.12) results in

$$\frac{\partial\omega'}{\partial t} + U\frac{\partial\omega'}{\partial x} + \frac{d\Omega}{dy}v' + \varepsilon\left(u'\frac{\partial\omega'}{\partial x} + v'\frac{\partial\omega'}{\partial y}\right) = \frac{1}{Re}\left(\frac{\partial^2\omega'}{\partial x^2} + \frac{\partial^2\omega'}{\partial y^2}\right), \tag{2.6.18a}$$

$$\omega' = -\left(\frac{\partial^2\psi'}{\partial x^2} + \frac{\partial^2\psi'}{\partial y^2}\right), \tag{2.6.18b}$$

$$u' = \frac{\partial\psi'}{\partial y}, \qquad v' = -\frac{\partial\psi'}{\partial x}. \tag{2.6.18c}$$

We shall use the above set of equations in our linear and nonlinear analysis of the critical layer.

2.6.1 Linear critical layer theory

Guided by (2.6.9) we scale y as

$$y = y_* + (\alpha Re)^{-1/3}Y, \tag{2.6.19}$$

and assume $Y = O(1)$ in the critical layer. The form of the asymptotic expansion of the stream function in the critical layer may be determined by re-expanding the 'outer solution' (2.6.6)–(2.6.8) in terms of the inner variable (2.6.19). For our purposes it is sufficient to represent $\phi_1(y)$ by the leading term in the sum in (2.6.7):

$$\phi_1 = y - y_* + \cdots.$$

Then, using (2.6.19), we can write

$$\phi_1 = (\alpha Re)^{-1/3}Y + \cdots. \tag{2.6.20}$$

Similarly, retaining the leading-order term in the sum in (2.6.8) we find that

$$\phi_2 = 1 + (\alpha Re)^{-1/3}\ln(\alpha Re)\left\{-\frac{1}{3}\sigma Y\right\} + (\alpha Re)^{-1/3}\left\{\sigma Y\ln|Y|\right\} + \cdots. \tag{2.6.21}$$

Now we substitute (2.6.20), (2.6.21) into (2.6.6) and then into (2.6.2). This yields 'the inner expansion of the outer solution':

$$\psi' = B^{\pm}e^{i\alpha\xi} + (\alpha Re)^{-1/3}\ln(\alpha Re)\left\{\left[-\frac{1}{3}B^{\pm}\sigma Y\right]e^{i\alpha\xi}\right\}$$

$$+ (\alpha Re)^{-1/3}\left\{\left[B^{\pm}\sigma Y\ln|Y| + A^{\pm}Y\right]e^{i\alpha\xi}\right\} + \cdots, \tag{2.6.22}$$

where $\xi = x - ct$ is the phase variable.

Guided by (2.6.22) we write the asymptotic expansion of ψ' in the critical layer in the form

$$\psi' = \psi_0'(\xi, Y) + (\alpha Re)^{-1/3} \ln(\alpha Re)\, \psi_{1L}'(\xi, Y) + (\alpha Re)^{-1/3} \psi_1'(\xi, Y) + \cdots, \quad (2.6.23)$$

where the functions ψ_0', ψ_{1L}', and ψ_1' have to satisfy the following matching conditions:

$$\left.\begin{aligned} \psi_0' &= B^{\pm} e^{i\alpha\xi} + \cdots, \\ \psi_{1L}' &= \left[-\frac{1}{3} B^{\pm} \sigma Y\right] e^{i\alpha\xi} + \cdots, \\ \psi_1' &= \left[B^{\pm}\sigma Y \ln|Y| + A^{\pm} Y\right] e^{i\alpha\xi} + \cdots \end{aligned}\right\} \quad \text{as} \quad Y \to \pm\infty. \quad (2.6.24)$$

Corresponding to (2.6.23), the asymptotic expansion for the vorticity perturbation function ω' is written as

$$\omega' = (\alpha Re)^{2/3} \omega_0'(\xi, Y) + (\alpha Re)^{1/3} \ln(\alpha Re)\, \omega_{1L}'(\xi, Y)$$
$$+ (\alpha Re)^{1/3} \omega_1'(\xi, Y) + \cdots. \quad (2.6.25)$$

We now substitute (2.6.23) and (2.6.25) together with (2.6.19) and (2.6.5) into the Navier–Stokes equations (2.6.18) where, to study the linear stability of the flow we set $\varepsilon = 0$. The equations for the leading-order terms ψ_0', ω_0' are

$$\tilde{\lambda} Y \frac{\partial \omega_0'}{\partial \xi} = \alpha \frac{\partial^2 \omega_0'}{\partial Y^2}, \qquad \omega_0' = -\frac{\partial^2 \psi_0'}{\partial Y^2}. \quad (2.6.26)$$

These have to be solved subject to the matching conditions

$$\psi_0' = B^{\pm} e^{i\alpha\xi} + \cdots \quad \text{as} \quad Y \to \pm\infty. \quad (2.6.27)$$

Guided by (2.6.27) we seek ψ_0' and ω_0' in the normal-mode form

$$\psi_0'(\xi, Y) = e^{i\alpha\xi} \breve{\psi}_0(Y), \qquad \omega_0'(\xi, Y) = e^{i\alpha\xi} \breve{\omega}_0(Y),$$

which turns (2.6.26), (2.6.27) into

$$\frac{d^2 \breve{\omega}_0}{dY^2} - i\tilde{\lambda} Y \breve{\omega}_0 = 0, \qquad \breve{\omega}_0 = -\frac{d^2 \breve{\psi}_0}{dY^2}, \quad (2.6.28a)$$

$$\breve{\psi}_0 \to B^{\pm} \quad \text{as} \quad Y \to \pm\infty. \quad (2.6.28b)$$

It is easily seen that the substitution of the second equation in (2.6.28a) into the first one leads to the equation (2.4.8) the properties of which have been already studied. We found that the solution of (2.4.8) that does not grow exponentially as $Y \to \pm\infty$ is simply a linear function of Y; see (2.4.11). Correspondingly, we shall write the solution to (2.6.28a) in the form

$$\breve{\psi}_0 = B + CY, \quad (2.6.29)$$

where B and C are constants. It immediately follows from (2.6.28b) that $C = 0$, which means that $\breve{\psi}_0$ remains unchanged across the critical layer and

$$B^+ = B^- = B.$$

Thus, we can conclude that

$$\psi_0' = Be^{i\alpha\xi}, \qquad \omega_0' = 0. \tag{2.6.30}$$

The logarithmic terms in (2.6.23), (2.6.25) are analysed in the same way. We find that in the critical layer

$$\psi_{1L}' = -\frac{1}{3}B\sigma Y e^{i\alpha\xi}, \qquad \omega_{1L}' = 0. \tag{2.6.31}$$

It remains to consider the functions ψ_1', ω_1'. The equations for these functions are found to be

$$\tilde{\lambda}Y\frac{\partial\omega_1'}{\partial\xi} + \tilde{\mu}Y^2\frac{\partial\omega_0'}{\partial\xi} + 2\tilde{\mu}\frac{\partial\psi_0'}{\partial\xi} = \alpha\frac{\partial^2\omega_1'}{\partial Y^2}, \qquad \omega_1' = -\frac{\partial^2\psi_1'}{\partial Y^2}. \tag{2.6.32a}$$

The boundary conditions for (2.6.32a) are given in (2.6.24):

$$\psi_1' = \left[B\sigma Y\ln|Y| + A^{\pm}Y\right]e^{i\alpha\xi} + \cdots \quad \text{as} \quad Y \to \pm\infty. \tag{2.6.32b}$$

We seek the solution of the boundary-value problem (2.6.32) in the normal-mode form

$$\psi_1' = e^{i\alpha\xi}\breve{\psi}_1(Y), \qquad \omega_1' - e^{i\alpha\xi}\breve{\omega}_1(Y). \tag{2.6.33}$$

Substitution of (2.6.33) together with (2.6.30) into (2.6.32) results in

$$\frac{d^2\breve{\omega}_1}{dY^2} - i\tilde{\lambda}Y\breve{\omega}_1 = i2\tilde{\mu}B, \qquad \breve{\omega}_1 = -\frac{d^2\breve{\psi}_1}{dY^2}, \tag{2.6.34a}$$

$$\breve{\psi}_1 = B\sigma Y\ln|Y| + A^{\pm}Y + \cdots \quad \text{as} \quad Y \to \pm\infty. \tag{2.6.34b}$$

Consider the first equation in (2.6.34a). If we introduce a new independent variable $Z = (i\tilde{\lambda})^{1/3}Y$, then this equation assumes the form of an inhomogeneous Airy equation

$$\frac{d^2\breve{w}_1}{dZ^2} - Z\breve{w}_1 = \mathcal{D}, \tag{2.6.35}$$

with constant $\mathcal{D} = 2i(i\tilde{\lambda})^{-2/3}\tilde{\mu}B$. Two complementary solutions of the homogeneous part of (2.6.35) are $Ai(Z)$ and $Bi(Z)$. We seek a particular solution in the form of a Laplace transform,[14]

$$\breve{w}_{1p}(Z) = \int_C F(\zeta)e^{\zeta Z}\,d\zeta, \tag{2.6.36}$$

where C is some contour (to be specified below) in the complex ζ-plane.

[14]For an alternative technique, see Problem 3 in Exercises 15.

Differentiating (2.6.36) with respect to Z twice, we have

$$\frac{d^2 \breve{w}_{1p}}{dZ^2} = \int\limits_C \zeta^2 F(\zeta) e^{\zeta Z} \, d\zeta. \tag{2.6.37}$$

Integrating (2.6.36) by parts gives

$$\breve{w}_{1p}(Z) = \int\limits_C F(\zeta) e^{\zeta Z} \, d\zeta = \frac{1}{Z} e^{\zeta Z} F(\zeta) \Big|_C - \frac{1}{Z} \int\limits_C \frac{dF}{d\zeta} e^{\zeta Z} \, d\zeta. \tag{2.6.38}$$

We now substitute (2.6.37), (2.6.38) into (2.6.35), which yields

$$\int\limits_C \left[\zeta^2 F(\zeta) + \frac{dF}{d\zeta} \right] e^{\zeta Z} \, d\zeta - e^{\zeta Z} F(\zeta) \Big|_C = \mathcal{D}.$$

This equation may be satisfied by setting

$$\frac{dF}{d\zeta} + \zeta^2 F = 0, \tag{2.6.39}$$

and

$$e^{\zeta Z} F(\zeta) \Big|_C = -\mathcal{D}. \tag{2.6.40}$$

The general solution of (2.6.39) is written as

$$F(\zeta) = \kappa e^{-\zeta^3/3}, \tag{2.6.41}$$

where κ is an arbitrary coefficient. We choose the contour C to start from $\zeta = 0$ and to extend to infinity along a ray that lies in one of the three unshaded sectors in Figure 2.8 where the real part of ζ^3 is positive, which makes $F(\infty) = 0$. We see that the condition (2.6.40) is satisfied provided that $\kappa = \mathcal{D}$. We can conclude that

$$\breve{w}_{1p} = \mathcal{D} \int\limits_C e^{-\zeta^3/3} e^{\zeta Z} \, d\zeta. \tag{2.6.42}$$

Now our task will be to find the asymptotic behaviour of (2.6.42) at large values of $|Y|$. To perform this task it is convenient to choose the integration contour C to coincide with the ray C^+ if $\tilde{\lambda} > 0$, and with the ray C^- if $\tilde{\lambda} < 0$; see Figure 2.8. The corresponding change of the integration variable

$$\zeta = \begin{cases} e^{-i\frac{2}{3}\pi} \tilde{\lambda}^{-1/3} s & \text{if } \tilde{\lambda} > 0, \\ e^{i\frac{2}{3}\pi} |\tilde{\lambda}|^{-1/3} s & \text{if } \tilde{\lambda} < 0, \end{cases}$$

allows us to express (2.6.42) in the form

$$\breve{w}_{1p} = -iB\sigma \mathrm{sgn}(\tilde{\lambda}) \int\limits_0^\infty f(s) \, e^{-i\,\mathrm{sgn}(\tilde{\lambda})Ys} \, ds, \tag{2.6.43}$$

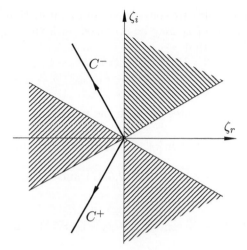

Fig. 2.8: Complex ζ-plane.

where

$$f(s) = e^{-s^3/3|\tilde{\lambda}|}.\tag{2.6.44}$$

Now we substitute (2.6.43) into the second equation in (2.6.34a), and integrate once with respect to Y from $Y = 0$. We obtain

$$\frac{d\breve{\psi}_{1p}}{dY} = -B\sigma I(Y) + \frac{d\breve{\psi}_1}{dY}\bigg|_{Y=0},\tag{2.6.45}$$

where

$$I(Y) = \int_0^\infty \frac{f(s)}{s}\left(e^{-i\,\mathrm{sgn}(\tilde{\lambda})Ys} - 1\right)ds.\tag{2.6.46}$$

In order to determine the behaviour in the far-field of the critical layer, we need to examine the integral I as $Y \to \pm\infty$. To facilitate this, we first split I up in the following way:

$$I(Y) = \sum_{n=1}^4 I_n,\tag{2.6.47}$$

where

$$\left.\begin{aligned}
I_1 &= \int_0^\infty \frac{f(s) - f(0)}{s}\, e^{i\eta s}\, ds, & I_2 &= f(0)\int_0^\infty \frac{\cos(\eta s) - \cos s}{s}\, ds, \\[2ex]
I_3 &= \int_0^\infty \frac{f(0)\cos s - f(s)}{s}\, ds, & I_4 &= -i\,f(0)\int_0^\infty \frac{\sin(\eta s)}{s}\, ds
\end{aligned}\right\}\tag{2.6.48}$$

with $\eta = \mathrm{sgn}(\tilde{\lambda})Y$.

All the integrals in (2.6.48) are convergent, and we note that I_3 is independent of Y. The asymptotic behaviour of I_1 may be obtained with the help of the integration by parts technique.[15] We find that

$$I_1 = -i\frac{f'(0)}{\eta} + O(\eta^{-2}) \quad \text{as} \quad \eta \to \pm\infty,$$

while I_2 and I_4 are standard integrals with the values

$$I_2 = -f(0)\ln|\eta|, \qquad I_4 = -\frac{i\pi}{2}f(0)\mathrm{sgn}(\eta).$$

Substituting these results back into (2.6.47) and using the definition (2.6.44) of $f(s)$, we find that

$$I(Y) = -f(0)\left[\ln|Y| + \frac{i\pi}{2}\mathrm{sgn}(\tilde{\lambda}Y)\right] + I_3 + O(Y^{-1}) \quad \text{as} \quad |Y| \to \infty. \tag{2.6.49}$$

We see that the particular solution we constructed does not grow exponentially at large values of $|Y|$. Therefore, if we return to the general solution

$$\breve{\omega}_1 = C_1 Ai(Z) + C_2 Bi(Z) + \breve{\omega}_{1p}$$

of (2.6.34a), then to avoid exponential growth of $\breve{\omega}_1$, we have to set $C_1 = C_2 = 0$. Thus, the particular solution we have constructed represents the required solution of (2.6.34a).

Using (2.6.49) in (2.6.45) shows that the matching condition (2.6.34b) is satisfied provided

$$A^+ - A^- = iB\sigma\pi\,\mathrm{sgn}(\tilde{\lambda}).$$

Thus, if one has the solution for the inviscid flow above the critical layer, then this solution can be adopted for the inviscid flow below the critical layer by a simple replacement

$$\ln(y - y_*) \implies \ln(y_* - y) - i\pi\,\mathrm{sgn}(\tilde{\lambda}). \tag{2.6.50}$$

We have used this result already (without proof) in Sections 2.4 and 2.5.

2.6.2   Nonlinear critical layer theory

Benney and Bergeron (1969) and independently Davis (1969) posed the question of what happens if we allow nonlinearity to affect the jump (2.6.50). Although to leading order the solution outside the critical layer remains unaffected, there is a significant change within the critical layer. It is found that the jump condition (2.6.50) is altered to

$$\underbrace{\ln(y - y_*)}_{y>y_*} \implies \underbrace{\ln(y_* - y) - i\theta(\varepsilon)\,\mathrm{sgn}(\tilde{\lambda})}_{y<y_*}, \tag{2.6.51}$$

where $\theta(\varepsilon)$ depends on the amplitude ε of the disturbance. In Benney and Bergeron (1969), it was assumed the phase shift θ is zero, but later in Haberman (1972) it was demonstrated how θ could vary between 0 and π depending on the perturbation size ε.

---

[15]See Section 1.1.2 in Part 2 of this book series.

We shall now derive the modified jump condition (2.6.51). As was already mentioned, we expect the solution outside the critical layer to remain unaffected by the nonlinearity. This means that 'the inner expansion of the outer solution' (2.6.22) remains unchanged, suggesting that the solution in the critical layer should still be written in the form of the asymptotic expansions (2.6.23), (2.6.25). Moreover, we have seen in Section 2.6.1 that the jump condition (2.6.50) is defined by the equations (2.6.32) for the third terms ψ_1', ω_1' in (2.6.23), (2.6.25). The role of the first two terms proved to be 'passive', which allows us to use the solutions (2.6.30), (2.6.31) for ψ_0', ψ_{1L}', ω_0', and ω_{1L}' as given by the linear theory. When dealing with nonlinear perturbations, we shall normalize the amplitude of the perturbations ε by choosing $B = 1$, and then ψ_0' and ψ_{1L}' may be written as

$$\left.\begin{aligned}
\psi_0' &= Be^{i\alpha\xi} + (c.c.) = e^{i\alpha\xi} + e^{-i\alpha\xi} = 2\cos(\alpha\xi), \\
\psi_{1L}' &= -\frac{2}{3}\sigma Y \cos(\alpha\xi).
\end{aligned}\right\} \tag{2.6.52}$$

Substituting (2.6.52) into (2.6.23) and taking into account that $\omega_0' = \omega_{1L}' = 0$ in (2.6.25) we seek the solution for the nonlinear critical layer in the form

$$\left.\begin{aligned}
\psi' &= 2\cos(\alpha\xi) + (\alpha Re)^{-1/3}\ln(\alpha Re)\left[-\frac{2}{3}B\sigma Y \cos(\alpha\xi)\right] \\
&\qquad\qquad\qquad + (\alpha Re)^{-1/3}\psi_1'(\xi, Y) + \cdots, \\
\omega' &= (\alpha Re)^{1/3}\omega_1'(\xi, Y) + \cdots,
\end{aligned}\right\} \tag{2.6.53}$$

where

$$\xi = x - ct, \qquad y = y_* + (\alpha Re)^{-1/3}Y. \tag{2.6.54}$$

Now we need to substitute (2.6.53), (2.6.54), together with (2.6.5) into the Navier–Stokes equations (2.6.18). We find that the inertial terms in the equation (2.6.18a) are

$$\frac{\partial\omega'}{\partial t} + U\frac{\partial\omega'}{\partial x} + \frac{d\Omega}{dy}v' = \tilde{\lambda}Y\frac{\partial\omega_1'}{\partial\xi} - 4\tilde{\mu}\alpha\sin(\alpha\xi) + \cdots. \tag{2.6.55}$$

The nonlinear terms are calculated as

$$\varepsilon\left(u'\frac{\partial\omega'}{\partial x} + v'\frac{\partial\omega'}{\partial y}\right) = \varepsilon(\alpha Re)^{2/3}2\alpha\frac{\partial\omega_1'}{\partial Y}\sin(\alpha\xi) + \cdots, \tag{2.6.56}$$

and the viscous terms are

$$\frac{1}{Re}\left(\frac{\partial^2\omega'}{\partial x^2} + \frac{\partial^2\omega'}{\partial y^2}\right) = \alpha\frac{\partial^2\omega_1'}{\partial Y^2} + \cdots. \tag{2.6.57}$$

As expected, the inertial (2.6.55) and viscous (2.6.57) are same order quantities. The order of magnitude of the nonlinear terms (2.6.56) depends on ε. If $\varepsilon \ll (\alpha Re)^{-2/3}$, then the nonlinear terms may be disregarded, which leads to the linear critical layer theory presented in Section 2.6.1. To study the effects of nonlinearity, we set

$$\varepsilon = (\alpha Re)^{-2/3}\varepsilon_0,$$

where ε_0 is an order one amplitude parameter. Then in place of (2.6.32a) we will have

$$\tilde{\lambda}Y\frac{\partial\omega_1'}{\partial\xi} - \left(4\tilde{\mu} - 2\varepsilon_0\frac{\partial\omega_1'}{\partial Y}\right)\alpha\sin(\alpha\xi) = \alpha\frac{\partial^2\omega_1'}{\partial Y^2}, \qquad \omega_1' = -\frac{\partial^2\psi_1'}{\partial Y^2}. \qquad (2.6.58a)$$

Setting $B = 1$ and recalling that $\sigma = 2\tilde{\mu}/\tilde{\lambda}$ we have the matching condition (2.6.32b) in the form

$$\psi_1' = \left[B\sigma Y\ln|Y| + A^{\pm}Y\right]e^{i\alpha\xi} + (c.c.) + \cdots$$

$$= \left[\frac{2\tilde{\mu}}{\tilde{\lambda}}Y\ln|Y| + A^{\pm}Y\right]2\cos(\alpha\xi) + \cdots \quad \text{as} \quad Y \to \pm\infty. \qquad (2.6.58b)$$

The transformation of the variables

$$\omega_1' = \frac{2\tilde{\mu}}{\varepsilon_0}\left[Y - \sqrt{\frac{2\varepsilon_0}{|\tilde{\lambda}|}}Q(\xi^*,Z)\right], \qquad \xi = \frac{1}{\alpha}\xi^*, \qquad Y = \sqrt{\frac{2\varepsilon_0}{|\tilde{\lambda}|}}Z$$

renders (2.6.58) in the following canonical form

$$Z\frac{\partial Q}{\partial\xi^*} + \text{sgn}(\tilde{\lambda})\sin(\xi^*)\frac{\partial Q}{\partial Z} = \gamma_c\frac{\partial^2 Q}{\partial Z^2}, \qquad (2.6.59)$$

$$Q = Z + \text{sgn}(\tilde{\lambda})\frac{\cos\xi^*}{Z} + \cdots \quad \text{as} \quad Z \to \pm\infty. \qquad (2.6.60)$$

The important parameter γ_c in (2.6.59) is given by

$$\gamma_c = \frac{1}{\tilde{\lambda}}\left(\frac{|\tilde{\lambda}|}{2\varepsilon_0}\right)^{3/2},$$

and controls the extent of the nonlinearity within the critical layer.

We shall see that the solution of (2.6.59) requires an additional term $2H^{\pm}$ in (2.6.60), that is

$$Q = Z + 2H^{\pm} + \text{sgn}(\tilde{\lambda})\frac{\cos\xi^*}{Z} + \cdots \quad \text{as} \quad Z \to \pm\infty. \qquad (2.6.61)$$

The need for this term is not immediately apparent from examining the solution outside the critical layer. The fact is that this term represents a correction to the mean flow arising from nonlinear interactions inside the critical layer. We shall assume $H^{\pm}$ to be constant, and refer to $H^+ - H^-$ as the *mean vorticity jump*. After the solution in the critical layer is constructed, one can revisit the outer flow and refine the solution there accordingly.

The boundary-value problem (2.6.59), (2.6.61) requires a numerical solution; we will return to this presently. To see how the nonlinear jump (2.6.51) emerges, it is convenient to introduce the 'normalized' stream function ψ^* such that

$$Q = \frac{\partial^2 \psi^*}{\partial Z^2}.$$

Integrating (2.6.61), we express the result in the form

$$\frac{\partial \psi^*}{\partial Z} = \frac{1}{2} Z^2 + 2H^{\pm} Z + \mathrm{sgn}(\tilde{\lambda}) \Re\{e^{i\xi^*} \ln|Z|\}$$
$$+ U^{\pm}(\xi^*) + \cdots \quad \text{as} \quad Z \to \pm\infty. \qquad (2.6.62)$$

The contribution $U^{\pm}$ corresponds to a jump in the velocity across the critical layer. Since this quantity is periodic in ξ^*, we can express it as a Fourier series

$$U^+ - U^- = \frac{1}{2} a_0 + \sum_{n=1}^{\infty} \left(a_n \cos n\xi^* + b_n \sin n\xi^* \right). \qquad (2.6.63)$$

Notice that the first three terms in (2.6.62) are even in ξ^* while the fourth term, $U^{\pm}(\xi^*)$, breaks the symmetry. Hence, we define the phase shift θ to be the coefficient of the highest symmetry-breaking harmonic in (2.6.63), namely, $\theta = -b_1$. The latter may be found by making use of the standard Fourier integral

$$\theta = -\frac{1}{\pi} \int_0^{2\pi} (U^+ - U^-) \sin\xi^* \, d\xi^*.$$

Using (2.6.63) in (2.6.62), the jump of $\partial\psi^*/\partial Z$ across the critical layer is calculated as

$$\left[\frac{\partial \psi^*}{\partial Z} \right]_-^+ = 2(H^+ - H^-)Z - \theta \sin\xi^*$$
$$+ \frac{1}{2} a_0 + \sum_{n=1}^{\infty} a_n \cos n\xi^* + \sum_{n=2}^{\infty} b_n \sin n\xi^*. \qquad (2.6.64)$$

This is equivalent to the logarithmic term in (2.6.62) effectively experiencing a jump of the form

$$\underbrace{\ln Z}_{Z>0} \quad \Longrightarrow \quad \underbrace{\ln|Z| - i\theta\,\mathrm{sgn}(\tilde{\lambda})}_{Z<0}, \qquad (2.6.65)$$

which is the condition stated in (2.6.51).

It is also possible to derive a relation between the phase shift θ and the mean vorticity jump $H^+ - H^-$. To demonstrate this, it is convenient to rewrite the critical layer equation (2.6.59) in the form

$$Z \frac{\partial Q}{\partial \xi^*} + \mathrm{sgn}(\tilde{\lambda}) \sin(\xi^*) \frac{\partial^3 \psi^*}{\partial Z^3} = \gamma_c \frac{\partial^2 Q}{\partial Z^2}.$$

Integrating with respect to ξ^* from 0 to 2π and using the fact that Q is a periodic function with period 2π we obtain

$$\text{sgn}(\tilde{\lambda}) \int_0^{2\pi} \sin(\xi^*) \frac{\partial^3 \psi^*}{\partial Z^3}\, d\xi^* = \gamma_c \int_0^{2\pi} \frac{\partial^2 Q}{\partial Z^2}\, d\xi^*.$$

Next, we integrate twice with respect to Z and then consider the jump in the quantities across the critical layer. This leads us to

$$\text{sgn}(\tilde{\lambda}) \int_0^{2\pi} \sin(\xi^*) \left[\frac{\partial \psi^*}{\partial Z}\right]_-^+ d\xi^* = \gamma_c \int_0^{2\pi} [Q]_-^+\, d\xi^*. \qquad (2.6.66)$$

The integral on the left-hand side of (2.6.66) is calculated using (2.6.64). We have

$$\int_0^{2\pi} \sin(\xi^*) \left[\frac{\partial \psi^*}{\partial Z}\right]_-^+ d\xi^* = -\pi\theta. \qquad (2.6.67)$$

Turning to the right-hand side of (2.6.66) we use the equation (2.6.61) according to which $[Q]_-^+ = 2(H^+ - H^-)$. We see that

$$\int_0^{2\pi} [Q]_-^+\, d\xi^* = 4\pi(H^+ - H^-). \qquad (2.6.68)$$

It remains to substitute (2.6.67) and (2.6.68) into (2.6.66) and solve the resulting equation for θ. We find that the *phase shift* is related to the *mean vorticity jump* by the relation

$$\theta = -4\gamma_c(H^+ - H^-)\,\text{sgn}(\tilde{\lambda}). \qquad (2.6.69)$$

For order one values of γ_c, to determine the vorticity jump $H^+ - H^-$, the equation (2.6.59) has to be solved numerically subject to the conditions (2.6.61), with the phase shift θ following from (2.6.69). The results of the calculations are displayed in Figure 2.9 for $\tilde{\lambda} > 0$, and show, as expected, that in the large γ_c limit, θ tends to its linear value of π while the mean vorticity jump $H^+ - H^-$ tends to zero. In the nonlinear limit $\gamma_c \to 0$ we see that the phase shift tends to zero, whereas the mean vorticity jump remains finite.[16] It is important to note also that the neutral wave numbers, phase speeds, and also disturbance amplitudes depend on the phase shift as shown by Benney and Bergeron (1969), Haberman (1972), and Smith and Bodonyi (1982*b*).

As the parameter γ_c becomes ever closer to zero, the effects of viscosity become ever more subtle, and the determination of the asymptotically small phase shift is more involved. Nevertheless, the basic principle of the balancing of the viscosity-induced phase shifts across the critical layer and the near-wall Stokes layer still underpins the

---

[16]These limits are examined in more detail in Exercises 9.

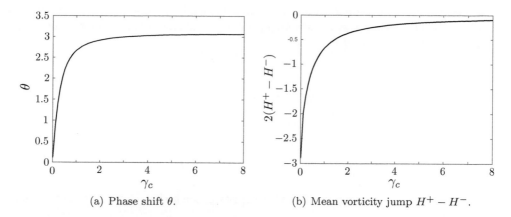

(a) Phase shift θ.   (b) Mean vorticity jump $H^+ - H^-$.

Fig. 2.9: A plot of the phase shift θ and the jump in mean vorticity $H^+ - H^-$ versus parameter γ_c.

structure and it this balancing which determines the small but finite amplitude of the neutral modes in this strongly nonlinear regime. One particular example where these ideas have been successfully applied concerns the Hagen–Poiseuille flow through a circular pipe which is well known to be linearly stable at all Reynolds numbers. Using strongly nonlinear critical layer theory, Smith and Bodonyi (1982a) proposed a nonlinear amplitude-dependent neutral mode structure involving spiral waves that has subsequently been found to agree quantitatively with full Navier–Stokes computations provided the Reynolds number is sufficiently large; see Deguchi and Walton (2013).

The analysis of the critical layer has shown that the vorticity disturbance becomes very large in the vicinity of the critical layer. In our case the singularity was smoothed out by re-introducing the viscous terms which were neglected everywhere except in the critical layer. In our analysis we have assumed that the disturbance amplitude does not grow or decay in space or time and in fact the critical layer is referred to as an *equilibrium critical layer*. The ideas can be extended to allow for a more general situation in which the disturbance is allowed to develop temporally and spatially. Hickernell (1984) demonstrated how slow unsteadiness or spatial growth could be incorporated into the critical layer. In Chapter 4 the ideas of weakly nonlinear theory are studied leading to the famous Landau–Stuart equation. Hickernell (1984) showed that unsteadiness and weak nonlinearity gives rise to an integro-differential equation describing the nonlinear development of the disturbance amplitude, quite different from the Landau–Stuart equation, and this is primarily because of the dynamics inside the critical layer. A discussion of these more complex ideas and applications to flows such as shear layers and cross-flow vortices may be found in the review articles and papers by Stewartson (1981), Maslowe (1986), Gajjar and Cole (1989), Gajjar (1996), Wu (2004).

The type of neutral mode investigated in this chapter, involving the interaction of the critical layer with the wall layer, can be regarded as a form of *self-sustaining process* in which wave interactions within the critical layer induce a distortion to the mean flow which itself sustains the neutral waves. These types of processes are much more prevalent in three-dimensional flows and typically produce strongly subcritical instabilities. The theory of self-sustaining processes and their application to shear flows is the focus of Chapter 5 of this volume.

Exercises 9

1. *Frobenius solution for the Rayleigh equation*

   Consider Rayleigh's equation for the basic flow $U(y)$

$$\mathcal{L}\phi \equiv \frac{d^2\phi}{dy^2} - \left(\alpha^2 + \frac{U''}{U-c}\right)\phi = 0, \qquad (2.6.70)$$

   where c is real. We suppose that there exists y_* such that $U(y_*) = c$, so that near $y = y_*$

$$U(y) = c + U'(y_*)(y - y_*) + \frac{1}{2}U''(y_*)(y - y_*)^2 + \cdots .$$

   (a) Introduce the function

$$q(y) = \alpha^2 + \frac{U''}{U-c}$$

   and show that near $y = y_*$ this function can be written in the form

$$q(y) = \sum_{n=0}^{\infty} q_n(y - y_*)^{n-1} = \frac{q_0}{y - y_*} + q_1 + \cdots ,$$

   where $q_0 = U''(y_*)/U'(y_*)$. This demonstrates that $y = y_*$ is a regular singular point of Rayleigh's equation.

   (b) Look for a solution to (2.6.70) in Frobenius form

$$\phi = \sum_{n=0}^{\infty} a_n(y - y_*)^{n+\gamma}, \qquad (2.6.71)$$

   and show that

$$\begin{aligned} \mathcal{L}\phi = {} & a_0\gamma(\gamma - 1)(y - y_*)^{\gamma-2} \\ & + \big[a_1(\gamma + 1)\gamma - a_0 q_0\big](y - y_*)^{\gamma-1} \\ & + \sum_{n=1}^{\infty}\bigg[a_{n+1}(n + 1 + \gamma)(n + \gamma) - \sum_{k=0}^{n} q_k a_{n-k}\bigg](y - y_*)^{n+\gamma-1}. \quad (2.6.72) \end{aligned}$$

   Deduce from the first term in (2.6.72) that if $a_0 \neq 0$ the indicial equation is

$$\gamma(\gamma - 1) = 0,$$

   with the solutions $\gamma = 1$ and $\gamma = 0$.

(c) Show that one complementary solution to (2.6.70)

$$\phi_1 = \sum_{n=0}^{\infty} a_n(y - y_*)^{n+1}$$

is obtained by taking $\gamma = 1$ and with $a_0 = 1$ the recurrence relation for determining a_{n+1} for $n \geq 0$ is

$$a_{n+1}(n+\gamma)(n+\gamma+1) = \sum_{k=0}^{n} q_k a_{n-k}. \qquad (2.6.73)$$

(d) Now take $\gamma = 0$. By considering the coefficients of $(y - y_*)^{-2}$ and $(y - y_*)^{-1}$ in (2.6.72), show that

$$0 \cdot a_0 = 0, \qquad 0 \cdot a_1 = q_0 a_0. \qquad (2.6.74)$$

This means that a second linearly independent solution of the form

$$\phi_2 = \sum_{n=0}^{\infty} a_n(y - y_*)^n \qquad (2.6.75)$$

can be found if $q_0 = U''(y_*)/U'(y_*) = 0$. Explain why we can take $a_0 = 1$, $a_1 = 0$ in (2.6.75) without loss of generality.

If $U''(y_*) \neq 0$ then deduce from (2.6.74) that a second linearly independent solution of Frobenius form (2.6.71) with $\gamma = 0$ cannot be found.

(e) To find a second linearly independent solution of (2.6.70) for the case $q_0 \neq 0$ let

$$\bar{\phi}(y; \gamma) = \sum_{n=0}^{\infty} a_n(\gamma)(y - y_*)^{n+\gamma}, \qquad (2.6.76)$$

where $a_n(\gamma)$ is given by the recurrence relation (2.6.73) with $a_0(\gamma)$ arbitrary. Show that

$$\mathcal{L}\left(\left.\frac{\partial \bar{\phi}}{\partial \gamma}\right|_{\gamma=1}\right) = \frac{a_0}{y - y_*}. \qquad (2.6.77)$$

To perform this task, first compute $\partial(\mathcal{L}\bar{\phi})/\partial\gamma$ using (2.6.72) for $\mathcal{L}\bar{\phi}$, and then use the fact that $\partial(\mathcal{L}\bar{\phi})/\partial\gamma = \mathcal{L}(\partial\bar{\phi}/\partial\gamma)$.

Hint: Recall that $d(x^\gamma)/d\gamma = x^\gamma \ln x$.

Now introduce the function

$$\phi_2 = \left.\frac{\partial \bar{\phi}}{\partial \gamma}\right|_{\gamma=1} + \sum_{n=0}^{\infty} b_n(y - y_*)^n, \qquad (2.6.78)$$

where the coefficients b_n are unknown. Show using (2.6.72), (2.6.77) that

$$\mathcal{L}(\phi_2) = \frac{a_0 - d_0}{y - y_*} + \sum_{n=1}^{\infty} \left[n(n+1)b_{n+1} - d_n\right](y - y_*)^{n-1},$$

where $d_n = \sum_{k=0}^{n} q_k b_{n-k}$. Hence deduce that by choosing the coefficients such that

$$b_0 = 1, \quad a_0 = d_0 = q_0,$$

$$b_{n+1} = \frac{1}{n(n+1)} \sum_{k=0}^{n} q_k b_{n-k}, \quad n \geq 1$$

we can ensure that $\mathcal{L}(\phi_2) = 0$ and hence ϕ_2 is a second linearly independent solution of the Rayleigh equation (2.6.70). The constant b_1 is arbitrary. Explain why we can take $b_1 = 0$ without loss of generality.

Finally demonstrate, by explicit differentiation of (2.6.76), that

$$\frac{\partial \bar{\phi}}{\partial \gamma} = \sum_{n=0}^{\infty} \frac{da_n}{d\gamma}(y - y_*)^{n+\gamma} + \ln(y - y_*) \sum_{n=0}^{\infty} a_n(y - y_*)^{n+\gamma}.$$

Substituting this expression into (2.6.78), show that ϕ_2 takes the form of (2.6.8).

2. *An inviscid nonlinear neutral mode at $O(1)$ wave number*

Consider the solution of the Rayleigh equation

$$(U - c)\left(\frac{d^2 \check{v}}{dy^2} - \alpha^2 \check{v}\right) = \frac{d^2 U}{dy^2}\check{v}, \tag{2.6.79}$$

subject to the boundary conditions

$$\check{v} = 0 \quad \text{at} \quad y = 0; \qquad \check{v} \to 0 \quad \text{as} \quad y \to \infty,$$

for the *asymptotic suction boundary layer* profile[17]

$$U(y) = 1 - e^{-y}, \qquad (0 \leq y < \infty).$$

We will seek neutral solutions with α and c real and assume the presence of a nonlinear critical layer at $y = y_*$ where $c = U(y_*)$.

(a) Show that in terms of the variable

$$\bar{y} = y - y_*,$$

the Rayleigh equation can be rewritten as

$$\frac{d^2 \check{v}}{d\bar{y}^2} - \left(\alpha^2 - \frac{e^{-\bar{y}}}{1 - e^{-\bar{y}}}\right)\check{v} = 0. \tag{2.6.80}$$

What are the boundary conditions in terms of this new variable?

---

[17]See Problem 3 in Exercises 2 in Part 3 of this book series.

(b) For the case $\alpha = 3/4$ it can be shown that the general solution of (2.6.80) for $\bar{y} > 0$ takes the form

$$\breve{v} = e^{-5\bar{y}/4}\left(e^{2\bar{y}} + 2e^{\bar{y}} - 3\right)\left[A + B\ln\left(\frac{1 - e^{-\bar{y}/2}}{1 + e^{-\bar{y}/2}}\right)\right] + 2Be^{-3\bar{y}/4}\left(e^{\bar{y}} - 3\right),$$

where A and B are arbitrary constants. Demonstrate that in order that $\breve{v} \to 0$ as $\bar{y} \to \infty$ we require $A = 0$. Explain how we could have anticipated the existence of the logarithmic singularity at $\bar{y} = 0$.

(c) Write down the continuation of the solution for $\bar{y} < 0$, assuming the existence of a nonlinear critical layer with zero phase shift at $\bar{y} = 0$, and show that

$$\breve{v} \to 0 \quad \text{as} \quad \bar{y} \to -\infty.$$

Deduce that for this value of the wave number α, the critical layer resides close to the boundary-layer edge. Why would this mode not exist if the critical layer were linear and viscous in nature?

3. *An inviscid nonlinear neutral mode at large wave number*

Return to the Rayleigh problem (2.6.79), but this time assume that we have a 'model' boundary-layer profile of the form

$$U(y) = 1 - e^{-y^2}, \qquad (0 \le y < \infty).$$

Again, seek neutral solutions with α and c real, and assume a strongly nonlinear critical layer at $y = y_*$, with zero phase shift, but now consider the large wave number limit $\alpha \gg 1$.

(a) Show that the Rayleigh equation for this particular velocity profile can be written in the form

$$\frac{d^2\breve{v}}{dy^2} - \left(\alpha^2 - \frac{(4y^2 - 2)e^{-(y^2 - y_*^2)}}{1 - e^{-(y^2 - y_*^2)}}\right)\breve{v} = 0. \qquad (2.6.81)$$

(b) Seek a solution with the critical layer in the far-field by supposing

$$y_* = \alpha Y_*,$$

with Y_* an order one quantity to be determined. Consider a new $O(1)$ variable Y such that

$$y = \alpha Y_* + \alpha^{-1}Y.$$

Show that to leading order, equation (2.6.81) becomes

$$\frac{d^2\breve{v}}{dY^2} - \left(1 - \frac{4Y_*^2 e^{-2Y_*Y}}{1 - e^{-2Y_*Y}}\right)\breve{v} = 0.$$

(c) Setting $\bar{y} = 2Y_*Y$, put this equation into the form

$$\frac{d^2\breve{v}}{d\bar{y}^2} - \left(\frac{1}{4Y_*^2} - \frac{e^{-\bar{y}}}{1 - e^{-\bar{y}}}\right)\breve{v} = 0,$$

and show that the accompanying boundary conditions are

$$\breve{v} \to 0 \quad \text{as} \quad \bar{y} \to \pm\infty.$$

(d) By comparing this equation with (2.6.80), show that in the large wave number limit there is a neutral mode in the free stream with

$$c = 1 - e^{-4\alpha^2/9} + \cdots \quad \text{as} \quad \alpha \to \infty.$$

As in Problem 2, explain why this mode would not exist if the critical layer were linear and viscous in nature.

4. *A cat's eye critical layer*

Consider the critical layer equation (2.6.59) for $\tilde{\lambda} > 0$ with the parameter γ_c set to zero so that the vorticity Q satisfies

$$Z\frac{\partial Q}{\partial \xi} + \sin \xi \frac{\partial Q}{\partial Z} = 0. \tag{2.6.82}$$

Here, for simplicity of notations, we have replaced ξ^* with ξ.

(a) Use the method of characteristics to show that the general solution of (2.6.82) is

$$Q = G(\eta), \tag{2.6.83}$$

where G is an arbitrary function of the variable

$$\eta = \frac{1}{2}Z^2 + \cos \xi. \tag{2.6.84}$$

(b) Sketch curves of constant η in the (ξ, Z)-plane. Show that they have a 'cat's eye' shape and are closed for $\eta < 1$.

(c) Invoke the Prandtl–Batchelor theorem (see Problem 1 in Exercises 6 in Part 3 of this book series) to deduce that G is equal to a constant (G_0, say) for $\eta < 1$.

5. *Critical layer properties in the nonlinear limit*

Again consider the critical layer equation (2.6.59) with $\tilde{\lambda} > 0$, together with the matching conditions (2.6.61), but now consider the strongly nonlinear limit $\gamma_c \to 0$.

(a) Consider an expansion of the form

$$Q = Q_0 + \gamma_c Q_1 + \cdots,$$

where Q_0 is the solution for Q in (2.6.83). Derive the equation to be satisfied by Q_1. Make a transformation of coordinates from (ξ, Z) to (ξ, η), where η is given in (2.6.84), and show that

$$\frac{\partial Q_1}{\partial \xi} = \pm\sqrt{2}\frac{\partial}{\partial \eta}\left(\sqrt{\eta - \cos \xi}\, G'(\eta)\right),$$

where the partial derivative on the left-hand side is evaluated at fixed η.

(b) Apply the condition that Q_1 is a periodic function of ξ with period 2π, and deduce that

$$G'(\eta)I(\eta) = D^{\pm}, \quad I(\eta) = \int\limits_{0}^{2\pi}\sqrt{\eta - \cos \xi}\, d\xi,$$

with $D^{\pm}$ being constants in the upper and lower regions where $\eta > 1$.

(c) Apply the outer condition (2.6.61) to deduce the values of $D^{\pm}$ and hence show that the solution satisfying $G = G_0$ on $\eta = 1$ can be expressed as

$$G = G_0 \pm \sqrt{2}\pi \int\limits_{1}^{\eta} \frac{ds}{I(s)}.$$

(d) Again, from consideration of the outer boundary condition (2.6.61), show that the mean vorticity jump is given by

$$H^+ - H^- = \sqrt{2}\pi \left\{ -\frac{1}{\pi} + \int\limits_{1}^{\infty} \left[\frac{1}{I(\eta)} - \frac{1}{2\pi\sqrt{\eta}} \right] d\eta \right\}.$$

Hence deduce from equation (2.6.69) that the phase shift $\theta = O(\gamma_c)$ as $\gamma_c \to 0$.

6. *The viscous limit of the nonlinear critical layer equation*

Consider the solution of the critical layer equation

$$Z\frac{\partial Q}{\partial \xi} + \sin \xi \frac{\partial Q}{\partial Z} = \gamma_c \frac{\partial^2 Q}{\partial Z^2},$$

subject to the far-field conditions

$$Q = Z + 2H^{\pm} + \frac{\cos \xi}{Z} + \cdots \quad \text{as} \quad Z \to \pm\infty,$$

in the viscous limit $\gamma_c \to \infty$.

(a) First, set $Q = Z + \tilde{Q}$ and show that

$$Z\frac{\partial \tilde{Q}}{\partial \xi} + \sin \xi \frac{\partial \tilde{Q}}{\partial Z} - \gamma_c \frac{\partial^2 \tilde{Q}}{\partial Z^2} = -\sin \xi, \qquad (2.6.85)$$

$$\tilde{Q} = 2H^{\pm} + \frac{\cos \xi}{Z} + \cdots \quad \text{as} \quad Z \to \pm\infty.$$

(b) Next, show that, in order to maintain a balance between the first and last terms on the left-hand side of (2.6.85), we require $Z = O(\gamma_c^{1/3})$. In view of this, set

$$Z = \gamma_c^{1/3}\tilde{z},$$

and seek a solution of the form

$$\tilde{Q} = \gamma_c^{-1/3}\tilde{Q}_1(\xi, \tilde{z}) + \gamma_c^{-1}\tilde{Q}_2(\xi, \tilde{z}) + \cdots.$$

Derive the equation to be satisfied by $\tilde{Q}_1$ and, guided by equation (2.6.34a) and solution (2.6.43), show that the appropriate solution is

$$\tilde{Q}_1 = \frac{1}{2}i\,e^{i\xi} \int\limits_{0}^{\infty} e^{-\tau^3/3}e^{-i\tau\tilde{z}}\,d\tau + (c.c.).$$

(c) Write down the equation and matching conditions to be satisfied by $\widetilde{Q}_2$, and deduce that $\widetilde{Q}_{20}$, the ξ-independent part of $\widetilde{Q}_2$, satisfies

$$\frac{d^2\widetilde{Q}_{20}}{d\tilde{z}^2} = \frac{1}{2}\int_0^\infty \tau e^{-\tau^3/3}\sin(\tau\tilde{z})\,d\tau, \qquad (2.6.86)$$

$$\widetilde{Q}_{20} = 2\widetilde{H}^\pm + \cdots \quad \text{as} \quad \tilde{z}\to\infty,$$

where $\widetilde{H}^\pm = \gamma_c H^\pm$.

(d) By integrating the equation (2.6.86) twice, and choosing the value of $d\widetilde{Q}_{20}/d\tilde{z}$ on $\tilde{z}=0$ appropriately, show that

$$\widetilde{Q}_{20}(\tilde{z}) = \widetilde{Q}_{20}(0) - \frac{1}{2}\int_0^\infty e^{-\tau^3/3}\frac{\sin(\tau\tilde{z})}{\tau}\,d\tau.$$

Using the imaginary part of the result (2.6.49), deduce that

$$\widetilde{H}^+ - \widetilde{H}^- = -\pi/4. \qquad (2.6.87)$$

(e) Substituting (2.6.87) into the relationship (2.6.69) between the vorticity jump and the phase shift, show that

$$\theta \to \pi \quad \text{as} \quad \gamma_c \to \infty,$$

thereby demonstrating that the linear value of the phase shift is recovered in the viscous limit.

3
Boundary-Layer Receptivity

Receptivity theory is a branch of fluid dynamics that deals with the interaction of the boundary layer with external perturbations. The importance of this interaction has been revealed by various experimental observations, including routine wind tunnel tests. It was observed that when the same aerodynamic body, say, a model of an aircraft wing, was tested in different wind tunnels, then the laminar-turbulent transition did not take place at the same position on the wing surface despite the similarity parameters of the flow, the Reynolds and Mach numbers, being the same in the tunnels. It is now generally accepted that the difference in position of the transition can be attributed to a number of factors including the difference in the quality of the flow in the test section, the level of turbulence in the oncoming flow, acoustic noise in the test section, and smoothness of the wind tunnel walls. The 'quieter' the wind tunnel the longer the boundary layer stays laminar (see, for example, Schneider, 2001).

It should be noted that the level of atmospheric turbulence in real flight conditions is significantly lower not only compared to standard wind tunnels but also compared to specially designed *low turbulence wind tunnels*. This makes the prediction of laminar-turbulent transition through wind tunnel tests a rather difficult proposition, and hence increases the importance of the theoretical studies.

Clearly, laminar-turbulent transition cannot be understood without taking into account the interaction of the boundary layer with the surrounding perturbation field. The analysis of the possible forms that interaction can take is the subject of receptivity theory. For a passenger airplane in a cruise flight, the flow past the wing represents an example of what is referred to as *weak turbulence flow*. In such flows, the laminar-turbulent transition follows the so-called *classical scenario* when the transition is caused by the production and amplification of the instability modes of the boundary layer. In two-dimensional boundary layers, which form the focus of this chapter, the instability assumes the form of Tollmien–Schlichting waves. Transition starts with the transformation of free-stream noise into Tollmien–Schlichting waves (see region 1 in

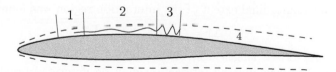

Fig. 3.1: Classical transition scenario; 1 receptivity stage, 2 linear amplification, 3 nonlinear region (which is normally rather short), 4 turbulent boundary layer.

Figure 3.1). Receptivity theory serves to describe this process; the main objective being to find the initial amplitude of the Tollmien–Schlichting waves. In the case of low free-stream turbulence, the generated Tollmien–Schlichting waves are weak and cannot lead to immediate transition to turbulence. They have to amplify in the boundary layer before triggering the nonlinear effects, characteristic of the turbulent flow. Obviously, the extent of the linear amplification region (region 2 in Figure 3.1) depends on the initial amplitude of the generated Tollmien–Schlichting waves, and in typical flight conditions it appears to be much longer than both the receptivity region and the nonlinear region (region 3 in Figure 3.1). This means that an accurate prediction of the position of laminar-turbulent transition is impossible without proper analysis of the receptivity process.

If we return to Figure 1.14, then we can see that most 'dangerous' are the Tollmien–Schlichting waves that are generated near the lower branch of the neutral curve, as they have a significant space to grow in the boundary layer before the upper branch is reached. Keeping this in mind we shall perform the receptivity analysis in this chapter based on the triple-deck theory which describes the Tollmien–Schlichting waves near the lower branch of the neutral curve (Section 2.3). We shall start with the 'vibrating ribbon' problem to which Terent'ev (1981) first applied the triple-deck theory.

3.1 Terent'ev's Problem

Here we present a mathematical model of the classical experiments of Schubauer and Skramstad (1948), where the Tollmien–Schlichting waves were generated by a vibrating ribbon installed a small distance above the plate surface; see Figure 1.13. To simplify the mathematical analysis of the flow, Terent'ev (1981) 'substituted' the ribbon by a short section of the body surface that performed harmonic oscillations in the direction perpendicular to the wall. In practice this can be arranged by making a narrow slit in the plate and covering it with a rubber membrane. The length of the vibrating section and frequency of the oscillations are chosen according to (2.3.56), which allows us to model the flow using a triple-deck description.

The *triple-deck theory*, in its present form, was first formulated for steady flows. This was performed independently by Neiland (1969) and Stewartson and Williams (1969) in their studies of the self-induced separation of the boundary layer in supersonic flow and by Stewartson (1969) and Messiter (1970) who analysed the incompressible fluid flow near the trailing edge of a flat plate. In both flows the boundary layer was found to come into interaction with the inviscid part of the flow. Using arguments unrelated to those in Section 2.3 these authors demonstrated that the interaction region has a three-tiered structure shown in Figure 2.4 with the scales given by (2.3.56). In particular, the longitudinal extent of the interaction region was found to be

$$\Delta \hat{x} = O(LRe^{-3/8}), \tag{3.1.1}$$

where L is a characteristic length scale of the body.

The process of viscous-inviscid interaction may be described briefly as follows.[1] We start with the near-wall viscous layer, region 3. The characteristic thickness of

---

[1] For a detailed discussion of triple-deck theory and its applications to steady flows, the reader is referred to Chapters 2–4 in Part 3 of this book series.

this region is an $O(LRe^{-5/8})$ quantity, so that it occupies an $O(Re^{-1/8})$ portion of the boundary layer near the body surface, where the fluid velocity $\hat{u}$ is an $O(Re^{-1/8})$ quantity relative to the free-stream velocity V_∞, that is

$$\frac{\hat{u}}{V_\infty} \sim Re^{-1/8}. \tag{3.1.2}$$

Due to the relatively slow motion of the fluid in this layer the flow exhibits high sensitivity to pressure variations. Even a small increase or decrease of the pressure causes a significant deceleration/acceleration of fluid particles. As a result the flow filaments change their thickness leading to a deformation of the streamlines. This process is known as the *displacement effect of the boundary layer.*

In region 2 occupying the main part of the boundary layer, $\hat{u}/V_\infty \sim 1$. The flow in this tier is less sensitive to the pressure variations. It does not produce any noticeable contribution to the displacement effect of the boundary layer, which means that all the streamlines in the middle tier are parallel to each other and carry the deformation produced by the displacement effect of the viscous sublayer to the upper tier.

Finally, the upper tier (region 1) is situated in the potential flow region outside the boundary layer. It serves to 'convert' the deformations in the form of the streamlines into perturbations of the pressure. These are then transmitted through the main part of the boundary layer back to the viscous sublayer. This process is self-sustained, and is referred to as the *viscous-inviscid interaction* or *free interaction*.

Following the pioneering works of Neiland (1969), Stewartson and Williams (1969), Stewartson (1969), and Messiter (1970) many researchers were involved in the development of the theory, and it became clear that viscous-inviscid interaction plays a key role in a wide variety of fluid-dynamic phenomena. An exposition of applications of the theory to different forms of the boundary-layer separation may be found, for example, in monographs by Sychev *et al.* (1998) and Neiland *et al.* (2008).

The above-mentioned works were devoted to steady or quasi-steady flows. Schneider (1974) was the first to show how the triple-deck theory may be extended to unsteady flows. He noticed that most sensitive to unsteadiness is the near-wall region 3, where the fluid velocity is relatively small. Using (3.1.1) and (3.1.2), he deduced that the flow in region 3 becomes unsteady when the characteristic time of the flow variations

$$\Delta\hat{t} \sim \frac{\Delta\hat{x}}{\hat{u}} \sim \frac{L}{V_\infty} Re^{-1/4}. \tag{3.1.3}$$

The flow in regions 1 and 2 remains quasi-steady.

Triple-deck theory was first applied to the stability analysis of the boundary-layer flow by Smith (1979*a*,*b*) and Zhuk and Ryzhov (1980). They were able to confirm that in subsonic flows, the unsteady triple-deck theory does describe the Tollmien–Schlichting waves at and near the lower branch of the neutral stability curve. In fact these studies reproduced the dispersion equation (2.3.34) first published by Lin (1946). Of course, Lin's initial assumption that the flow perturbations could be described by the Orr–Sommerfeld equation precluded nonparallel or nonlinear effects to be a part of his analysis. These restrictions can be lifted once the full version of triple-deck theory is applied; see Smith (1979*a*,*b*).

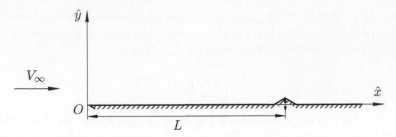

Fig. 3.2: Flow layout.

3.1.1 Problem formulation

Let us consider a two-dimensional incompressible fluid flow past a flat plate with a vibrating ribbon installed on the surface of the plate a distance L from its leading edge; see Figure 3.2. We shall assume that the velocity vector upstream of the plate is parallel to its surface. We shall also assume that the fluid density ρ and the kinematic viscosity ν are constant throughout the flow field. Then the governing Navier–Stokes equations can be written as

$$\left.\begin{aligned}
\frac{\partial \hat{u}}{\partial \hat{t}} + \hat{u}\frac{\partial \hat{u}}{\partial \hat{x}} + \hat{v}\frac{\partial \hat{u}}{\partial \hat{y}} &= -\frac{1}{\rho}\frac{\partial \hat{p}}{\partial \hat{x}} + \nu\left(\frac{\partial^2 \hat{u}}{\partial \hat{x}^2} + \frac{\partial^2 \hat{u}}{\partial \hat{y}^2}\right), \\
\frac{\partial \hat{v}}{\partial \hat{t}} + \hat{u}\frac{\partial \hat{v}}{\partial \hat{x}} + \hat{v}\frac{\partial \hat{v}}{\partial \hat{y}} &= -\frac{1}{\rho}\frac{\partial \hat{p}}{\partial \hat{y}} + \nu\left(\frac{\partial^2 \hat{v}}{\partial \hat{x}^2} + \frac{\partial^2 \hat{v}}{\partial \hat{y}^2}\right), \\
\frac{\partial \hat{u}}{\partial \hat{x}} + \frac{\partial \hat{v}}{\partial \hat{y}} &= 0.
\end{aligned}\right\} \tag{3.1.4}$$

Here $(\hat{x}, \hat{y})$ are Cartesian coordinates with $\hat{x}$ measured along the plate's surface from the leading edge O. The velocity components in these coordinates are denoted as $(\hat{u}, \hat{v})$, and the pressure as $\hat{p}$. As before, the 'hat' is used to indicate that the corresponding variables are dimensional.

The non-dimensional variables are introduced as

$$\left.\begin{aligned}
\hat{u} = V_\infty u, \qquad \hat{v} = V_\infty v, \qquad \hat{p} = p_\infty + \rho V_\infty^2 p, \\
\hat{x} = Lx, \qquad \hat{y} = Ly, \qquad \hat{t} = \frac{L}{V_\infty}t.
\end{aligned}\right\} \tag{3.1.5}$$

Here V_∞ is the modulus of the free-stream velocity vector and p_∞ is the pressure in the unperturbed flow upstream of the plate.

Substitution of (3.1.5) into the Navier–Stokes equations (3.1.4) renders them in the non-dimensional form

$$\frac{\partial u}{\partial t} + u\frac{\partial u}{\partial x} + v\frac{\partial u}{\partial y} = -\frac{\partial p}{\partial x} + \frac{1}{Re}\left(\frac{\partial^2 u}{\partial x^2} + \frac{\partial^2 u}{\partial y^2}\right), \tag{3.1.6a}$$

$$\frac{\partial v}{\partial t} + u\frac{\partial v}{\partial x} + v\frac{\partial v}{\partial y} = -\frac{\partial p}{\partial y} + \frac{1}{Re}\left(\frac{\partial^2 v}{\partial x^2} + \frac{\partial^2 v}{\partial y^2}\right), \tag{3.1.6b}$$

$$\frac{\partial u}{\partial x} + \frac{\partial v}{\partial y} = 0. \tag{3.1.6c}$$

Here Re is the Reynolds number:

$$Re = \frac{V_\infty L}{\nu}.$$

We shall study the behaviour of the flow through the asymptotic analysis of the Navier–Stokes equations (3.1.6) under the assumption that $Re \to \infty$.

3.1.2 Boundary layer before the vibrating ribbon

We start with the boundary layer upstream of the vibrating ribbon. The flow in this region is expected to be steady, and is described by Prandtl's classical boundary-layer theory. According to this theory, the velocity components and the pressure are represented by the asymptotic expansions

$$u = U_0(x, Y) + \cdots, \quad v = Re^{-1/2}V_0(x, Y) + \cdots, \quad p = Re^{-1}P_0(x, Y) + \cdots, \quad (3.1.7)$$

where

$$Y = Re^{1/2}y. \tag{3.1.8}$$

Substitution of (3.1.7) and (3.1.8) into the Navier–Stokes equations (3.1.6) results in

$$U_0 \frac{\partial U_0}{\partial x} + V_0 \frac{\partial U_0}{\partial Y} = \frac{\partial^2 U_0}{\partial Y^2}, \tag{3.1.9a}$$

$$\frac{\partial U_0}{\partial x} + \frac{\partial V_0}{\partial Y} = 0. \tag{3.1.9b}$$

When solving these equations one needs to pose the following boundary conditions. First, at the leading edge of the plate the flow is still unaffected by the presence of the boundary layer, and we can write:

$$U_0 = 1 \quad \text{at} \quad x = 0, \ Y \in (0, \infty). \tag{3.1.10a}$$

Second, the matching condition with the inviscid flow above the boundary layer reads

$$U_0 = 1 \quad \text{at} \quad Y = \infty, \ x \in [0, 1). \tag{3.1.10b}$$

Finally we need to impose the no-slip conditions on the plate surface:

$$U_0 = V_0 = 0 \quad \text{at} \quad Y = 0, \ x \in [0, 1). \tag{3.1.10c}$$

This is the same problem that was discussed in Section 1.3.1, but now it is written in non-dimensional form. We know that it admits a self-similar solution (1.3.2)–(1.3.4b). However, for the receptivity analysis we do not need to know the precise velocity distribution in the boundary layer. We only need to know that the solution is smooth near the position $(x = 1)$ of the vibrating ribbon, and can be represented by the Taylor expansion

$$U_0(x, Y) = U_{00}(Y) + (x - 1)U_{01}(Y) + \cdots , \qquad (3.1.11)$$

where $U_{00}(Y)$ represents the velocity profile immediately before the vibrating ribbon. Near the body surface it behaves as

$$U_{00}(Y) = \lambda Y + \cdots \quad \text{as} \quad Y \to 0, \qquad (3.1.12)$$

with λ being a positive skin friction parameter. At the outer edge of the boundary layer

$$U_{00}(Y) = 1 \quad \text{at} \quad Y = \infty. \qquad (3.1.13)$$

The theory to be developed here is applicable to any boundary layer for which equations (3.1.11), (3.1.12), and (3.1.13) hold.

3.1.3   Triple-deck region

Now we turn our attention to the flow in the vicinity of the vibrating section of the wall. Guided by (3.1.1) and (3.1.3) we write the equation for this section in the form

$$y_w(t, x) = Re^{-5/8} f(x_*) \cos(\omega_* t_*), \qquad (3.1.14)$$

where

$$x_* = \frac{x - 1}{Re^{-3/8}}, \qquad t_* = \frac{t}{Re^{-1/4}}. \qquad (3.1.15)$$

The $O(Re^{-5/8})$ amplitude of the vibrations is chosen to 'fit' into the lower tier (region 3) of the triple-deck structure; see Figure 3.3.

Viscous sublayer (region 3)

We begin the flow analysis in the triple-deck region with the viscous sublayer (region 3). The solution of the Navier–Stokes equations (3.1.6) in region 3 is sought in the form

$$u(t, x, y; Re) = Re^{-1/8} U_*(t_*, x_*, Y_*) + \cdots , \qquad (3.1.16a)$$

$$v(t, x, y; Re) = Re^{-3/8} V_*(t_*, x_*, Y_*) + \cdots , \qquad (3.1.16b)$$

$$p(t, x, y; Re) = Re^{-1/4} P_*(t_*, x_*, Y_*) + \cdots . \qquad (3.1.16c)$$

Here all the arguments are assumed to be order one quantities; t_* and x_* are defined by (3.1.15), and Y_* is obtained by scaling of the lateral coordinate y with the characteristic thickness of region 3:

$$Y_* = \frac{y}{Re^{-5/8}}. \qquad (3.1.17)$$

The form of the asymptotic expansion (3.1.16a) of the longitudinal velocity component u follows directly from (3.1.2). To find the scaling coefficient $Re^{-3/8}$ in the

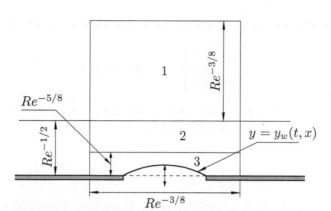

Fig. 3.3: The flow in the vicinity of the vibrator.

asymptotic expansion (3.1.16b) of the lateral velocity component v, the following arguments are used. As in classical boundary-layer theory, the two terms in the continuity equation (3.1.6c) should be in balance with one another. Approximating the derivatives in this equation by finite differences, we have

$$\frac{u}{\Delta x} \sim \frac{v}{y}. \tag{3.1.18}$$

If we now solve (3.1.18) for v and use the fact that $\Delta x \sim Re^{-3/8}$ and $y \sim Re^{-5/8}$, then we will have

$$v \sim u\frac{y}{\Delta x} \sim Re^{-1/8}\frac{Re^{-5/8}}{Re^{-3/8}} \sim Re^{-3/8}.$$

As far as the asymptotic expansion (3.1.16c) of the pressure p is concerned, its form may be predicted by balancing the following two terms in the longitudinal momentum equation (3.1.6a):

$$u\frac{\partial u}{\partial x} \sim \frac{\partial p}{\partial x},$$

which leads to the scalings

$$p \sim u^2 \sim Re^{-1/4}.$$

Now we need to deduce the equations governing the flow in region 3. This is done by substituting (3.1.16), (3.1.15), and (3.1.17) into the Navier–Stokes equations (3.1.6) and setting $Re \to \infty$. We find that

$$\frac{\partial U_*}{\partial t_*} + U_*\frac{\partial U_*}{\partial x_*} + V_*\frac{\partial U_*}{\partial Y_*} = -\frac{\partial P_*}{\partial x_*} + \frac{\partial^2 U_*}{\partial Y_*^2}, \tag{3.1.19a}$$

$$\frac{\partial P_*}{\partial Y_*} = 0, \tag{3.1.19b}$$

$$\frac{\partial U_*}{\partial x_*} + \frac{\partial V_*}{\partial Y_*} = 0. \tag{3.1.19c}$$

The next step is to formulate the boundary conditions for equations (3.1.19). We start with the no-slip conditions on the vibrating section of the wall. If we assume that each element of the wall moves vertically up and down, then we will have

$$\left.\begin{array}{l} u = 0, \\ v = \dfrac{\partial y_w}{\partial t} \end{array}\right\} \quad \text{at} \quad y = y_w(t, x). \tag{3.1.20}$$

To express these conditions in terms of the variables of region 3, we need to substitute (3.1.16a), (3.1.16b), and (3.1.14) into (3.1.20). This yields

$$\left.\begin{array}{l} U_* = 0, \\ V_* = -\omega_* f(x_*) \sin(\omega_* t_*) \end{array}\right\} \quad \text{at} \quad Y_* = f(x_*) \cos(\omega_* t_*). \tag{3.1.21}$$

Since equation (3.1.19a) is parabolic in x_*, it requires a boundary condition for U_* upstream of the vibrating section of the wall as well as at the outer edge of region 3. The former is deduced through the matching with the solution (3.1.7) in the boundary layer upstream of the interaction region. It follows from (3.1.15), (3.1.8), and (3.1.17) that the independent variables in the Blasius boundary layer before the ribbon and in region 3 are related as

$$x = 1 + Re^{-3/8} x_*, \qquad Y = Re^{-1/8} Y_*. \tag{3.1.22}$$

Using (3.1.22), the first equation in (3.1.7) may be expressed in the form

$$u = U_0(1 + Re^{-3/8} x_*, Re^{-1/8} Y_*) + \cdots . \tag{3.1.23}$$

At large values of the Reynolds number, the first argument of U_0 in (3.1.23) is close to one, and the second tends to zero, which makes (3.1.11) and (3.1.12) applicable, reducing (3.1.23) to

$$u = Re^{-1/8} \lambda Y_* + \cdots . \tag{3.1.24}$$

It remains to compare (3.1.24) with the asymptotic expansion (3.1.16a) of the longitudinal velocity u in region 3, and we can see that the sought boundary condition reads

$$U_* = \lambda Y_* + \cdots \quad \text{as} \quad x_* \to -\infty. \tag{3.1.25}$$

Finally, we need to study the asymptotic behaviour of the solution of equations (3.1.19) at the outer edge of the viscous sublayer, that is, in the limit $Y_* \to \infty$. For this purpose it is convenient to introduce the stream function $\Psi_*(t_*, x_*, Y_*)$ such that

$$U_* = \frac{\partial \Psi_*}{\partial Y_*}, \qquad V_* = -\frac{\partial \Psi_*}{\partial x_*}. \tag{3.1.26}$$

Let us seek the asymptotic expansion of Ψ_* at the outer edge of region 3 in the form

$$\Psi_*(t_*, x_*, Y_*) = F(t_*, x_*) Y_*^{\alpha} + \cdots \quad \text{as} \quad Y_* \to \infty, \tag{3.1.27}$$

where α is a parameter to be found. Substitution of (3.1.27) into (3.1.26) yields

$$U_* = \alpha F Y_*^{\alpha-1} + \cdots , \qquad V_* = -\frac{\partial F}{\partial x_*} Y_*^{\alpha} + \cdots . \tag{3.1.28}$$

Using (3.1.28), one can calculate the three terms on the left-hand side of equation (3.1.19a) and the viscous term on the right-hand side. These are:

$$\frac{\partial U_*}{\partial t_*} = \alpha \frac{\partial F}{\partial t_*} Y_*^{\alpha-1} + \cdots, \tag{3.1.29a}$$

$$U_* \frac{\partial U_*}{\partial x_*} = \alpha^2 F \frac{\partial F}{\partial x_*} Y_*^{2\alpha-2} + \cdots, \tag{3.1.29b}$$

$$V_* \frac{\partial U_*}{\partial Y_*} = -\alpha(\alpha-1) F \frac{\partial F}{\partial x_*} Y_*^{2\alpha-2} + \cdots, \tag{3.1.29c}$$

$$\frac{\partial^2 U^*}{\partial Y_*^2} = \alpha(\alpha-1)(\alpha-2) F Y_*^{\alpha-3} + \cdots. \tag{3.1.29d}$$

If we assume, subject to subsequent confirmation, that $\alpha > 1$, then the convective terms (3.1.29b), (3.1.29c) will dominate over the rest of the terms in equation (3.1.19a) including the one representing the pressure gradient, which remains finite as $Y_* \to \infty$. This reduces (3.1.19a) to

$$F \frac{\partial F}{\partial x_*} = 0.$$

The initial condition for this equation is obtained by substituting the first equation in (3.1.28) into (3.1.25). We see that

$$F \Big|_{x_*=-\infty} = \begin{cases} \lambda/\alpha & \text{if } \alpha = 2, \\ 0 & \text{if } \alpha \neq 2. \end{cases}$$

Clearly, a non-trivial solution exists only if $\alpha = 2$, in which case $F = \lambda/2$, and (3.1.27) assumes the form

$$\Psi^*(x_*, Y_*) = \frac{1}{2}\lambda Y_*^2 + \cdots \quad \text{as} \quad Y_* \to \infty.$$

To find the next order term we write:

$$\Psi_*(t_*, x_*, Y_*) = \frac{1}{2}\lambda Y_*^2 + A_*(t_*, x_*) Y_*^\alpha + \cdots \quad \text{as} \quad Y_* \to \infty. \tag{3.1.30}$$

Here, to ensure that the second term in (3.1.30) is small compared to the first term, we have to assume that $\alpha < 2$. Substitution of (3.1.30) into (3.1.26) and then into equation (3.1.19a) results in

$$(\alpha - 1) \frac{\partial A_*}{\partial x_*} = 0. \tag{3.1.31}$$

It further follows from (3.1.25) that

$$A_* \Big|_{x_*=-\infty} = 0. \tag{3.1.32}$$

We see that a non-trivial solution for $A_*(t_*, x_*)$ exists only if $\alpha = 1$. The function $A_*(t_*, x_*)$ remains arbitrary in the framework of the asymptotic analysis of equations (3.1.19) at large values of Y_*. We, of course, expect this function to be found as a part of the solution of the problem as a whole.

Setting $\alpha = 1$ in (3.1.30) and using equations (3.1.26) we can conclude that at the outer edge of region 3

$$\left.\begin{aligned} U_* &= \lambda Y_* + A_*(t_*, x_*) + \cdots, \\ V_* &= -\frac{\partial A_*}{\partial x_*} Y_* + \cdots \end{aligned}\right\} \qquad \text{as} \quad Y_* \to \infty. \qquad (3.1.33)$$

The physical interpretation of the function $A_*(t_*, x_*)$ may be clarified by calculating the streamline slope angle

$$\vartheta = \arctan \frac{v}{u}. \qquad (3.1.34)$$

Remember that in region 3, the velocity components u and v are given by the asymptotic expansions (3.1.16a) and (3.1.16b). Substituting (3.1.33) into (3.1.16a) and (3.1.16b), and then into (3.1.34), we see that at the outer edge of region 3

$$\vartheta = \arctan \frac{v}{u} = Re^{-1/4} \frac{V^*}{U^*}\bigg|_{Y_* \to \infty} + \cdots = Re^{-1/4} \left(-\frac{1}{\lambda} \frac{\partial A_*}{\partial x_*} \right) + \cdots. \qquad (3.1.35)$$

In view of this, the function $A_*(t_*, x_*)$ is termed the *displacement function*.

Main part of the boundary layer (region 2)

Region 2, the middle tier of the triple-deck structure (see Figure 3.3), is a continuation of the Blasius boundary layer into the triple-deck region. Therefore, the thickness of region 2 is estimated as $y = O(Re^{-1/2})$. The form of the asymptotic expansions of the fluid-dynamic functions in this region may be predicted by analysing the solution behaviour in the overlap region that lies between regions 3 and 2. Substitution of (3.1.33) into (3.1.16a) and (3.1.16b) shows that the 'outer expansion of the inner solution' has the form

$$u = Re^{-1/8} \lambda Y_* + Re^{-1/8} A_*(t_*, x_*) + \cdots, \qquad v = Re^{-3/8} \left(-\frac{\partial A_*}{\partial x_*} Y_* \right) + \cdots.$$

The above expansions can be expressed in terms of the variables of region 2. For this purpose the second equation in (3.1.22) is used. We find that the 'inner expansion of the outer solution' is written as

$$u = \lambda Y + Re^{-1/8} A_*(t_*, x_*) + \cdots, \qquad v = Re^{-1/4} \left(-\frac{\partial A_*}{\partial x_*} Y \right) + \cdots. \qquad (3.1.36)$$

This suggests that the solution in region 2 should be sought in the form

$$\left.\begin{aligned} u(t, x, y; Re) &= U_{00}(Y) + Re^{-1/8} \tilde{U}_1(t_*, x_*, Y) + \cdots, \\ v(t, x, y, Re) &= Re^{-1/4} \tilde{V}_1(t_*, x_*, Y) + \cdots, \\ p(t, x, y; Re) &= Re^{-1/4} \tilde{P}_1(t_*, x_*, Y) + \cdots. \end{aligned}\right\} \qquad (3.1.37)$$

The leading-order term $U_{00}(Y)$ in the asymptotic expansion of $u(t, x, y; Re)$ coincides with the velocity profile (3.1.11) in the Blasius boundary layer immediately before the triple-deck region. According to (3.1.12)

$$U_{00} = \lambda Y + \cdots \quad \text{as} \quad Y \to 0. \tag{3.1.38}$$

It further follows from (3.1.36) that

$$\left. \begin{aligned} \widetilde{U}_1 &= A_*(t_*, x_*) + \cdots, \\ \widetilde{V}_1 &= -\frac{\partial A_*}{\partial x_*} Y + \cdots \end{aligned} \right\} \quad \text{as} \quad Y \to 0. \tag{3.1.39}$$

Substitution of (3.1.37) into the Navier–Stokes equations (3.1.6) results in

$$U_{00}(Y)\frac{\partial \widetilde{U}_1}{\partial x_*} + \widetilde{V}_1 U'_{00}(Y) = 0, \tag{3.1.40a}$$

$$\frac{\partial \widetilde{P}_1}{\partial Y} = 0, \tag{3.1.40b}$$

$$\frac{\partial \widetilde{U}_1}{\partial x_*} + \frac{\partial \widetilde{V}_1}{\partial Y} = 0. \tag{3.1.40c}$$

To solve these equations we eliminate $\partial \widetilde{U}_1/\partial x_*$ from (3.1.40a) with the help of (3.1.40c). This results in

$$U_{00}(Y)\frac{\partial \widetilde{V}_1}{\partial Y} - \widetilde{V}_1 U'_{00}(Y) = 0. \tag{3.1.41}$$

Dividing both terms in (3.1.41) by U_{00}^2, we have

$$\frac{1}{U_{00}(Y)}\frac{\partial \widetilde{V}_1}{\partial Y} - \widetilde{V}_1\frac{U'_{00}}{U_{00}^2} = 0,$$

or, equivalently,

$$\frac{\partial}{\partial Y}\left(\frac{\widetilde{V}_1}{U_{00}}\right) = 0.$$

We see that the ratio $\widetilde{V}_1/U_{00}$ is a function of t_* and x_* only, say $G(t_*, x_*)$:

$$\frac{\widetilde{V}_1}{U_{00}} = G(t_*, x_*). \tag{3.1.42}$$

Setting $Y \to 0$ in (3.1.42) and using (3.1.38) and (3.1.39), we find that

$$G(t_*, x_*) = -\frac{1}{\lambda}\frac{\partial A_*}{\partial x_*}. \tag{3.1.43}$$

It remains to substitute (3.1.43) back into (3.1.42), and we can conclude that in region 2

$$\frac{\widetilde{V}_1}{U_{00}} = -\frac{1}{\lambda}\frac{\partial A_*}{\partial x_*}. \tag{3.1.44}$$

Now we can find the streamline slope angle ϑ. For this purpose we return to the asymptotic expansions (3.1.37) of the velocity components u and v in region 2 and use the equation (3.1.44). We see that

$$\vartheta = \arctan \frac{v}{u} = Re^{-1/4} \frac{\widetilde{V_1}}{U_{00}} + \cdots = Re^{-1/4}\left(-\frac{1}{\lambda}\frac{\partial A_*}{\partial x_*} \right) + \cdots . \qquad (3.1.45)$$

Notice that ϑ does not depend on Y, which means that region 2 does not contribute to the displacement effect of the boundary layer.

Upper tier (region 1)

The upper tier lies in the inviscid flow outside the boundary layer; see Figure 3.3. The flow in this region can be described in the framework of thin aerofoil theory.[2] To start the flow analysis in region 1, we first need to determine its dimensions. Of course, the longitudinal extent of region 1 coincides with that of the interaction region as a whole: $\Delta x \sim Re^{-3/8}$. To determine the lateral size, we use the fact that any inviscid potential flow of an incompressible fluid is described by Laplace's equation

$$\frac{\partial^2 \varphi}{\partial x^2} + \frac{\partial^2 \varphi}{\partial y^2} = 0$$

for the velocity potential φ.[3] The principle of least degeneration, applied to this equation, suggests that $y \sim \Delta x$. Consequently, the asymptotic analysis of the Navier–Stokes equations (3.1.6) in region 1 should be performed assuming that

$$y_* = \frac{y}{Re^{-3/8}} = O(1). \qquad (3.1.46)$$

To predict the form of the asymptotic expansion of the fluid-dynamic functions in region 1, we note that if the flow in this region was not affected by the displacement of the boundary layer, then we would have

$$u = 1, \qquad v = 0, \qquad p = 0, \qquad (3.1.47)$$

that is, the velocity components u and v, and the pressure p would assume their free-stream values. Equation (3.1.45) shows that the displacement effect of the boundary layer leads to $O(Re^{-1/4})$ perturbations in (3.1.47), which means that the solution in region 1 should be sought in the form

$$\left.\begin{aligned}
u(t,x,y;Re) &= 1 + Re^{-1/4} u_*(t_*,x_*,y_*) + \cdots, \\
v(t,x,y;Re) &= Re^{-1/4} v_*(t_*,x_*,y_*) + \cdots, \\
p(t,x,y;Re) &= Re^{-1/4} p_*(t_*,x_*,y_*) + \cdots.
\end{aligned}\right\} \qquad (3.1.48)$$

Substitution of (3.1.48) into the Navier–Stokes equations (3.1.6) yields the following set of equations

---

[2]See Section 2.1 in Part 2 of this book series.
[3]See Section 3.2 in Part 1.

$$\frac{\partial u_*}{\partial x_*} = -\frac{\partial p_*}{\partial x_*}, \qquad \frac{\partial v_*}{\partial x_*} = -\frac{\partial p_*}{\partial y_*}, \qquad \frac{\partial u_*}{\partial x_*} + \frac{\partial v_*}{\partial y_*} = 0, \qquad (3.1.49)$$

governing the flow in region 1.

Now we need to formulate the boundary conditions for equations (3.1.49). It follows from (3.1.48) that in region 1 the streamline slope angle is calculated as

$$\vartheta = \arctan \frac{v}{u} = Re^{-1/4} v_* + \cdots . \qquad (3.1.50)$$

Comparing (3.1.50) with (3.1.45) we see that the condition of matching of ϑ in regions 1 and 2 reads

$$v_* \Big|_{y_*=0} = -\frac{1}{\lambda} \frac{\partial A_*}{\partial x_*}. \qquad (3.1.51)$$

It should be supplemented with the condition of the attenuation of the perturbations far from the vibrating section of the plate surface:

$$\left. \begin{aligned} u_* &\to 0, \\ v_* &\to 0, \\ p_* &\to 0 \end{aligned} \right\} \quad \text{as} \quad x_*^2 + y_*^2 \to \infty. \qquad (3.1.52)$$

The set of equations (3.1.49) can be easily reduced to a single equation for the pressure p_*. To perform this task, we first eliminate $\partial u_*/\partial x_*$ from the first and third equations in (3.1.49). As a result, we have

$$\frac{\partial v_*}{\partial y_*} = \frac{\partial p_*}{\partial x_*}, \qquad \frac{\partial v_*}{\partial x_*} = -\frac{\partial p_*}{\partial y_*}. \qquad (3.1.53)$$

Then v_* is eliminated by cross-differentiation of (3.1.53), leading to Laplace's equation for p_*:

$$\frac{\partial^2 p_*}{\partial x_*^2} + \frac{\partial^2 p_*}{\partial y_*^2} = 0. \qquad (3.1.54)$$

The boundary condition (3.1.51) is converted into a condition for the pressure p_* by setting $y_* = 0$ in the second equation in (3.1.49) and using (3.1.51). This leads to

$$\frac{\partial p_*}{\partial y_*} \Big|_{y_*=0} = \frac{1}{\lambda} \frac{\partial^2 A_*}{\partial x_*^2}. \qquad (3.1.55)$$

The condition of attenuation of the perturbations with distance from the 'vibrator', namely,

$$p_* \to 0 \quad \text{as} \quad x_*^2 + y_*^2 \to \infty \qquad (3.1.56)$$

closes the boundary-value problem (3.1.54)–(3.1.56).

3.1.4 Viscous-inviscid interaction problem

The results of the analysis we have just presented may be summarized as follows. We demonstrated that, if the vibrating section of the wall can be expressed by equation (3.1.14), then the region of viscous-inviscid interaction forms in the vicinity of this section. It has a three-tiered structure, being composed of the viscous sublayer (region 3), the main part of the boundary layer (region 2), and the inviscid potential flow (region 1) situated outside the boundary layer. The interaction takes place between the lower and upper tiers, with the middle tier playing a passive role in the interaction process. The flow in the lower tier is described by equations (3.1.19) subject to the boundary conditions (3.1.21), (3.1.25), and (3.1.33). If the pressure distribution P_* were given, as in the case of Prandtl's classical boundary-layer theory, then equations (3.1.19) could be solved for the velocity components U_*, V_*, and, as a part of the solution, the displacement function A_* could be found using (3.1.33). However, in the interactive flow considered here, neither the pressure P_* nor the displacement function A_* is known in advance. Instead, a relationship between P_* and A_* can be found by solving the boundary-value problem (3.1.54)–(3.1.56) for the upper tier. Considered together, the lower tier problem and the upper tier problem constitute the *viscous-inviscid interaction problem*. Before solving it, we shall transform this problem into a *canonical form* by excluding the parameter λ. This is done by performing the following affine transformations of the variables in region 3:

$$\left. \begin{aligned} t_* &= \lambda^{-3/2}\bar{t}, & x_* &= \lambda^{-5/4}\bar{x}, & Y_* &= \lambda^{-3/4}\bar{Y}, \\ U_* &= \lambda^{1/4}\bar{U}, & V_* &= \lambda^{3/4}\bar{V}, & P_* &= \lambda^{1/2}\bar{P}, \\ A_* &= \lambda^{1/4}\bar{A}, & f &= \lambda^{-3/4}\bar{f}, & \omega_* &= \lambda^{3/2}\omega. \end{aligned} \right\} \tag{3.1.57}$$

As a result, the boundary-value problem (3.1.19), (3.1.21), (3.1.25), (3.1.33) for the lower tier assumes the form

$$\frac{\partial \bar{U}}{\partial \bar{t}} + \bar{U}\frac{\partial \bar{U}}{\partial \bar{x}} + \bar{V}\frac{\partial \bar{U}}{\partial \bar{Y}} = -\frac{\partial \bar{P}}{\partial \bar{x}} + \frac{\partial^2 \bar{U}}{\partial \bar{Y}^2}, \tag{3.1.58a}$$

$$\frac{\partial \bar{U}}{\partial \bar{x}} + \frac{\partial \bar{V}}{\partial \bar{Y}} = 0, \tag{3.1.58b}$$

$$\left. \begin{aligned} \bar{U} &= 0, \\ \bar{V} &= -\omega\bar{f}(\bar{x})\sin(\omega\bar{t}) \end{aligned} \right\} \quad \text{at} \quad \bar{Y} = \bar{f}(\bar{x})\cos(\omega\bar{t}), \tag{3.1.58c}$$

$$\bar{U} = \bar{Y} + \cdots \qquad \text{as} \quad \bar{x} \to -\infty, \tag{3.1.58d}$$

$$\bar{U} = \bar{Y} + \bar{A}(\bar{t}, \bar{x}) + \cdots \quad \text{as} \quad \bar{Y} \to \infty. \tag{3.1.58e}$$

Corresponding to (3.1.57), we transform the variables in region 1 as

$$y_* = \lambda^{-5/4}\bar{y}, \qquad p_* = \lambda^{1/2}\bar{p}.$$

This allows us to express the boundary-value problem (3.1.54)–(3.1.56) in the form

$$\left.\begin{array}{l} \dfrac{\partial^2 \bar{p}}{\partial \bar{x}^2} + \dfrac{\partial^2 \bar{p}}{\partial \bar{y}^2} = 0, \\[3mm] \dfrac{\partial \bar{p}}{\partial \bar{y}} = \dfrac{\partial^2 \bar{A}}{\partial \bar{x}^2} \quad \text{at} \quad \bar{y} = 0, \\[3mm] \bar{p} \to 0 \quad \text{as} \quad \bar{x}^2 + \bar{y}^2 \to \infty. \end{array}\right\} \tag{3.1.59}$$

The solutions in region 1 and 3 are related to one another through the equation:

$$\bar{P} = \bar{p}\Big|_{\bar{y}=0}. \tag{3.1.60}$$

3.1.5 Linear problem

The viscous-inviscid interaction problem (3.1.58)–(3.1.60) can be linearized and solved analytically in the case of weak perturbations. Such a situation occurs when the amplitude of the wall vibrations is small compared to the thickness of the viscous region 3, that is

$$\bar{f}(\bar{x}) = \varepsilon F(\bar{x}), \tag{3.1.61}$$

where ε is a small parameter.

Let us start by setting $\varepsilon = 0$. In this case the vibrating section of the wall becomes motionless and forms a single flat surface together with the rest of the plate. Clearly, no perturbations to the Blasius flow can be expected in these conditions, and the solution of (3.1.58)–(3.1.60) proves to be

$$\bar{U} = \bar{Y}, \quad \bar{V} = \bar{P} = \bar{A} = \bar{p} = 0. \tag{3.1.62}$$

If ε is small but non-zero, then the solution in the lower tier (region 3) should be sought in the form

$$\left.\begin{array}{ll} \bar{U} = \bar{Y} + \varepsilon U'(\bar{t}, \bar{x}, \bar{Y}) + \cdots, & \bar{V} = \varepsilon V'(\bar{t}, \bar{x}, \bar{Y}) + \cdots, \\[2mm] \bar{P} = \varepsilon P'(\bar{t}, \bar{x}) + \cdots, & \bar{A} = \varepsilon A'(\bar{t}, \bar{x}) + \cdots. \end{array}\right\} \tag{3.1.63}$$

Substitution of (3.1.63) and (3.1.61) into (3.1.58) yields for the $O(\varepsilon)$ terms

$$\frac{\partial U'}{\partial \bar{t}} + \bar{Y}\frac{\partial U'}{\partial \bar{x}} + V' = -\frac{\partial P'}{\partial \bar{x}} + \frac{\partial^2 U'}{\partial \bar{Y}^2}, \tag{3.1.64a}$$

$$\frac{\partial U'}{\partial \bar{x}} + \frac{\partial V'}{\partial \bar{Y}} = 0, \tag{3.1.64b}$$

$$\left.\begin{array}{l} U' = -F(\bar{x})\cos(\omega \bar{t}), \\[2mm] V' = -\omega F(\bar{x})\sin(\omega \bar{t}) \end{array}\right\} \quad \text{at} \quad \bar{Y} = 0, \tag{3.1.64c}$$

$$U' = 0 \qquad\qquad \text{at} \quad \bar{x} = -\infty, \tag{3.1.64d}$$

$$U' = A'(\bar{t}, \bar{x}) \qquad \text{at} \quad \bar{Y} = \infty. \tag{3.1.64e}$$

In particular, the condition for U' in (3.1.64c) is deduced in the following way. Substitution of (3.1.61) and of the first of equations (3.1.63) into the condition for $\bar{U}$ in (3.1.58c) results in

$$\varepsilon F(\bar{x})\cos(\omega\bar{t}) + \varepsilon U'\big[\bar{t},\bar{x},\varepsilon F(\bar{x})\cos(\omega\bar{t})\big] + \cdots = 0. \tag{3.1.65}$$

Since the third argument of U' is small, we can use the Taylor expansion

$$U'\big[\bar{t},\bar{x},\varepsilon F(\bar{x})\cos(\omega\bar{t})\big] = U'(\bar{t},\bar{x},0) + \frac{\partial U'}{\partial \bar{Y}}(\bar{t},\bar{x},0)\,\varepsilon F(\bar{x})\cos(\omega\bar{t}) + \cdots. \tag{3.1.66}$$

Substituting (3.1.66) into (3.1.65) and disregarding the $O(\varepsilon^2)$ terms, we arrive at the conclusion that the function U' does, indeed, satisfy the condition

$$U'(\bar{t},\bar{x},0) = -F(\bar{x})\cos(\omega\bar{t}).$$

If the harmonic oscillations of the vibrating section of the wall persist for long enough, then the response of the flow is expected to be periodic with the same frequency ω. Keeping this in mind we shall seek the solution of the boundary-value problem (3.1.64) in the form

$$\left.\begin{aligned} U'(\bar{t},\bar{x},\bar{Y}) &= e^{i\omega\bar{t}}U^\star(\bar{x},\bar{Y}) + (c.c.), \qquad V'(\bar{t},\bar{x},\bar{Y}) = e^{i\omega\bar{t}}V^\star(\bar{x},\bar{Y}) + (c.c.),\\ P'(\bar{t},\bar{x}) &= e^{i\omega\bar{t}}P^\star(\bar{x}) + (c.c.), \qquad\quad\; A'(\bar{t},\bar{x}) = e^{i\omega\bar{t}}A^\star(\bar{x}) + (c.c.), \end{aligned}\right\} \tag{3.1.67}$$

where $(c.c.)$ stands for the complex conjugate of the term to its left.

Substitution of (3.1.67) into (3.1.64) results in

$$i\omega U^\star + \bar{Y}\frac{\partial U^\star}{\partial \bar{x}} + V^\star = -\frac{dP^\star}{d\bar{x}} + \frac{\partial^2 U^\star}{\partial \bar{Y}^2}, \tag{3.1.68a}$$

$$\frac{\partial U^\star}{\partial \bar{x}} + \frac{\partial V^\star}{\partial \bar{Y}} = 0, \tag{3.1.68b}$$

$$\left.\begin{aligned} U^\star &= -\frac{1}{2}F(\bar{x}),\\ V^\star &= \frac{i}{2}\omega F(\bar{x}) \end{aligned}\right\} \quad \text{at}\quad \bar{Y}=0, \tag{3.1.68c}$$

$$U^\star \to 0 \qquad\qquad \text{as}\quad \bar{x}\to -\infty, \tag{3.1.68d}$$

$$U^\star = A^\star(\bar{x}) \qquad \text{at}\quad \bar{Y}=\infty. \tag{3.1.68e}$$

Similarly, we represent the pressure $\bar{p}$ in the upper tier (region 1) in the form

$$\bar{p}(\bar{t},\bar{x},\bar{y}) = \varepsilon\, e^{i\omega\bar{t}}p^\star(\bar{x},\bar{y}) + (c.c.),$$

and then the boundary-value problem (3.1.59) may be written as

$$\frac{\partial^2 p^\star}{\partial \bar{x}^2} + \frac{\partial^2 p^\star}{\partial \bar{y}^2} = 0, \tag{3.1.69a}$$

$$\frac{\partial p^\star}{\partial \bar{y}} = \frac{\partial^2 A^\star}{\partial \bar{x}^2} \quad \text{at}\quad \bar{y}=0, \tag{3.1.69b}$$

$$p^\star \to 0 \qquad \text{as}\quad \bar{x}^2 + \bar{y}^2 \to \infty. \tag{3.1.69c}$$

Finally, equation (3.1.60) is written in the new variables as

$$P^\star = p^\star \Big|_{\bar{y}=0}. \tag{3.1.70}$$

Notice that the coefficients in all the equations and boundary conditions in (3.1.68)–(3.1.70) are independent of $\bar{x}$. This allows us to solve the viscous-inviscid interaction problem (3.1.68)–(3.1.70) by making use of Fourier transforms with respect to the $\bar{x}$-coordinate. We start with the upper tier (region 1). The Fourier transform $\breve{p}(\alpha; \bar{y})$ of function $p^\star(\bar{x}, \bar{y})$ is defined as

$$\breve{p}(\alpha; \bar{y}) = \int_{-\infty}^{\infty} p^\star(\bar{x}, \bar{y}) \, e^{-i\alpha\bar{x}} \, d\bar{x}. \tag{3.1.71}$$

Applying a Fourier transform to equation (3.1.69a) and boundary conditions (3.1.69b), (3.1.69c) we have

$$-\alpha^2 \breve{p} + \frac{d^2\breve{p}}{d\bar{y}^2} = 0, \tag{3.1.72a}$$

$$\frac{d\breve{p}}{d\bar{y}} = -\alpha^2 \breve{A}(\alpha) \quad \text{at} \quad \bar{y} = 0, \tag{3.1.72b}$$

$$\breve{p} = 0 \qquad \qquad \text{at} \quad \bar{y} = \infty. \tag{3.1.72c}$$

Here $\breve{A}(\alpha)$ is the Fourier transform of the function $A^\star(\bar{x})$.

The general solution of equation (3.1.72a) is written as

$$\breve{p} = C_1 e^{\alpha\bar{y}} + C_2 e^{-\alpha\bar{y}}.$$

By definition, the parameter α in the Fourier transform (3.1.71) is real, and may assume positive or negative values. Therefore, the solution that satisfies boundary condition (3.1.72c) should be written as

$$\breve{p} = \begin{cases} C_2 e^{-\alpha\bar{y}} & \text{if } \alpha > 0, \\ C_1 e^{\alpha\bar{y}} & \text{if } \alpha < 0, \end{cases}$$

or, equivalently,

$$\breve{p} = C e^{-|\alpha|\bar{y}}. \tag{3.1.73}$$

The constant C is easily found from boundary condition (3.1.72b):

$$C = |\alpha| \breve{A}(\alpha). \tag{3.1.74}$$

Substituting (3.1.74) back into (3.1.73) we have the solution in region 1 in the form

$$\breve{p} = |\alpha| \breve{A}(\alpha) e^{-|\alpha|\bar{y}}. \tag{3.1.75}$$

In order to find the pressure in region 3, we express equation (3.1.70) in terms of Fourier transforms as

$$\breve{P} = \breve{p} \Big|_{\bar{y}=0}, \tag{3.1.76}$$

and use (3.1.75) on the right-hand side of (3.1.76). We then find that

$$\breve{P} = |\alpha| \breve{A}.$$

(3.1.77)

Let us now consider the boundary-value problem (3.1.68) describing the flow in the lower tier (region 3). Applying Fourier transforms to equations (3.1.68a), (3.1.68b) and boundary conditions (3.1.68c), (3.1.68e), we have

$$i\omega \breve{U} + i\alpha \bar{Y}\breve{U} + \breve{V} = -i\alpha \breve{P} + \frac{d^2 \breve{U}}{d\bar{Y}^2},$$

(3.1.78a)

$$i\alpha \breve{U} + \frac{d\breve{V}}{d\bar{Y}} = 0,$$

(3.1.78b)

$$\left.\begin{array}{l} \breve{U} = -\dfrac{1}{2}\breve{F}(\alpha), \\[2mm] \breve{V} = \dfrac{i}{2}\omega \breve{F}(\alpha) \end{array}\right\} \quad \text{at} \quad \bar{Y} = 0,$$

(3.1.78c)

$$\breve{U} = \breve{A}(\alpha) \qquad \text{at} \quad \bar{Y} = \infty.$$

(3.1.78d)

The boundary-value problem (3.1.78), (3.1.77) is solved in the following steps. We start with the momentum equation (3.1.78a). Differentiation of this equation with respect to $\bar{Y}$ yields

$$i(\omega + \alpha \bar{Y})\frac{d\breve{U}}{d\bar{Y}} + i\alpha \breve{U} + \frac{d\breve{V}}{d\bar{Y}} = \frac{d^3 \breve{U}}{d\bar{Y}^3}.$$

(3.1.79)

Now we use the continuity equation (3.1.78b) to eliminate $d\breve{V}/d\bar{Y}$, which turns (3.1.79) into the following equation for $\breve{U}$:

$$i(\omega + \alpha \bar{Y})\frac{d\breve{U}}{d\bar{Y}} = \frac{d^3 \breve{U}}{d\bar{Y}^3}.$$

(3.1.80)

Since (3.1.80) is a third-order differential equation, it requires an additional boundary condition. The latter may be obtained by setting $\bar{Y} = 0$ in (3.1.78a). Using (3.1.78c) and (3.1.77), we find that

$$\left.\frac{d^2 \breve{U}}{d\bar{Y}^2}\right|_{\bar{Y}=0} = i\alpha \breve{P} = i\alpha |\alpha| \breve{A}.$$

(3.1.81)

If we introduce a new independent variable

$$z = z_0 + \theta \bar{Y},$$

with

$$z_0 = \frac{\omega}{\alpha}\theta, \qquad \theta = (i\alpha)^{1/3},$$

then equation (3.1.80) transforms into the Airy equation

$$\frac{d^3 \breve{U}}{dz^3} - z\frac{d\breve{U}}{dz} = 0$$

(3.1.82a)

for the derivative $d\breve{U}/dz$.

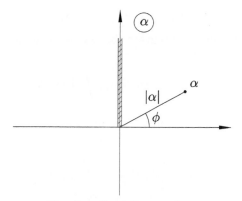

Fig. 3.4: Complex α-plane.

The boundary conditions (3.1.78c), (3.1.78d), and (3.1.81) are written in terms of the new variable z as

$$\check{U} = -\frac{1}{2}\check{F}(\alpha) \qquad \text{at} \quad z = z_0, \qquad\qquad (3.1.82b)$$

$$\frac{d^2\check{U}}{dz^2} = (i\alpha)^{1/3}|\alpha|\check{A} \qquad \text{at} \quad z = z_0, \qquad\qquad (3.1.82c)$$

$$\check{U} = \check{A} \qquad\qquad\qquad \text{at} \quad z = \infty. \qquad\qquad (3.1.82d)$$

Thus far, the Fourier transform parameter α has been assumed real, but for future purposes, we need to extend the analysis into the complex plane. We shall do this in the same way as in Section 2.3.2. We make a branch cut along the imaginary positive semi-axis (see Figure 3.4) and, for α written in the exponential form, $\alpha = |\alpha|e^{i\phi}$, define an analytic branch of $(i\alpha)^{1/3}$ as

$$(i\alpha)^{1/3} = \left(e^{i\pi/2}|\alpha|e^{i\phi}\right)^{1/3} = |\alpha|^{1/3}e^{i(\pi/6+\phi/3)}. \qquad\qquad (3.1.83)$$

The general solution of equation (3.1.82a) is written as

$$\frac{d\check{U}}{dz} = C_1 Ai(z) + C_2 Bi(z),$$

where $Ai(z)$ and $Bi(z)$ are two complementary solutions of the Airy equation. It follows from (2.3.12) that, for any α in the complex plane in Figure 3.4, the Airy function $Ai(z)$ tends to zero while $Bi(z)$ grows exponentially as $z = z_0 + (i\alpha)^{1/3}\check{Y}$ tends to infinity. Clearly, the boundary condition (3.1.82d) can only be satisfied if $d\check{U}/dz$ decays at large z. Therefore, we have to set $C_2 = 0$, and we can write

$$\frac{d\check{U}}{dz} = C_1 Ai(z). \qquad\qquad (3.1.84)$$

Substitution of (3.1.84) into (3.1.82c) gives a first equation relating the constant C_1 with the Fourier transform $\check{A}$ of the displacement function:

$$C_1 Ai'(z_0) = (i\alpha)^{1/3}|\alpha|\breve{A}. \tag{3.1.85}$$

To deduce a second relation, we integrate (3.1.84) with the initial condition (3.1.82b). This gives

$$\breve{U} = -\frac{1}{2}\breve{F}(\alpha) + C_1 \int\limits_{z_0}^{z} Ai(\zeta)\,d\zeta. \tag{3.1.86}$$

Now we set $z \to \infty$ in (3.1.86), and use boundary condition (3.1.82d), which yields the desired second relation between C_1 and $\breve{A}$:

$$\breve{A} = -\frac{1}{2}\breve{F}(\alpha) + C_1 \int\limits_{z_0}^{\infty} Ai(z)\,dz. \tag{3.1.87}$$

It remains to eliminate C_1 from (3.1.85) and (3.1.87), leading us to the solution of the viscous-inviscid interaction problems in terms of the Fourier transform of the displacement function:

$$\breve{A} = \frac{\frac{1}{2}\breve{F}(\alpha)Ai'(z_0)}{(i\alpha)^{1/3}|\alpha| \int\limits_{z_0}^{\infty} Ai(z)\,dz - Ai'(z_0)}. \tag{3.1.88}$$

Now that $\breve{A}$ is known, any other function can be easily found. In particular, the Fourier transform of the pressure is obtained by substitution of (3.1.88) into (3.1.77):

$$\breve{P} = \frac{\frac{1}{2}\breve{F}(\alpha)|\alpha|Ai'(z_0)}{(i\alpha)^{1/3}|\alpha| \int\limits_{z_0}^{\infty} Ai(z)\,dz - Ai'(z_0)}. \tag{3.1.89}$$

We are ready now to return back from Fourier to physical variables. To study the behaviour of the perturbations produced in the flow by the wall vibrations, it is sufficient to consider the pressure, which is calculated by applying the inverse Fourier transform:

$$P^{\star}(\bar{x}) = \frac{1}{2\pi} \int\limits_{-\infty}^{\infty} \breve{P}(\alpha)e^{i\alpha\bar{x}}\,d\alpha. \tag{3.1.90}$$

Substituting (3.1.89) into (3.1.90) and then into the equation for $P'(t,x)$ in (3.1.67), we see that the pressure perturbations in the lower tier (region 3) are given by

$$P'(\bar{t},\bar{x}) = \frac{e^{i\omega\bar{t}}}{2\pi} \int\limits_{-\infty}^{\infty} \frac{\frac{1}{2}\breve{F}(\alpha)|\alpha|Ai'(z_0)\,e^{i\alpha\bar{x}}}{(i\alpha)^{1/3}|\alpha| \int\limits_{z_0}^{\infty} Ai(z)\,dz - Ai'(z_0)}\,d\alpha + (c.c.)$$

$$= \Re\left\{ \frac{e^{i\omega\bar{t}}}{2\pi} \int\limits_{-\infty}^{\infty} \frac{\breve{F}(\alpha)|\alpha|Ai'(z_0)\,e^{i\alpha\bar{x}}}{(i\alpha)^{1/3}|\alpha| \int\limits_{z_0}^{\infty} Ai(z)\,dz - Ai'(z_0)}\,d\alpha \right\}, \tag{3.1.91}$$

where the integration variable α is assumed real, and $\Re$ stands for the real part of the expression in the curly brackets.

3.1.6   Receptivity coefficient

The distribution of the pressure $P'(\bar{t},\bar{x})$ along the wall may be, of course, found by means of a numerical calculation of the integral in (3.1.91); see Terent'ev (1981). However, in receptivity theory, the main interest is in the development of the perturbations downstream of their 'source', which may be studied through the asymptotic analysis of (3.1.91) at large values of $\bar{x}$. To perform such an analysis, it is convenient to consider an analytical extension of the integrand into the upper-half of the complex α-plane, and alter the contour of integration in a way we will explain later. When deforming the integration contour we need to identify the poles of the integrand. These are given by the dispersion equation

$$(i\alpha)^{1/3}|\alpha| \int\limits_{z_0}^{\infty} Ai(z)\,dz - Ai'(z_0) = 0, \qquad z_0 = \frac{i\omega}{(i\alpha)^{2/3}}. \tag{3.1.92}$$

This equation was studied in Section 2.3.3, and we know that there are an infinite (countable) number of roots of (3.1.92). The position of each root in the complex α-plane depends on the frequency ω. The trajectories of the first five roots, as ω changes from zero to infinity, are shown in Figure 2.3. All the roots originate at $\omega = 0$ from the coordinate origin, and all of them, except the first one, remain in the second quadrant for all $\omega \in (0,\infty)$. The behaviour of the first root is different. It stays in the second quadrant until the frequency reaches its critical value, $\omega_* \simeq 2.29797$, and then it crosses the real axis at the point $\alpha_* \simeq -1.00049$ and remains in the third quadrant for all $\omega \in (\omega_*,\infty)$. This root represents the Tollmien–Schlichting wave, and our task is to determine its amplitude.

Let us consider a value of the frequency ω smaller than the critical value.[4] Then all the roots of (3.1.92) lie in the second quadrant of the complex α-plane, as shown in Figure 3.5. Remember that when introducing an analytical branch of the function $(i\alpha)^{1/3}$ we made a branch cut along the positive imaginary axis in the α-plane. Also, the analytical continuation of $|\alpha|$ requires the branch cut to be extended to the entire imaginary axis. Therefore, we shall split the integration interval in (3.1.91) into two parts, the negative real semi-axis and positive real semi-axis, denoted by C_- and C_+ in Figure 3.5. We have

$$P'(\bar{t},\bar{x}) = \Re\left\{\frac{e^{i\omega\bar{t}}}{2\pi}\left[\mathcal{N}(\bar{x};\omega) + \mathcal{M}(\bar{x};\omega)\right]\right\}, \tag{3.1.93}$$

where

$$\mathcal{N}(\bar{x};\omega) = \int\limits_{-\infty}^{0} \frac{\alpha\breve{F}(\alpha)Ai'(z_0)e^{i\alpha\bar{x}}}{(i\alpha)^{1/3}\alpha\int\limits_{z_0}^{\infty}Ai(z)\,dz + Ai'(z_0)}\,d\alpha, \tag{3.1.94}$$

$$\mathcal{M}(\bar{x};\omega) = \int\limits_{0}^{\infty} \frac{\alpha\breve{F}(\alpha)Ai'(z_0)e^{i\alpha\bar{x}}}{(i\alpha)^{1/3}\alpha\int\limits_{z_0}^{\infty}Ai(z)\,dz - Ai'(z_0)}\,d\alpha. \tag{3.1.95}$$

---

[4]This restriction will be lifted in Section 3.2.

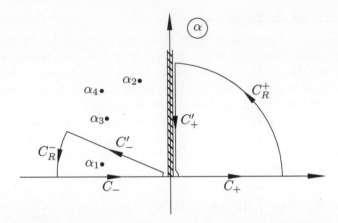

Fig. 3.5: Deformation of the contour of integration in (3.1.94) and (3.1.95).

When calculating the integral in (3.1.94) we close the contour of integration by adding to C_- a ray C'_- and a circular arc C_R^- of a large radius R. If the ray C'_- is chosen such that only one root, α_1, finds itself inside the combined contour, then, using the residue theorem, we will have

$$\mathcal{N}(\bar{x};\omega) = 2\pi i \frac{\alpha \breve{F}(\alpha) Ai'(z_0) e^{i\alpha\bar{x}}}{\frac{4}{3}(i\alpha)^{1/3} \int\limits_{z_0}^{\infty} Ai(z)\,dz - \frac{2}{3}(z_0/\alpha)[i(i\alpha)^{4/3} + z_0] Ai(z_0)}\Bigg|_{\alpha=\alpha_1}$$

$$- \int\limits_{C'_-} \frac{\alpha \breve{F}(\alpha) Ai'(z_0) e^{i\alpha\bar{x}}}{(i\alpha)^{1/3}\alpha \int\limits_{z_0}^{\infty} Ai(z)\,dz + Ai'(z_0)}\, d\alpha. \qquad (3.1.96)$$

Here it is taken into account that, according to Jordan's lemma, the integral along C_R^- tends to zero as $R \to \infty$.

Let us now assume that $\bar{x} \to \infty$. The integral along C'_- in (3.1.96) is a Laplace type integral. Its asymptotic behaviour may be determined with the help of Watson's lemma.[5] According to this lemma, at large values of $\bar{x}$ the 'region of dominant contribution' for this integral is represented by a small section of C'_- lying near the point $\alpha = 0$. Since $z_0 = i\omega/(i\alpha)^{2/3}$ tends to infinity in the limit

$$\alpha \to 0, \qquad \omega = O(1),$$

we can use asymptotic formulae (2.3.43) for the derivative and the integral of the Airy function. We see that as $z_0 \to \infty$

$$\frac{Ai'(z_0)}{(i\alpha)^{1/3}\alpha \int\limits_{z_0}^{\infty} Ai(z)\,dz} = -\frac{z_0}{(i\alpha)^{1/3}\alpha} \gg 1, \qquad (3.1.97)$$

---

[5]See Section 1.2 in Part 2 of this book series.

which allows us to disregard the first term in the denominator in (3.1.96). We have

$$\int_{C'_-} \frac{\alpha \breve{F}(\alpha) Ai'(z_0) e^{i\alpha\bar{x}}}{(i\alpha)^{1/3}\alpha \int\limits_{z_0}^{\infty} Ai(z)\,dz + Ai'(z_0)}\,d\alpha = \breve{F}(0) \int\limits_0^{\infty} \alpha e^{i\alpha\bar{x}}\,d\alpha = -\frac{\breve{F}(0)}{\bar{x}^2} + \cdots. \quad (3.1.98)$$

The integral (3.1.95) is analysed in the same way, except that there are no roots of the dispersion equation (3.1.92) inside the closed contour composed of C_+, C_R^+ and C'_+; see Figure 3.5. By Jordan's lemma, the integral along the arc C_R^+ tends to zero as the arc radius R tends to infinity, and the integral along C'_+ can be evaluated using Watson's lemma. We find that

$$\mathcal{M}(\bar{x};\omega) = \frac{\breve{F}(0)}{\bar{x}^2} + \cdots \quad \text{as} \quad \bar{x} \to \infty. \quad (3.1.99)$$

It remains to substitute (3.1.98) into (3.1.96) and then together with (3.1.99) into (3.1.93). We find that downstream of the vibrating section of the wall

$$P'(\bar{t},\bar{x}) = \Re\left\{ \mathcal{K}(\omega)\breve{F}(\alpha_1)e^{i(\omega\bar{t}+\alpha_1\bar{x})} + \frac{\breve{F}(0)e^{i\omega\bar{t}}}{\pi\bar{x}^2} + \cdots \right\} \quad \text{as} \quad \bar{x} \to \infty. \quad (3.1.100)$$

Here

$$\mathcal{K}(\omega) = \frac{i\alpha Ai'(z_0)}{\frac{4}{3}(i\alpha)^{1/3}\int\limits_{z_0}^{\infty} Ai(z)\,dz - \frac{2}{3}(z_0/\alpha)\left[i(i\alpha)^{4/3} + z_0\right]Ai(z_0)}\Bigg|_{\alpha=\alpha_1} \quad (3.1.101)$$

is a universal function of the frequency ω, which does not depend on the shape of the wall vibrations. It is termed the *receptivity coefficient*. The results of numerical calculation of $\mathcal{K}(\omega)$ are displayed in Figure 3.6. We represent the receptivity coefficient as $\mathcal{K} = |\mathcal{K}|e^{i\theta}$. Figure 3.6(a) shows the modulus $|\mathcal{K}|$, and Figure 3.6(b) the argument θ of the receptivity coefficient.

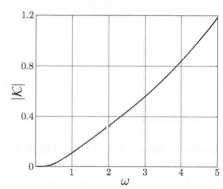

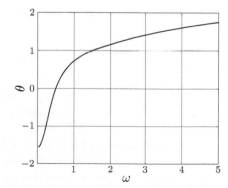

(a) Modulus of the receptivity coefficient. (b) Argument of the receptivity coefficient.

Fig. 3.6: The receptivity coefficient $\mathcal{K} = |\mathcal{K}|e^{i\theta}$ as a function of the frequency ω.

It should be noted that the second term on the right-hand side of (3.1.100) decreases with $\bar{x}$ with a decay rate $O(\bar{x}^{-2})$ that is independent of the frequency ω. The behaviour of the first term is different. It decreases with $\bar{x}$ for subcritical values of the frequency ($\omega < \omega_*$) but when $\omega \to \omega_* - 0$, the first root α_1 approaches the real axis (see Figure 3.5), and the decay rate (being equal to the imaginary part of α_1) tends to zero. Thus, downstream of the vibrating section of the wall, a Tollmien–Schlichting wave forms with the amplitude $\mathcal{K}(\omega)\check{F}(\alpha_1)$.

We shall conclude with the following remark. If one is interested in the behaviour of the perturbations around the vibrating section of the wall at finite values of $\bar{x}$, and chooses the ray C'_- to coincide with the left-hand side of the branch cut along the imaginary semi-axis, then the contributions from the rest of the roots of the dispersion equation (3.1.92) should be added in (3.1.100). Considered together they constitute the so-called *discrete spectrum* of perturbations, while the integral along the branch cut represents the *continuous spectrum*.

Exercises 10

1. In Section 2.3, the dispersion equation (2.3.34) was deduced by means of asymptotic analysis of the Orr–Sommerfeld equation. Here your task is to obtain (2.3.34) using the equations of triple-deck theory as a starting point. Remember that a dispersion equation describes natural oscillations of a flow that is free of external perturbations. Therefore, when writing the equations (3.1.58) for the lower tier, you need to set $\bar{f} = 0$ to ensure that you are dealing with the boundary layer on a surface free of a roughness, thereby obtaining

$$
\left.
\begin{aligned}
\frac{\partial \bar{U}}{\partial \bar{t}} + \bar{U}\frac{\partial \bar{U}}{\partial \bar{x}} + \bar{V}\frac{\partial \bar{U}}{\partial \bar{Y}} &= -\frac{d\bar{P}}{d\bar{x}} + \frac{\partial^2 \bar{U}}{\partial \bar{Y}^2}, \\
\frac{\partial \bar{U}}{\partial \bar{x}} + \frac{\partial \bar{V}}{\partial \bar{Y}} &= 0, \\
\bar{U} = \bar{V} = 0 \qquad &\text{at} \quad \bar{Y} = 0, \\
\bar{U} = \bar{Y} + \bar{A}(\bar{t}, \bar{x}) + \cdots \qquad &\text{as} \quad \bar{Y} \to \infty.
\end{aligned}
\right\}
\tag{3.1.102}
$$

The equations (3.1.59) for the upper tier preserve their form:

$$
\left.
\begin{aligned}
\frac{\partial^2 \bar{p}}{\partial \bar{x}^2} + \frac{\partial^2 \bar{p}}{\partial \bar{y}^2} &= 0, \\
\frac{\partial \bar{p}}{\partial \bar{y}} = \frac{\partial^2 \bar{A}}{\partial \bar{x}^2} \qquad &\text{at} \quad \bar{y} = 0, \\
\bar{p} \to 0 \qquad &\text{as} \quad \bar{x}^2 + \bar{y}^2 \to \infty.
\end{aligned}
\right\}
\tag{3.1.103}
$$

Using equations (3.1.102), (3.1.103) perform linear stability analysis of the flow. You may do this in the following steps:

(a) Confirm, by direct substitution into (3.1.102), that in the lower tier, the steady basic flow is given by

$$
\bar{U} = \bar{Y}, \qquad \bar{V} = \bar{P} = \bar{A} = 0,
$$

with the corresponding solution in the upper tier being

$$\bar{p} = 0.$$

(b) Introduce small perturbations to the basic flow, namely, seek the solution for the lower tier in the form

$$\left.\begin{array}{ll}\bar{U} = \bar{Y} + \varepsilon U'(\bar{t}, \bar{x}, \bar{Y}) + \cdots, & \bar{V} = \varepsilon V'(\bar{t}, \bar{x}, \bar{Y}) + \cdots, \\ \bar{P} = \varepsilon P'(\bar{t}, \bar{x}) + \cdots, & \bar{A} = \varepsilon A'(\bar{t}, \bar{x}) + \cdots,\end{array}\right\} \qquad (3.1.104)$$

with the pressure in the upper tier

$$\bar{p} = \varepsilon p'(\bar{t}, \bar{x}, \bar{y}) + \cdots. \qquad (3.1.105)$$

Assume that the perturbation amplitude parameter ε is small. Substitute (3.1.104) and (3.1.105) into (3.1.102) and (3.1.103), respectively. When doing so, disregard the $O(\varepsilon^2)$ terms, which will allow you to deduce the linearized equations of the viscous-inviscid interaction theory.

(c) Represent the solution of the linear equations for the lower tier in the simple wave form

$$\begin{array}{ll}U'(\bar{t}, \bar{x}, \bar{Y}) = e^{i(\omega \bar{t} + \alpha \bar{x})} \breve{U}(\bar{Y}), & V'(\bar{t}, \bar{x}, \bar{Y}) = e^{i(\omega \bar{t} + \alpha \bar{x})} \breve{V}(\bar{Y}), \\ P'(\bar{t}, \bar{x}) = e^{i(\omega \bar{t} + \alpha \bar{x})} \breve{P}, & A'(\bar{t}, \bar{x}) = e^{i(\omega \bar{t} + \alpha \bar{x})} \breve{A},\end{array}$$

and show that

$$\left.\begin{array}{ll} i(\omega + \alpha \bar{Y})\breve{U} + \breve{V} = -i\alpha \breve{P} + \dfrac{d^2 \breve{U}}{d\bar{Y}^2}, \\[2mm] i\alpha \breve{U} + \dfrac{d\breve{V}}{d\bar{Y}} = 0, \\[2mm] \breve{U} = \breve{V} = 0 & \text{at} \quad \bar{Y} = 0, \\[1mm] \breve{U} = \breve{A} & \text{at} \quad \bar{Y} = \infty. \end{array}\right\} \qquad (3.1.106)$$

Similarly, write the pressure in the upper tier as

$$p'(\bar{t}, \bar{x}, \bar{y}) = e^{i(\omega \bar{t} + \alpha \bar{x})} \breve{p}(\bar{y}),$$

and show that the function $\breve{p}(\bar{y})$ satisfies the boundary-value problem

$$\left.\begin{array}{ll} -\alpha^2 \breve{p} + \dfrac{d^2 \breve{p}}{d\bar{y}^2} = 0, \\[2mm] \dfrac{d\breve{p}}{d\bar{y}} = -\alpha^2 \breve{A} & \text{at} \quad \bar{y} = 0, \\[2mm] \breve{p} = 0 & \text{at} \quad \bar{y} = \infty. \end{array}\right\} \qquad (3.1.107)$$

(d) After solving problem (3.1.107), deduce that the pressure in the lower tier is given by

$$\breve{P} = \breve{p}\Big|_{\bar{y}=0} = |\alpha|\breve{A}. \qquad (3.1.108)$$

(e) Finally, consider the lower tier problem (3.1.106) and perform similar manipulations to those applied to the boundary-value problem (3.1.78) Your task
is to show that a non-trivial solution to (3.1.106), (3.1.108) can only exist if

$$(i\alpha)^{1/3}|\alpha| \int\limits_{z_0}^{\infty} Ai(z)\,dz - Ai'(z_0) = 0, \qquad z_0 = \frac{i\omega}{(i\alpha)^{2/3}}.$$

3.2   Initial-Value Problem

In Section 3.1 we studied the generation of a Tollmien–Schlichting wave by a 'vibrator'
which consists of a short section of the body surface performing harmonic oscillations.
When analysing the flow, we assumed that the perturbations induced in the boundary layer were periodic in time with a frequency ω less than the critical value ω_*.
This restriction enabled us to solve the receptivity problem with the help of Fourier
transforms. Remember that the classical Fourier transform is only applicable to functions that tend to zero both upstream ($\bar{x} \to -\infty$) and downstream ($\bar{x} \to \infty$). The
latter requirement is only satisfied if the generated Tollmien–Schlichting wave decays
downstream, which happens when $\omega < \omega_*$.

It should be noted that the Fourier transforms method can still be formally applied
for the case of a supercritical frequency ($\omega > \omega_*$), but it produces a solution with an
unexpected behaviour. What one expects is that there should be a growing Tollmien–
Schlichting wave in the boundary layer behind the vibrator. However, once ω exceeds
its critical value ω_*, the first root α_1 of the dispersion equation (3.1.92) moves into
the lower half-plane in Figure 3.5, and the first term in (3.1.96) simply disappears.
Thus the flow downstream of the vibrator appears to be free of Tollmien–Schlichting
waves. Somewhat miraculously, a Tollmien–Schlichting wave now emerges upstream of
the vibrator instead. The amplitude of this wave may be calculated in the usual way
by introducing a suitable closed contour for the integral in (3.1.94) and applying the
residue theorem. We form the closed contour by combining the original integration
contour C_- with a ray C_I along the negative imaginary semi-axis and a quarter-circle
C_R of large radius R; see Figure 3.7. For $\bar{x} < 0$, the integral along C_R obeys Jordan's
lemma, while the integral along C_I may be estimated using Watson's lemma. Taking
the residue at α_1, we find that

$$P'(\bar{t},\bar{x}) = -\Re\left\{ \mathcal{K}(\omega)\breve{F}(\alpha_1)e^{i(\omega\bar{t}+\alpha_1\bar{x})} + \frac{\breve{F}(0)e^{i\omega\bar{t}}}{\pi(-\bar{x})^2} + \cdots \right\} \quad \text{as} \quad \bar{x} \to -\infty, \quad (3.2.1)$$

where $\mathcal{K}(\omega)$ is given by (3.1.101). Notice that, compared to (3.1.100), there is a minus
sign in (3.2.1). This is due to the fact that now the contour of integration is traversed
in the clockwise direction.

This 'formal solution' describes the following physical situation. Suppose that there
is a Tollmien–Schlichting wave growing in the boundary layer, and we want to suppress
its development which would allow us to delay laminar-turbulent transition. With this
aim in mind, a vibrator may be installed on the body surface to generate a second
Tollmien–Schlichting wave in anti-phase with the first one. Then the two waves will

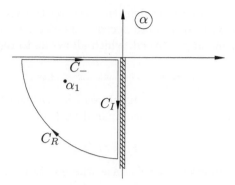

Fig. 3.7: Deformation of the integration contour in (3.1.94) for a supercritical frequency $\omega > \omega_*$.

cancel each other, and the boundary layer downstream of the vibrator will be free of Tollmien–Schlichting waves.

Of course, we can always add to this solution a Tollmien–Schlichting wave as the latter represents an eigen-solution of the triple-deck equations. If we choose the amplitude of the added wave according to (3.1.100), then the Tollmien–Schlichting wave (3.2.1) upstream of the vibrator will disappear, and we will return to the original problem of the generation of a Tollmien–Schlichting wave by the vibrator. Thus the solution given by (3.1.100), (3.1.101) is applicable not only for subcritical but also for supercritical frequency ω.

In this section, we shall confirm this result by solving the corresponding initial-value problem.[6]

3.2.1   Problem formulation

Here we shall consider the case where the boundary layer is initially free of perturbations, but then at time $t = 0$ a vibrator, made of a short section of the wall, starts to oscillate, and our task will be to study the flow response. We shall assume, as before, that the length of the vibrator is an $O(Re^{-3/8})$ quantity; the frequency of its oscillations is assumed to be $O(Re^{1/4})$ and their amplitude is $O(Re^{-5/8})$. Under these conditions, the flow analysis in the previous section leading to the formulation of the viscous-inviscid interaction problem (3.1.58)–(3.1.60) remains valid, except the vibrator equation is now written as[7]

$$\bar{Y} = \begin{cases} 0 & \text{if } \bar{t} \leq 0, \\ \bar{f}(\bar{x})\sin(\omega\bar{t}) & \text{if } \bar{t} > 0, \end{cases} \qquad (3.2.2)$$

and we have to add to (3.1.58c), (3.1.58d), and (3.1.58e) the initial condition

$$\bar{U} = \bar{Y} \quad \text{at} \quad \bar{t} = 0. \qquad (3.2.3)$$

---

[6]The original study of the initial-value problem was performed by Terent'ev (1984, 1987).

[7]Note that we replaced $\cos(\omega_* t_*)$ in (3.1.14) by $\sin(\omega\bar{t})$. This is to make (3.2.2) continuous at $\bar{t} = 0$.

Of course, the perturbations induced in the flow by the vibrator cannot propagate in the boundary layer with an infinite speed. Therefore, for all finite $\bar{t}$, the flow far from the vibrator remains unperturbed, which allows us to replace (3.1.58d) with the boundary condition

$$\bar{U} = \bar{Y} + \cdots \quad \text{as} \quad \bar{x} \to \pm\infty. \tag{3.2.4}$$

To linearize the equations of viscous-inviscid interaction theory, we shall assume that the vibrator amplitude is small compared with the thickness of the lower tier, that is,

$$\bar{f}(\bar{x}) = \varepsilon F(\bar{x}),$$

where ε is a small parameter. In this case, the solution in the lower tier may be represented in the form

$$\left. \begin{array}{ll} \bar{U} = \bar{Y} + \varepsilon U'(\bar{t}, \bar{x}, \bar{Y}) + \cdots, & \bar{V} = \varepsilon V'(\bar{t}, \bar{x}, \bar{Y}) + \cdots, \\ \bar{P} = \varepsilon P'(\bar{t}, \bar{x}) + \cdots, & \bar{A} = \varepsilon A'(\bar{t}, \bar{x}) + \cdots. \end{array} \right\} \tag{3.2.5}$$

Substitution of (3.2.5) into (3.1.58), (3.2.3), and (3.2.4) yields

$$\frac{\partial U'}{\partial \bar{t}} + \bar{Y}\frac{\partial U'}{\partial \bar{x}} + V' = -\frac{\partial P'}{\partial \bar{x}} + \frac{\partial^2 U'}{\partial \bar{Y}^2}, \tag{3.2.6a}$$

$$\frac{\partial U'}{\partial \bar{x}} + \frac{\partial V'}{\partial \bar{Y}} = 0, \tag{3.2.6b}$$

$$U' = 0 \qquad\qquad \text{at} \quad \bar{t} = 0, \tag{3.2.6c}$$

$$\left. \begin{array}{l} U' = -F(\bar{x})\sin(\omega \bar{t}), \\ V' = \omega F(\bar{x})\cos(\omega \bar{t}) \end{array} \right\} \quad \text{at} \quad \bar{Y} = 0, \tag{3.2.6d}$$

$$U' \to 0 \qquad\qquad \text{as} \quad \bar{x} \to \pm\infty, \tag{3.2.6e}$$

$$U' = A'(\bar{t}, \bar{x}) \qquad \text{at} \quad \bar{Y} = \infty. \tag{3.2.6f}$$

The pressure in the upper tier is sought in the form

$$\bar{p} = \varepsilon p'(\bar{t}, \bar{x}, \bar{y}) + \cdots,$$

which, being substituted into (3.1.59), results in

$$\left. \begin{array}{l} \dfrac{\partial^2 p'}{\partial \bar{x}^2} + \dfrac{\partial^2 p'}{\partial \bar{y}^2} = 0, \\[2mm] \dfrac{\partial p'}{\partial \bar{y}} = \dfrac{\partial^2 A'}{\partial \bar{x}^2} \quad \text{at} \quad \bar{y} = 0, \\[2mm] p' \to 0 \quad \text{as} \quad \bar{x}^2 + \bar{y}^2 \to \infty. \end{array} \right\} \tag{3.2.7}$$

The solutions in the upper and lower tiers are related to one another through the equation:

$$P' = p' \big|_{\bar{y}=0}. \tag{3.2.8}$$

3.2.2 Numerical solution of the linear problem

To see how the perturbations behave in the boundary layer, we shall start with a numerical solution of the linearized viscous-inviscid interaction (3.2.6)–(3.2.8). We shall use for this purpose the following simple approach. When dealing with the lower tier, we apply to (3.2.6) a Fourier transform with respect to $\bar{x}$. In particular, the Fourier transform of the longitudinal velocity U' is defined as

$$\breve{U}(\bar{t}, \alpha, \bar{Y}) = \int_{-\infty}^{\infty} U'(\bar{t}, \bar{x}, \bar{Y}) e^{-i\alpha\bar{x}} \, d\bar{x}, \tag{3.2.9}$$

with similar transformations for the other functions. As a result, the lower tier problem becomes

$$\frac{\partial \breve{U}}{\partial \bar{t}} + i\alpha\bar{Y}\breve{U} + \breve{V} = -i\alpha\breve{P} + \frac{\partial^2 \breve{U}}{\partial \bar{Y}^2}, \tag{3.2.10a}$$

$$i\alpha\breve{U} + \frac{\partial \breve{V}}{\partial \bar{Y}} = 0, \tag{3.2.10b}$$

$$\breve{U} = 0 \qquad\qquad \text{at} \quad \bar{t} = 0, \tag{3.2.10c}$$

$$\left.\begin{aligned}\breve{U} &= -\breve{F}(\alpha)\sin(\omega\bar{t}),\\ \breve{V} &= \omega\breve{F}(\alpha)\cos(\omega\bar{t})\end{aligned}\right\} \quad \text{at} \quad \bar{Y} = 0, \tag{3.2.10d}$$

$$\breve{U} = \breve{A}(\bar{t}, \alpha) \qquad \text{at} \quad \bar{Y} = \infty. \tag{3.2.10e}$$

Here $\breve{F}(\alpha)$ is the Fourier transform of the vibrator shape function $F(\bar{x})$.

Applying Fourier transforms to the upper tier problem (3.2.7), we have

$$\left.\begin{aligned} -\alpha^2\breve{p} + \frac{\partial^2 \breve{p}}{\partial \bar{y}^2} &= 0,\\ \frac{\partial \breve{p}}{\partial \bar{y}} &= -\alpha^2\breve{A} \quad \text{at} \quad \bar{y} = 0,\\ \breve{p} &\to 0 \quad \text{as} \quad \bar{y} \to \infty.\end{aligned}\right\} \tag{3.2.11}$$

The boundary-value problem (3.2.11) is identical to (3.1.72); its solution is given by (3.1.75):

$$\breve{p} = |\alpha|\breve{A}(\bar{t}, \alpha) e^{-|\alpha|\bar{y}}. \tag{3.2.12}$$

Setting $\bar{y} = 0$ in (3.2.12) yields the pressure in the lower tier:

$$\breve{P} = |\alpha|\breve{A}(\bar{t}, \alpha). \tag{3.2.13}$$

When dealing with the lower tier problem (3.2.10) it is convenient to introduce the shear stress function

$$\tau = \frac{\partial \breve{U}}{\partial \bar{Y}}. \tag{3.2.14}$$

Differentiating (3.2.10a) with respect to $\bar{Y}$, and eliminating $\check{V}$ with the help of the continuity equation (3.2.10b), we find that τ satisfies the following equation:

$$\frac{\partial \tau}{\partial \bar{t}} + i\alpha \bar{Y} \tau = \frac{\partial^2 \tau}{\partial \bar{Y}^2}. \tag{3.2.15}$$

Now we need to formulate the initial and boundary conditions for (3.2.15). Obviously, the initial condition (3.2.10c) is written in terms of the shear stress function (3.2.14) as

$$\tau = 0 \quad \text{at} \quad \bar{t} = 0. \tag{3.2.16}$$

Differentiation of both sides of (3.2.10e) with respect to $\bar{Y}$ yields the first boundary condition:

$$\tau = 0 \quad \text{at} \quad \bar{Y} = \infty. \tag{3.2.17}$$

To deduce the second condition, we set $\bar{Y} = 0$ in (3.2.10a). Using (3.2.10d) and (3.2.13), we find that

$$\frac{\partial \tau}{\partial \bar{Y}} = i\alpha |\alpha| \check{A}(\bar{t}, \alpha) \quad \text{at} \quad \bar{Y} = 0. \tag{3.2.18}$$

Finally, we integrate (3.2.14) with the initial condition for $\check{U}$ in (3.2.10d). This leads to

$$\check{U} = -\check{F}(\alpha) \sin(\omega \bar{t}) + \int\limits_0^{\bar{Y}} \tau(\bar{t}, \alpha, \bar{Y}') \, d\bar{Y}'. \tag{3.2.19}$$

Setting $\bar{Y} = \infty$ in (3.2.19) and using condition (3.2.10e), we arrive at the conclusion that

$$\int\limits_0^\infty \tau(\bar{t}, \alpha, \bar{Y}) \, d\bar{Y} = \check{A}(\bar{t}, \alpha) + \check{F}(\alpha) \sin(\omega \bar{t}). \tag{3.2.20}$$

After eliminating $\check{A}(\bar{t}, \alpha)$ from (3.2.18) and (3.2.20), we have the following integro-differential condition on τ:

$$\frac{\partial \tau}{\partial \bar{Y}}\bigg|_{\bar{Y}=0} = i\alpha |\alpha| \int\limits_0^\infty \tau(\bar{t}, \alpha, \bar{Y}) \, d\bar{Y} - i\alpha |\alpha| \check{F}(\alpha) \sin(\omega \bar{t}). \tag{3.2.21}$$

Thus, our task is to solve equation (3.2.15) with initial condition (3.2.16), boundary condition (3.2.17) and integro-differential condition (3.2.21).

Numerical procedure

We start by introducing a mesh $\{\bar{t}_k, \bar{Y}_j\}$ with uniform spacing ΔT, ΔY in the $\bar{t}$- and $\bar{Y}$-directions, respectively:

$$\bar{t}_k = k \, \Delta T, \qquad k = 0, \, 1, \, 2, \, \ldots, \, K,$$
$$\bar{Y}_j = j \, \Delta Y, \qquad j = 0, \, 1, \, 2, \, \ldots, \, J.$$

Owing to the fact that equation (3.2.15) is parabolic, its solution may be constructed using a marching procedure where the calculations progress from one time line, $\bar{t} = \bar{t}_k$,

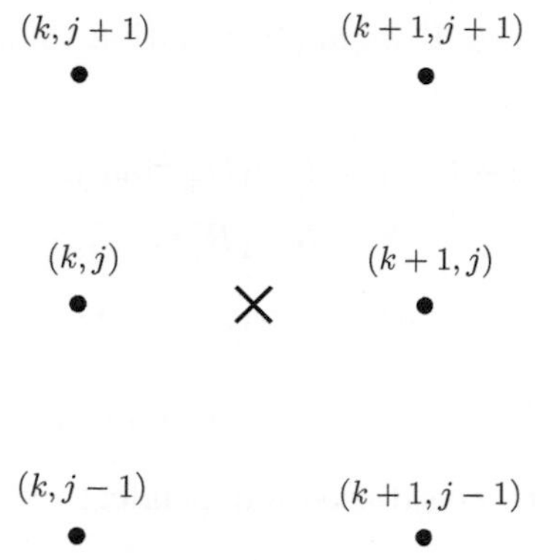

$(k, j+1)$ $(k+1, j+1)$

(k, j) $\times$ $(k+1, j)$

$(k, j-1)$ $(k+1, j-1)$

Fig. 3.8: Finite-differencing stencil for equation (3.2.15).

to the next one, $\bar{t} = \bar{t}_{k+1}$. Figure 3.8 shows the stencil employed for finite-differencing equation (3.2.15). The derivatives in the equation are represented by the formulae

$$\left. \begin{aligned} \frac{\partial \tau}{\partial \bar{t}} &= \frac{\tau_{k+1,j} - \tau_{k,j}}{\Delta T}, \\ \frac{\partial^2 \tau}{\partial \bar{Y}^2} &= \frac{1}{2}\left(\frac{\tau_{k,j+1} - 2\tau_{k,j} + \tau_{k,j-1}}{(\Delta Y)^2} + \frac{\tau_{k+1,j+1} - 2\tau_{k+1,j} + \tau_{k+1,j-1}}{(\Delta Y)^2} \right), \end{aligned} \right\} \qquad (3.2.22)$$

all being second-order accurate with respect to the central point in the stencil; the latter is shown by the cross in Figure 3.8.

Substitution of (3.2.22) into (3.2.15) results in the following set of algebraic equations:

$$a_j \tau_{k+1,j+1} + b_j \tau_{k+1,j} + c_j \tau_{k+1,j-1} = d_j, \qquad j = 1, 2, \ldots, J-1, \qquad (3.2.23)$$

where

$$a_j = \frac{1}{2(\Delta Y)^2}, \qquad b_j = -\frac{1}{(\Delta Y)^2} - \frac{1}{\Delta T} - \frac{i\alpha \Delta Y j}{2}, \qquad c_j = \frac{1}{2(\Delta Y)^2},$$

$$d_j = -\frac{\tau_{k,j+1} - 2\tau_{k,j} + \tau_{k,j-1}}{2(\Delta Y)^2} - \left(\frac{1}{\Delta T} - \frac{i\alpha \Delta Y j}{2} \right)\tau_{k,j}.$$

The set of equations (3.2.23) may be solved in various ways. One can use, for example, the Thomas algorithm. This is an elimination technique with the solution sought in the form

$$\tau_{k+1,j} = R_j \tau_{k+1,j-1} + Q_j, \qquad j = 1, 2, \ldots, J. \qquad (3.2.24)$$

The calculations are performed in two steps. First, the Thomas coefficients, R_j, Q_j, are calculated using the recurrence relations

$$R_j = -\frac{c_j}{b_j + a_j R_{j+1}}, \qquad Q_j = \frac{d_j - a_j Q_{j+1}}{b_j + a_j R_{j+1}}, \qquad j = J-1, \ldots, 1. \qquad (3.2.25)$$

These are deduced as follows. If we apply equation (3.2.24) to $\tau_{k+1,j+1}$, then we will have

$$\tau_{k+1,j+1} = R_{j+1}\tau_{k+1,j} + Q_{j+1} = R_{j+1}(R_j\tau_{k+1,j-1} + Q_j) + Q_{j+1}$$
$$= R_{j+1}R_j\tau_{k+1,j-1} + R_{j+1}Q_j + Q_{j+1}. \qquad (3.2.26)$$

Substituting (3.2.26) together with (3.2.24) into (3.2.23) results in

$$(a_jR_{j+1}R_j + b_jR_j + c_j)\tau_{k+1,j-1} + a_jR_{j+1}Q_j + b_jQ_j + a_jQ_{j+1} = d_j.$$

If we want this equation to be satisfied independent of the value of $\tau_{k+1,j-1}$, then we need to set

$$a_jR_{j+1}R_j + b_jR_j + c_j = 0, \qquad a_jR_{j+1}Q_j + b_jQ_j + a_jQ_{j+1} = d_j. \qquad (3.2.27)$$

Solving equations (3.2.27) for R_j and Q_j respectively, leads to (3.2.25).

To start the calculations in (3.2.25), we need to know R_J and Q_J. These are provided by boundary condition (3.2.17),

$$\tau_{k+1,J} = 0. \qquad (3.2.28)$$

Setting $j = J$ in (3.2.24) and substituting the result into (3.2.28), we have

$$R_J\tau_{k+1,J-1} + Q_J = 0.$$

This equation is satisfied independently of the value of $\tau_{k+1,J-1}$ provided that

$$R_J = Q_J = 0.$$

The second step in Thomas' procedure is the recurrent use of equation (3.2.24). It requires $\tau_{k+1,0}$ to be known. To find $\tau_{k+1,0}$ one needs to consider the integro-differential condition (3.2.21). It follows from (3.2.24) that $\tau_{k+1,j}$ is a linear function of $\tau_{k+1,0}$ and can be expressed in the form

$$\tau_{k+1,j} = A_j\tau_{k+1,0} + B_j, \qquad j = 0, 1, \ldots, J. \qquad (3.2.29)$$

Here

$$A_0 = 1, \qquad B_0 = 0,$$

and

$$\left.\begin{array}{l} A_j = R_jA_{j-1}, \\ B_j = R_jB_{j-1} + Q_j \end{array}\right\} \qquad j = 1, 2, \ldots, J. \qquad (3.2.30)$$

With known A_j, B_j, the derivative on the left-hand side of (3.2.21) is written as

$$\left.\frac{\partial\tau}{\partial\bar{Y}}\right|_{\bar{Y}=0} = \frac{4\tau_{k+1,1} - 3\tau_{k+1,0} - \tau_{k+1,2}}{2\Delta Y} = D_1\tau_{k+1,0} + D_2, \qquad (3.2.31)$$

with

$$D_1 = \frac{2A_1}{\Delta Y} - \frac{3}{2\Delta Y} - \frac{A_2}{2\Delta Y}, \qquad D_2 = \frac{2B_1}{\Delta Y} - \frac{B_2}{2\Delta Y}.$$

We can also express the integral in (3.2.21) as

$$\int_0^\infty \tau \, dY = G_1 \tau_{k+1,0} + G_2, \qquad (3.2.32)$$

where

$$G_1 = \sum_{j=1}^M (A_j + A_{j-1})\frac{\Delta Y}{2}, \qquad G_2 = \sum_{j=1}^M (B_j + B_{j-1})\frac{\Delta Y}{2}.$$

Substituting (3.2.31) and (3.2.32) into (3.2.21) and solving the resulting equation for $\tau_{k+1,0}$, we find that

$$\tau_{k+1,0} = \frac{i\alpha|\alpha|\left[G_2 - \check{F}(\alpha)\sin(\omega\bar{t}_{k+1})\right] - D_2}{D_1 - i\alpha|\alpha|G_1}. \qquad (3.2.33)$$

Now we can return to (3.2.29) and find the distribution of τ on a new time line $\bar{t}_{k+1}$.

At each time step, the Fourier transform of the pressure $\check{P}(\bar{t},\alpha)$ can be found using the equation

$$\check{P}(\bar{t},\alpha) = \frac{1}{i\alpha}\frac{\partial\tau}{\partial\bar{Y}}\bigg|_{\bar{Y}=0}, \qquad (3.2.34)$$

which is obtained by combining (3.2.18) with (3.2.13).

The calculations should be repeated for a suitable set of values of α to allow us to perform the inverse Fourier transform

$$P'(\bar{t},\bar{x}) = \frac{1}{2\pi}\int_{-\infty}^\infty \check{P}(\bar{t},\alpha)e^{i\alpha\bar{x}}\,d\alpha.$$

The results of the calculations are presented in Figures 3.9 and 3.10. When performing the calculations, we chose $F(\bar{x}) = e^{-\bar{x}^2}$, in which case $\check{F}(\alpha) = \sqrt{\pi}e^{-\alpha^2/4}$.

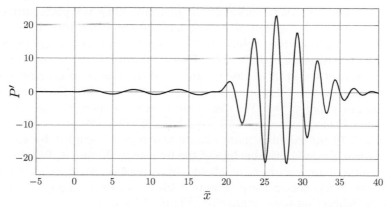

Fig. 3.9: Pressure perturbations at time $\bar{t} = 7$ for the vibrator frequency $\omega = 2.6$.

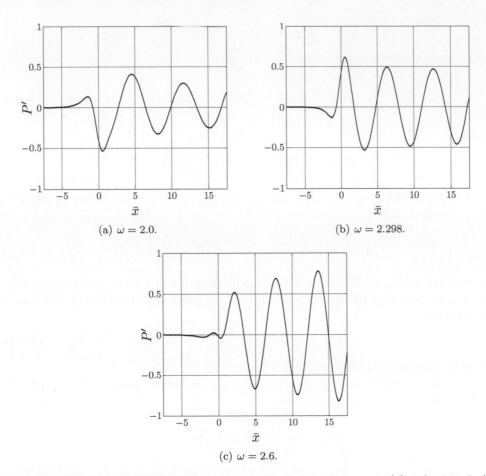

(a) $\omega = 2.0$.

(b) $\omega = 2.298$.

(c) $\omega = 2.6$.

Fig. 3.10: The Tollmien–Schlichting wave behind the vibrator at (a) subcritical, (b) critical, and (c) supercritical values of the frequency ω.

Figure 3.9 shows the distribution of the pressure along the plate surface at time $\bar{t} = 7.0$. The vibrator is centred at $\bar{x} = 0$, and we see a well established Tollmien–Schlichting wave immediately behind it. At time $\bar{t} = 7.0$ the Tollmien–Schlichting wave occupies the region up to $\bar{x} \approx 17.5$. Further downstream, we see a *wave packet*. The packet forms at the start of the vibrator, and then moves downstream spreading in the $\bar{x}$-direction. At $\bar{t} = 7.0$ it occupies the region $\bar{x} \in [17.5, 40]$. Further downstream, the flow remains unperturbed. We also observe that the perturbations decay rather fast upstream of the vibrator.

In Figure 3.10 we examine more closely the region occupied by the Tollmien–Schlichting wave. For subcritical values of the frequency (Figure 3.10a), the Tollmien–Schlichting wave decays with $\bar{x}$, as should be expected. At $\omega = \omega_*$ we observe a neutral Tollmien–Schlichting wave (Figure 3.10b), and for a supercritical frequency, the Tollmien–Schlichting wave is growing (Figure 3.10c).

Returning to Figure 3.9, we note that the amplitude associated with the wave packet is significantly larger than that of the Tollmien–Schlichting wave. In fact, we will show in the following section that the perturbations in the wave packet grow exponentially with time.

3.2.3   Analysis of the wave packet

We shall now try to make further progress in the analytical study of the perturbations. For this purpose, in addition to the Fourier transform (3.2.9) with respect to $\bar{x}$, we shall also apply a Laplace transform with respect to time $\bar{t}$. In particular, the Laplace transform $\check{\tau}(\sigma, \alpha, \bar{Y})$ of the shear stress function $\tau(\bar{t}, \alpha, \bar{Y})$ is defined as

$$\check{\tau}(\sigma, \alpha, \bar{Y}) = \int\limits_{0}^{\infty} \tau(\bar{t}, \alpha, \bar{Y}) e^{-\sigma \bar{t}} \, d\bar{t}.$$

Notice that, with initial condition (3.2.16), the Laplace transform of $\partial \tau / \partial \bar{t}$ is calculated using integration by parts as

$$\int\limits_{0}^{\infty} \frac{\partial \tau}{\partial \bar{t}} e^{-\sigma \bar{t}} \, d\bar{t} = \tau e^{-\sigma \bar{t}} \Big|_{0}^{\infty} + \sigma \int\limits_{0}^{\infty} \tau e^{-\sigma \bar{t}} \, d\bar{t} = \sigma \check{\tau}.$$

Hence, equation (3.2.15) is written in terms of Laplace transforms as

$$(\sigma + i\alpha \bar{Y})\check{\tau} = \frac{d^2 \check{\tau}}{d\bar{Y}^2}. \tag{3.2.35a}$$

Applying a Laplace transform to boundary condition (3.2.17) and to the integro-differential condition (3.2.21) we have

$$\check{\tau} = 0 \quad \text{at} \quad \bar{Y} = \infty, \tag{3.2.35b}$$

$$\frac{d\check{\tau}}{d\bar{Y}}\bigg|_{\bar{Y}=0} = i\alpha|\alpha| \int\limits_{0}^{\infty} \check{\tau}(\sigma, \alpha, \bar{Y}) \, d\bar{Y} - i\alpha|\alpha|\check{F}(\alpha)\frac{\omega}{\sigma^2 + \omega^2}. \tag{3.2.35c}$$

If we introduce a new independent variable

$$z = (i\alpha)^{1/3}\bar{Y} + z_0, \qquad z_0 = \frac{\sigma}{(i\alpha)^{2/3}}, \tag{3.2.36}$$

then the boundary-value problem (3.2.35) takes the form

$$\frac{d^2 \check{\tau}}{dz^2} - z\check{\tau} = 0, \tag{3.2.37a}$$

$$\check{\tau} = 0 \quad \text{at} \quad z = \infty, \tag{3.2.37b}$$

$$\frac{d\check{\tau}}{dz}\bigg|_{z=z_0} = (i\alpha)^{1/3}|\alpha| \int\limits_{z_0}^{\infty} \check{\tau} \, dz - (i\alpha)^{2/3}|\alpha|\check{F}(\alpha)\frac{\omega}{\sigma^2 + \omega^2}. \tag{3.2.37c}$$

Equation (3.2.37a) is the Airy equation. We can deal with it in the same way we did when analysing equation (3.1.82a) in Section 3.1. We make a branch cut along the positive imaginary semi-axis in the complex α-plane (see Figure 3.4) and define an analytic branch of the function $(i\alpha)^{1/3}$ with the help of equation (3.1.83). Then the solution of equation (3.2.37a) that satisfies condition (3.2.37b) is written as

$$\check{\tau} = C_1 Ai(z), \tag{3.2.38}$$

where $Ai(z)$ is the Airy function. The constant C_1 is determined by substituting (3.2.38) into (3.2.37c). We find that

$$C_1 = \frac{\omega}{\sigma^2 + \omega^2} \frac{(i\alpha)^{2/3}|\alpha|\check{F}(\alpha)}{(i\alpha)^{1/3}|\alpha| \int\limits_{z_0}^{\infty} Ai(z)\, dz - Ai'(z_0)}. \tag{3.2.39}$$

Now any of the other fluid-dynamic functions can be easily found. In particular, let us consider equation (3.2.34) for the Fourier transform $\check{P}(\check{t}, \alpha)$ of the pressure $P'(\check{t}, \bar{x})$. Taking Laplace transforms of both sides of (3.2.34), we have

$$\check{P}(\sigma, \alpha) = \frac{1}{i\alpha} \frac{d\check{\tau}}{d\bar{Y}}\bigg|_{\bar{Y}=0},$$

which can be written in terms of the new independent variable (3.2.36) as

$$\check{P}(\sigma, \alpha) = \frac{1}{(i\alpha)^{2/3}} \frac{d\check{\tau}}{dz}\bigg|_{z=z_0}. \tag{3.2.40}$$

It remains to substitute (3.2.39) into (3.2.38) and then into (3.2.40). We find that the Laplace/Fourier image of the pressure is given by

$$\check{P}(\sigma, \alpha) = \frac{\omega}{\sigma^2 + \omega^2} \frac{|\alpha| Ai'(z_0)\check{F}(\alpha)}{(i\alpha)^{1/3}|\alpha| \int\limits_{z_0}^{\infty} Ai(z)\, dz - Ai'(z_0)}. \tag{3.2.41}$$

Let us now return to physical variables. We first apply to (3.2.41) the inverse Laplace transform:

$$\check{P}(\check{t}, \alpha) = \frac{1}{2\pi i} \int\limits_{a-i\infty}^{a+i\infty} \check{P}(\sigma, \alpha) e^{\sigma \check{t}}\, d\sigma. \tag{3.2.42}$$

Here the integration is performed along a vertical line $\mathcal{L}$ in the complex σ-plane that crosses the real axis at $\Re(\sigma) = a$, where a is chosen such that all the singularities of (3.2.41) find themselves on the left-hand side of $\mathcal{L}$; see Figure 3.11. Clearly, in addition to two poles

$$\sigma = i\omega, \qquad \sigma = -i\omega$$

that lie on the imaginary axis, there are an infinite (countable) number of poles defined by the equation

$$(i\alpha)^{1/3}|\alpha| \int\limits_{z_0}^{\infty} Ai(z)\, dz - Ai'(z_0) = 0, \tag{3.2.43}$$

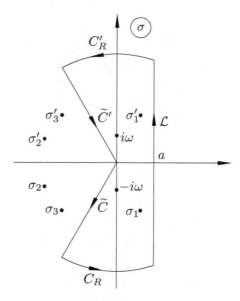

Fig. 3.11: Deformation of the contour of integration in (3.2.42).

where $z_0 = \sigma/(i\alpha)^{2/3}$. This is a standard dispersion equation of triple-deck theory, studied in detail in Section 2.3.3, but now it has to be treated differently. Since our intention is to use (3.2.42) in the inverse Fourier transform

$$P'(\bar{t}, \bar{x}) = \frac{1}{2\pi} \int_{-\infty}^{\infty} \check{P}(\bar{t}, \alpha) e^{i\alpha \bar{x}} \, d\alpha, \qquad (3.2.44)$$

we shall assume, to start with, that α is real.

The results of the numerical solution of (3.2.43) for real positive α are displayed in Figure 3.12. We see that all the roots of the dispersion equation (3.2.43) start (at $\alpha = 0$) from the coordinate origin, and all of them, except the first one, remain in the third quadrant of the complex σ-plane for all $\alpha \in [0, \infty)$. The first root stays in the third quadrant until α reaches the critical (neutral) value,

$$\alpha_* \simeq 1.00049. \qquad (3.2.45)$$

At $\alpha = \alpha_*$, this root crosses the imaginary axis at the point

$$\Im(\sigma) \simeq -2.20707. \qquad (3.2.46)$$

The real part of σ continues to increase with α until a maximum

$$\Re(\sigma)_{\max} \simeq 1.24055 \qquad (3.2.47)$$

is reached, which occurs at

$$\alpha_{\max} \simeq 2.7146. \qquad (3.2.48)$$

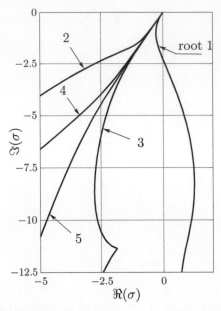

Fig. 3.12: The trajectories of the first five roots of (3.2.43) in the complex σ-plane as α increases from zero to infinity.

The calculations further show that there is a second local maximum of $\Re(\sigma)$ at a larger value of α, but it is smaller than (3.2.47) and does not influence the behaviour of the wave packet at large time $\bar{t}$.

To extend these results to negative α, one simply needs to mirror reflect the trajectories of the roots in Figure 3.12 in the real axis. In fact, the following statement is valid (see Problem 1 in Exercises 11): for each solution σ of the dispersion equation (3.2.43) for a positive value of α, there is a corresponding complex conjugate solution σ' for $\alpha' = -\alpha$.

Armed with this information, we return to the inverse Laplace transform (3.2.42). As was stated above, the integration in (3.2.42) is performed along a vertical line $\mathcal{L}$ that crosses the real axis at the point $\Re(\sigma) = a$. If we choose the quantity a to be larger than (3.2.47), then all the singularities of (3.2.41) will lie to the left of $\mathcal{L}$. When analysing the flow perturbations at large values of time $\bar{t}$, it is convenient to create a closed contour C by adding to $\mathcal{L}$ two circular arcs C_R and C'_R of large radius R, and two rays $\widetilde{C}$ and $\widetilde{C}'$ as shown in Figure 3.11. We choose the rays to lie in the second and third quadrants of the complex σ-plane, respectively, and draw them in such a way that all the roots of the dispersion equation (3.2.43), except σ_1 and σ'_1, find themselves to the left of $\widetilde{C}$ and $\widetilde{C}'$.

Applying the Residue theorem to the integral along the closed contour C, we have

$$\frac{1}{2\pi i} \oint_C \check{P}(\sigma, \alpha)\, e^{\sigma t}\, d\sigma = \mathcal{R}_1 + \mathcal{R}_2 + \mathcal{R}_3. \qquad (3.2.49)$$

Here $\mathcal{R}_1$ and $\mathcal{R}_2$ are the residues of $\check{P}(\sigma, \alpha)\, e^{\sigma t}$ at the points $\sigma = i\omega$ and $\sigma = -i\omega$ (see

Figure 3.11). They are given by

$$\mathcal{R}_1 = \frac{e^{i\omega\bar{t}}}{2i} \left. \frac{|\alpha|Ai'(z_0)\breve{F}(\alpha)}{(i\alpha)^{1/3}|\alpha|\int\limits_{z_0}^{\infty} Ai(z)\,dz - Ai'(z_0)} \right|_{z_0=i\omega/(i\alpha)^{2/3}}, \tag{3.2.50}$$

$$\mathcal{R}_2 = -\frac{e^{-i\omega\bar{t}}}{2i} \left. \frac{|\alpha|Ai'(z_0)\breve{F}(\alpha)}{(i\alpha)^{1/3}|\alpha|\int\limits_{z_0}^{\infty} Ai(z)\,dz - Ai'(z_0)} \right|_{z_0=-i\omega/(i\alpha)^{2/3}}, \tag{3.2.51}$$

respectively. The third term on the right-hand side of (3.2.49) is calculated differently according to whether α is positive or negative. If $\alpha > 0$, then it is given by the residue at the point $\sigma = \sigma_1(\alpha)$:

$$\mathcal{R}_3 = -\frac{\omega e^{\sigma_1(\alpha)\bar{t}}}{\sigma_1^2 + \omega^2} \left. \frac{(i\alpha)^{2/3}|\alpha|Ai'(z_0)\breve{F}(\alpha)}{\left[(i\alpha)^{1/3}|\alpha| + z_0\right]Ai(z_0)} \right|_{z_0=\sigma_1/(i\alpha)^{2/3}}. \tag{3.2.52}$$

If however $\alpha < 0$, then the residue should be taken at the complex conjugate point $\sigma = \sigma_1'$:

$$\mathcal{R}_3 = -\frac{\omega e^{\sigma_1'(\alpha)\bar{t}}}{\sigma_1'^2 + \omega^2} \left. \frac{(i\alpha)^{2/3}|\alpha|Ai'(z_0)\breve{F}(\alpha)}{\left[(i\alpha)^{1/3}|\alpha| + z_0\right]Ai(z_0)} \right|_{z_0=\sigma_1'/(i\alpha)^{2/3}}. \tag{3.2.53}$$

The left-hand side of (3.2.49) admits the following simplifications. Firstly, it may be shown (see Problem 2 in Exercises 11) that for any positive value of time $\bar{t}$, the integrals along C_R and C_R' vanish as $R \to \infty$. Secondly, the integrals along $\widetilde{C}$ and $\hat{C}'$ satisfy the conditions of Watson's lemma,[8] and prove to decay like $1/\bar{t}$ at large values of time $\bar{t}$; see Problem 3 in Exercises 11. The remaining integral along $\mathcal{L}$ represents the Fourier transform (3.2.42) of the pressure perturbation function $P'(\bar{t}, \bar{x})$. Thus, assuming that $\bar{t}$ is large, we can write

$$\breve{P}(\bar{t}, \alpha) = \mathcal{R}_1 + \mathcal{R}_2 + \mathcal{R}_3. \tag{3.2.54}$$

Let us now consider the right-hand side of equation (3.2.54). According to (3.2.50), (3.2.51), the first two terms, $\mathcal{R}_1$ and $\mathcal{R}_2$, are periodic functions of time $\bar{t}$. Meanwhile, the numerical results presented in Section 3.2.2 revealed the formation of the wave packet where the perturbations grew with $\bar{t}$. This growth should be attributed to the third term $\mathcal{R}_3$. Thus, to study the behaviour of the wave packet, we shall write

$$\breve{P}(\bar{t}, \alpha) = \mathcal{R}_3. \tag{3.2.55}$$

To return to physical variables, we need to apply the inverse Fourier transform (3.2.44) to equation (3.2.55). When performing this task, equation (3.2.53) should be used for negative α and equation (3.2.52) for positive α. We have

$$P'(\bar{t}, \bar{x}) = \int\limits_{-\infty}^{0} \breve{h}(\alpha)e^{\sigma_1'(\alpha)\bar{t}+i\alpha\bar{x}}\,d\alpha + \int\limits_{0}^{\infty} h(\alpha)e^{\sigma_1(\alpha)\bar{t}+i\alpha\bar{x}}\,d\alpha, \tag{3.2.56}$$

---

[8]See Section 1.2.2 in Part 2 of this book series.

where

$$\check{h}(\alpha) = -\frac{\omega \check{F}(\alpha)}{2\pi(\sigma_1'^2 + \omega^2)} \left. \frac{(i\alpha)^{2/3}|\alpha|Ai'(z_0)}{\left[(i\alpha)^{1/3}|\alpha| + z_0\right]Ai(z_0)} \right|_{z_0 = \sigma_1'/(i\alpha)^{2/3}},$$

$$h(\alpha) = -\frac{\omega \check{F}(\alpha)}{2\pi(\sigma_1^2 + \omega^2)} \left. \frac{(i\alpha)^{2/3}|\alpha|Ai'(z_0)}{\left[(i\alpha)^{1/3}|\alpha| + z_0\right]Ai(z_0)} \right|_{z_0 = \sigma_1/(i\alpha)^{2/3}}.$$

It can be noticed that the first integral in (3.2.56) is the complex conjugate of the second integral (see Problem 4 in Exercises 11). Hence, we can write

$$P'(\bar{t}, \bar{x}) = \int\limits_0^\infty h(\alpha) e^{\sigma_1(\alpha)\bar{t} + i\alpha\bar{x}} \, d\alpha + (c.c.). \tag{3.2.57}$$

Equation (3.2.57) serves to determine the pressure perturbations in the wave packet. As mentioned before, our main interest is in the behaviour of the perturbations at large values of $\bar{t}$. The corresponding asymptotic representation of the integral in (3.2.57) may be obtained with the help of the steepest descent method.[9] To employ this method, we need, first of all, to bring the integral in (3.2.57) to the form of a Laplace's type integral:

$$\int\limits_C h(z) e^{\lambda f(z)} dz. \tag{3.2.58}$$

Here C is an integration contour in the complex z-plane, $h(z)$ and $f(z)$ are analytic function of z, and λ is a real positive parameter, assumed large. When using the method of steepest descent, one needs to deform the contour of integration C so that it passes through a saddle point $z = \zeta_0$. The latter is defined by the equation

$$\left. \frac{df}{dz} \right|_{z = \zeta_0} = 0.$$

At large λ, the dominant contribution to (3.2.58) is given by a small vicinity of ζ_0, and takes the form[10]

$$\int\limits_C h(z) e^{\lambda f(z)} dz = \mathcal{J} \frac{e^{\lambda f(\zeta_0)}}{\sqrt{\lambda}} + \cdots \quad \text{as} \quad \lambda \to \infty.$$

Here the factor $\mathcal{J}$ depends on the value of the function $h(z)$ at the point $z = \zeta_0$ and on the second derivative of $f(z)$ at this point only.

Returning back to the integral in (3.2.57), we set

$$\bar{x} = D\bar{t}, \tag{3.2.59}$$

and assume that the parameter D is fixed as $\bar{t} \to \infty$. This allows us to represent the

---

[9]See Section 1.2.3 in Part 2 of this book series.

[10]See equation (1.2.35) on page 28 in Part 2 of this book series.

integral in (3.2.57) in the form

$$P'(\bar{t}, \bar{x}) = \int_0^\infty h(\alpha) e^{\bar{t} f(\alpha)} \, d\alpha + (c.c.), \tag{3.2.60}$$

where

$$f(\alpha) = \sigma_1(\alpha) + i\alpha D. \tag{3.2.61}$$

Next, we consider an analytic continuation of the integrand in (3.2.60) from the real axis into the complex α-plane. The saddle point is defined by the equation

$$\frac{df}{d\alpha} = \frac{d\sigma_1}{d\alpha} + iD = 0. \tag{3.2.62}$$

Recall that

$$\sigma_1 = (i\alpha)^{2/3} z_0 \tag{3.2.63}$$

is the first root of the dispersion equation (3.2.43), relating α and z_0. For numerical purposes it is convenient to convert (3.2.62) into a second equation for α and z_0. Differentiation of (3.2.63) with respect to α yields

$$\frac{d\sigma_1}{d\alpha} = \frac{2}{3} i(i\alpha)^{-1/3} z_0 + (i\alpha)^{2/3} \frac{dz_0}{d\alpha}. \tag{3.2.64}$$

Substituting (3.2.64) into (3.2.62) and solving the resulting equation for $dz_0/d\alpha$, we have

$$\frac{dz_0}{d\alpha} = -iD(i\alpha)^{-2/3} - \frac{2}{3} i(i\alpha)^{-1} z_0. \tag{3.2.65}$$

Now we differentiate the dispersion equation (3.2.43) with respect to α.[11] We find that

$$\frac{4}{3} (i\alpha)^{1/3} \int_{z_0}^\infty Ai(z) \, dz - \left[(i\alpha)^{1/3} \alpha + z_0 \right] Ai(z_0) \frac{dz_0}{d\alpha} = 0. \tag{3.2.66}$$

Substitution of (3.2.65) into (3.2.66) leads to the desired second equation relating z_0 and α:

$$\frac{4}{3} (i\alpha)^{1/3} \int_{z_0}^\infty Ai(z) \, dz + i \left[(i\alpha)^{1/3} \alpha + z_0 \right] \left[\frac{D}{(i\alpha)^{2/3}} + \frac{2}{3} \frac{z_0}{i\alpha} \right] Ai(z_0) = 0. \tag{3.2.67}$$

Considered together with the dispersion equation (3.2.43), it allows us to determine α and z_0 as functions of D. Once α and z_0 are found, equation (3.2.63) may be used to calculate σ_1.

---

[11] When performing this task, we take into account that originally the contour of integration in (3.2.60) coincided with the positive real semi-axis. Therefore, when dealing with an analytic extension of the integrand into the complex α-plane, we have to write $|\alpha| = \alpha$.

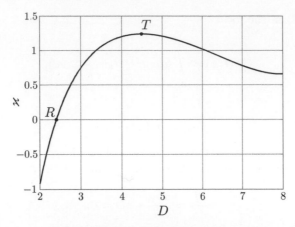

Fig. 3.13: Amplification rate $\varkappa$.

At large values of time $\bar{t}$ the pressure perturbations are represented by

$$P' = \mathcal{J}\frac{e^{\bar{t}f(\alpha)}}{\sqrt{\bar{t}}} + \cdots \quad \text{as} \quad \bar{t} \to \infty, \tag{3.2.68}$$

which shows that the amplification rate of the perturbations is given by the real part of (3.2.61):

$$\varkappa = \Re\{f\} = \Re\{\sigma_1\} - D\Im\{\alpha\}. \tag{3.2.69}$$

The position of the saddle point in the complex α-plane depends on D. For each D, equations (3.2.67), (3.2.43) can be solved numerically for α and z_0 using Newton iteration. The results of the calculations are displayed in Figure 3.13 where the amplification rate $\varkappa$ is plotted as a function of the parameter D.

We see that the wave packet (shown earlier in Figure 3.9) possesses the following properties. The width of the wave packet increases as a linear function of time. At each ray $\bar{x}/\bar{t} = D$ inside the packet, the amplitude of the perturbations grows according to (3.2.68). Point T in Figure 3.13 corresponds to the *centre of the wave packet*, where the amplification rate has its maximum $\varkappa_{\max} \simeq 1.24055$. This point moves downstream with speed

$$D_T = \bar{x}/\bar{t} \simeq 4.48913. \tag{3.2.70}$$

Point R in Figure 3.13 represents the rear end of the wave packet. At this point $\varkappa = 0$, and it moves downstream with speed

$$D_R \simeq 2.40585.$$

These results show that the flow considered is *convectively unstable*. While the perturbations in the wave packet grow with time, the wave packet as a whole, is swept downstream leaving a space for the Tollmien–Schlichting waves to be observed behind the vibrator (see Figure 3.10).

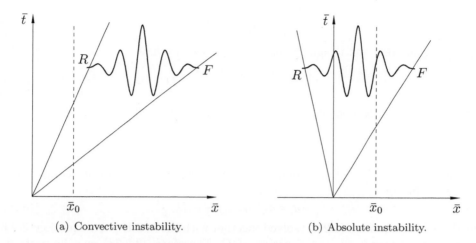

(a) Convective instability. (b) Absolute instability.

Fig. 3.14: Convective and absolute instabilities.

3.2.4 Convective and absolute instabilities

The flow considered above belongs to the class of *convectively unstable* flows. For such flows, a localized perturbation introduced at time $\bar{t} = 0$ leads to a formation of a wave packet that behaves as shown in Figure 3.14(a). The perturbations inside the wave packet show a monotonic growth, but both the rear, R, and front, F, ends of the packet move downstream, such that at any fixed point $\bar{x} = \bar{x}_0$, after a certain time, the flow returns to its unperturbed state. The other category of instability is the class of *absolutely unstable* flows. The behaviour of the wave packet in such flows is shown in Figure 3.14(b). In this case, any fixed point $\bar{x} = \bar{x}_0$ eventually finds itself inside the region of growing perturbations, which ultimately lead to laminar-turbulent transition.

3.2.5 Centre of the wave packet

Here we shall look at the centre of the wave packet more closely. We start with the dispersion equation (3.2.43). As we have seen, it establishes a relationship between α and σ, which can be expressed in the form

$$\sigma = \sigma(\alpha). \qquad (3.2.71)$$

This is a complex function, and it may be written in terms of two real equations for the real and imaginary parts of σ, respectively:

$$\sigma_r = \sigma_r(\alpha_r, \alpha_i), \qquad \sigma_i = \sigma_i(\alpha_r, \alpha_i). \qquad (3.2.72)$$

Since (3.2.71) is an analytic function, $\sigma_r(\alpha_r, \alpha_i)$ and $\sigma_i(\alpha_r, \alpha_i)$ satisfy the Cauchy–Riemann conditions

$$\frac{\partial \sigma_r}{\partial \alpha_r} = \frac{\partial \sigma_i}{\partial \alpha_i}, \qquad \frac{\partial \sigma_i}{\partial \alpha_r} = -\frac{\partial \sigma_r}{\partial \alpha_i}. \qquad (3.2.73)$$

Next, we consider the saddle point equation (3.2.62). It may be written in terms of the functions (3.2.72) as

$$\frac{\partial \sigma_r}{\partial \alpha_r} + i \frac{\partial \sigma_i}{\partial \alpha_r} + iD = 0. \tag{3.2.74}$$

Separating the real and imaginary parts in (3.2.74), we have

$$\frac{\partial \sigma_r}{\partial \alpha_r} = 0, \qquad \frac{\partial \sigma_i}{\partial \alpha_r} = -D. \tag{3.2.75}$$

Comparing the second equation in (3.2.73) with the second equation in (3.2.75), we see that

$$\frac{\partial \sigma_r}{\partial \alpha_i} = D. \tag{3.2.76}$$

Equation (3.2.69) for the amplification rate $\varkappa$ is recast in terms of the functions in (3.2.72) as follows:

$$\varkappa = \sigma_r(\alpha_r, \alpha_i) - \alpha_i D. \tag{3.2.77}$$

The dispersion equation (3.2.43) solved together with the saddle point equation (3.2.62) allows us to determine α as a function of D. Therefore, (3.2.77) may be written in more detail as

$$\varkappa = \sigma_r\big[\alpha_r(D), \alpha_i(D)\big] - D\alpha_i(D). \tag{3.2.78}$$

The maximum of $\varkappa$ is achieved at the centre of the wave packet (point T in Figure 3.13), where

$$\frac{d\varkappa}{dD} = \frac{\partial \sigma_r}{\partial \alpha_r} \frac{d\alpha_r}{dD} + \frac{\partial \sigma_r}{\partial \alpha_i} \frac{d\alpha_i}{dD} - \alpha_i - D\frac{d\alpha_i}{dD} = 0. \tag{3.2.79}$$

Clearly, the first term in (3.2.79) disappears in view of the first equation in (3.2.75). Using (3.2.76) for the second term, we arrive at the conclusion that at point T

$$\alpha_i = 0.$$

Thus, we see that at the centre of the wave packet T, the saddle point finds itself on the real axis in the complex α-plane. It then follows from (3.2.77) that at this point $\varkappa$ coincides with σ_r which is given by (3.2.47). The value of the parameter D at point T is referred to as the *group velocity* of the wave packet. It may be calculated using the second equation in (3.2.75):

$$D = -\left. \frac{\partial \sigma_i}{\partial \alpha_r} \right|_{\alpha_i = 0}. \tag{3.2.80}$$

To use this formula, one only needs to know the solution of the dispersion equation (3.2.43) for real positive values of α; see Figure 3.12. The derivative in (3.2.80) should be taken at the maximum point (3.2.48), which yields the value of D given by (3.2.70).

Exercises 11

1. Suppose that the dispersion equation

$$(i\alpha)^{1/3}|\alpha| \int_{z_0}^{\infty} Ai(z)\, dz - Ai'(z_0) = 0, \qquad z_0 = \frac{\sigma}{(i\alpha)^{2/3}} \tag{3.2.81}$$

   admits a solution, σ, at a real positive value of α. Prove that for $\alpha' = -\alpha$, there exists a solution σ' that is the complex conjugate of σ.

Suggestion: First, prove that changing the sign of α turns $(i\alpha)^{1/3}$ into its complex conjugate. To perform this task use equation (3.1.83), where ϕ should be chosen to be zero for positive α, and $\phi = -\pi$ for negative α. Second, take complex conjugates on both sides of equation (3.2.81). When performing this task, use (without proof) *Schwarz reflection principle*. The latter is formulated as follows. Suppose that $F(z)$ is a function of a complex variable z, analytic in a domain that lies in the upper half plane ($\Im(z) > 0$). Suppose further that $F(z)$ assumes real values on the real axis. Then an analytic continuation of $F(z)$ into the lower half plane is given by

$$F(\bar{z}) = \overline{F(z)},$$

where the 'overline' denotes the complex conjugate.

2. The Laplace/Fourier image of the pressure, obtained as part of the solution of the initial-value problem, is given by equation (3.2.41):

$$\check{P}(\sigma, \alpha) = \frac{\omega}{\sigma^2 + \omega^2} \frac{|\alpha| Ai'(z_0) \check{F}(\alpha)}{(i\alpha)^{1/3}|\alpha| \int\limits_{z_0}^{\infty} Ai(z)\,dz - Ai'(z_0)}, \qquad z_0 = \frac{\sigma}{(i\alpha)^{2/3}}. \quad (3.2.82)$$

Consider the integrals

$$\int\limits_{C_R} \check{P}(\sigma, \alpha) e^{\sigma \bar{t}}\,d\sigma, \qquad \int\limits_{C_R'} \check{P}(\sigma, \alpha) e^{\sigma \bar{t}}\,d\sigma \qquad (3.2.83)$$

along circular arcs C_R and C_R' in Figure 3.11. Assuming that $\bar{t} > 0$ and α is finite, show that for large $|\sigma|$, the integrand in (3.2.83) may be bounded as

$$\left| \check{P}(\sigma, \alpha) e^{\sigma \bar{t}} \right| < \frac{M}{|\sigma|^2},$$

where M is a positive constant. Hence, conclude that the integrals (3.2.83) tend to zero as the arcs' radius R tends to infinity.

Hint: You may use without proof the fact that at large values of z,

$$Ai'(z) = -\frac{1}{2\sqrt{\pi}} z^{1/4} e^{-\frac{2}{3} z^{3/2}} + \cdots, \qquad \int\limits_{z}^{\infty} Ai(\zeta)d\zeta = \frac{1}{2\sqrt{\pi}} z^{-3/4} e^{-\frac{2}{3} z^{3/2}} + \cdots.$$

3. Consider the integrals

$$\int\limits_{\widetilde{C}} \check{P}(\sigma, \alpha) e^{\sigma \bar{t}}\,d\sigma, \qquad \int\limits_{\widetilde{C}'} \check{P}(\sigma, \alpha) e^{\sigma \bar{t}}\,d\sigma \qquad (3.2.84)$$

along rays $\widetilde{C}$ and $\widetilde{C}'$; see Figure 3.11. Assume that $\bar{t} \to \infty$ and use Watson's lemma to show that integrals (3.2.84) may be estimated as $O(1/\bar{t})$ quantities.[12]

[12]For a discussion of Watson's lemma, the reader is referred to Section 1.2.2 in Part 2 of this book series.

Hint: Instead of formally applying Watson's lemma, one can use the fact that at large $\bar{t}$ the dominant contribution to integrals (3.2.84) is given by small sections of $\widetilde{C}$ and $\widetilde{C}'$ lying in a vicinity of the point $\sigma = 0$. This means that $\check{P}(\sigma, \alpha)$ may be substituted in (3.2.84) by its approximation obtained by setting $\sigma \to 0$ in (3.2.82).

4. Consider equation (3.2.56) for the pressure perturbations in the boundary layer:

$$P'(\bar{t}, \bar{x}) = \int\limits_{-\infty}^{0} \check{h}(\alpha) e^{\sigma_1'(\alpha)\bar{t} + i\alpha\bar{x}}\, d\alpha + \int\limits_{0}^{\infty} h(\alpha) e^{\sigma_1(\alpha)\bar{t} + i\alpha\bar{x}}\, d\alpha, \qquad (3.2.85)$$

where

$$\check{h}(\alpha) = -\frac{\omega \check{F}(\alpha)}{2\pi(\sigma_1'^2 + \omega^2)} \frac{(i\alpha)^{2/3}|\alpha| Ai'(z_0)}{\left[(i\alpha)^{1/3}|\alpha| + z_0\right] Ai(z_0)}\bigg|_{z_0 = \sigma_1'(\alpha)/(i\alpha)^{2/3}},$$

$$h(\alpha) = -\frac{\omega \check{F}(\alpha)}{2\pi(\sigma_1^2 + \omega^2)} \frac{(i\alpha)^{2/3}|\alpha| Ai'(z_0)}{\left[(i\alpha)^{1/3}|\alpha| + z_0\right] Ai(z_0)}\bigg|_{z_0 = \sigma_1(\alpha)/(i\alpha)^{2/3}}.$$

Your task is to prove that the first integral in (3.2.84) is the complex conjugate of the second integral. You may perform this task in the following steps:

(a) Introduce a new integration variable $\alpha' = -\alpha$ in the first integral on the right-hand side of (3.2.85), and show that equation (3.2.85) may be expressed in the form

$$P'(\bar{t}, \bar{x}) = \int\limits_{0}^{\infty} h'(-\alpha) e^{\sigma_1'(-\alpha)\bar{t} - i\alpha\bar{x}}\, d\alpha + \int\limits_{0}^{\infty} h(\alpha) e^{\sigma_1(\alpha)\bar{t} + i\alpha\bar{x}}\, d\alpha. \qquad (3.2.86)$$

(b) Using the definition of Fourier transform, show that for any real function $F(\bar{x})$, its Fourier transform $\check{F}(\alpha)$ possesses the following property: $\check{F}(-\alpha)$ is the complex conjugate of $\check{F}(\alpha)$.

(c) Now, using the suggestion in Problem 1, prove that $\check{h}(-\alpha)$ is the complex conjugate of $h(\alpha)$.

(d) Finally, keeping in mind that $\sigma_1'(-\alpha)$ is the complex conjugate of $\sigma_1(\alpha)$, prove that the first integral in (3.2.86) is the complex conjugate of the second integral.

3.3   Generation of Tollmien–Schlichting Waves by Sound

Of course, in real flight, there are no vibrating ribbons on the wing surface. Instead, the Tollmien–Schlichting waves are produced through the interaction of the boundary layer with disturbances in the oncoming flow. These disturbances are weak and, in the general case, may be represented as a superposition of acoustic waves, vorticity waves, and entropy waves (see Problem 1 in Exercises 12). The vorticity and entropy waves, considered together, constitute the so-called *free-stream turbulence*.

In this section, we shall consider the interaction of the boundary layer with acoustic waves. These are always present in the flow field around an aircraft, being produced

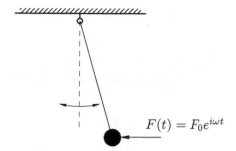

$$F(t) = F_0 e^{i\omega t}$$

Fig. 3.15: Pendulum resonance.

by the engines and emitted by turbulent boundary layers on the fuselage and rear sections of the wings. Experimental observations starting with Schubauer and Skramstad (1948) have shown that acoustic waves easily penetrate the boundary layer. However, in order to generate the Tollmien–Schlichting waves, they have to satisfy rather restrictive resonance conditions which were first formulated by Kachanov *et al.* (1982). Unlike in a simple mechanical system, say, a pendulum (see Figure 3.15), where the resonance is observed provided that the frequency ω of the external forcing is close to the natural frequency of the pendulum oscillations, in fluid flows effective transformation of external disturbances into instability modes of the boundary layer is only possible if in addition to the frequency, the wave number of the external perturbations is in tune with the natural internal oscillations of the boundary layer. This requirement is termed the *double-resonance principle*.

In the vibrating ribbon problem, considered in the previous section, the two resonance conditions are satisfied by simply choosing the frequency and the length of the vibrating part of the wall appropriately. The situation is more complex in the case of boundary-layer receptivity to acoustic noise. The theory of the generation of Tollmien–Schlichting waves by acoustic noise was developed by Ruban (1984) and Goldstein (1985). To satisfy the first resonance condition, they chose the frequency of the acoustic wave to be an $O(Re^{1/4})$ quantity, but since the speed of propagation of acoustic waves is finite, their wavelength appears to be $O(Re^{-1/4})$ long, which is much longer than the wavelength of the Tollmien–Schlichting wave. Hence, the acoustic wave alone is insufficient for Tollmien–Schlichting wave generation. To satisfy the resonance condition with respect to the wave number, the acoustic wave has to come into interaction with a wall roughness of which there are, of course, plenty on a real aircraft wing. An effective generation of the Tollmien–Schlichting wave is observed when the length of the roughness is an order $O(Re^{-3/8})$ quantity.

3.3.1 Problem formulation

Let us consider a subsonic flow of a perfect gas past a flat plate that is aligned with the mean velocity vector in the free stream; see Figure 3.16. We shall assume that small-amplitude acoustic waves are present in the flow. We shall further assume that there is a small roughness on the plate surface at distance L from the leading edge. For simplicity we restrict our analysis to a two-dimensional basic flow and perturbations. We use Cartesian coordinates $(\hat{x}, \hat{y})$, with $\hat{x}$ measured along the plate surface from its

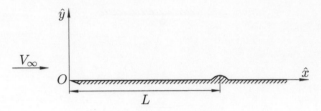

Fig. 3.16: Flow layout.

leading edge O, and $\hat{y}$ in the perpendicular direction. The velocity components in these coordinates are denoted by $(\hat{u}, \hat{v})$. We further denote the time by $\hat{t}$, the gas density by $\hat{\rho}$, pressure by $\hat{p}$, enthalpy by $\hat{h}$ and dynamic viscosity coefficient by $\hat{\mu}$. As usual, the 'hat' stands for dimensional variables. The non-dimensional variables are introduced as follows:

$$\left.\begin{aligned} &\hat{t} = \frac{L}{V_\infty}t, \quad \hat{x} = Lx, \quad \hat{y} = Ly, \quad \hat{u} = V_\infty u, \quad \hat{v} = V_\infty v, \\ &\hat{\rho} = \rho_\infty \rho, \quad \hat{p} = p_\infty + \rho_\infty V_\infty^2 p, \quad \hat{h} = V_\infty^2 h, \quad \hat{\mu} = \mu_\infty \mu, \end{aligned}\right\} \tag{3.3.1}$$

with V_∞, p_∞, ρ_∞ and μ_∞ being the dimensional free-stream velocity, pressure, density and viscosity, respectively.

In the non-dimensional variables, the Navier–Stokes equations are written as[13]

$$\rho\left(\frac{\partial u}{\partial t} + u\frac{\partial u}{\partial x} + v\frac{\partial u}{\partial y}\right) = -\frac{\partial p}{\partial x} + \frac{1}{Re}\left\{\frac{\partial}{\partial x}\left[\mu\left(\frac{4}{3}\frac{\partial u}{\partial x} - \frac{2}{3}\frac{\partial v}{\partial y}\right)\right]\right.$$
$$\left. + \frac{\partial}{\partial y}\left[\mu\left(\frac{\partial u}{\partial y} + \frac{\partial v}{\partial x}\right)\right]\right\}, \tag{3.3.2a}$$

$$\rho\left(\frac{\partial v}{\partial t} + u\frac{\partial v}{\partial x} + v\frac{\partial v}{\partial y}\right) = -\frac{\partial p}{\partial y} + \frac{1}{Re}\left\{\frac{\partial}{\partial y}\left[\mu\left(\frac{4}{3}\frac{\partial v}{\partial y} - \frac{2}{3}\frac{\partial u}{\partial x}\right)\right]\right.$$
$$\left. + \frac{\partial}{\partial x}\left[\mu\left(\frac{\partial u}{\partial y} + \frac{\partial v}{\partial x}\right)\right]\right\}, \tag{3.3.2b}$$

$$\rho\left(\frac{\partial h}{\partial t} + u\frac{\partial h}{\partial x} + v\frac{\partial h}{\partial y}\right) = \frac{\partial p}{\partial t} + u\frac{\partial p}{\partial x} + v\frac{\partial p}{\partial y}$$
$$+ \frac{1}{Re}\left\{\frac{1}{Pr}\left[\frac{\partial}{\partial x}\left(\mu\frac{\partial h}{\partial x}\right) + \frac{\partial}{\partial y}\left(\mu\frac{\partial h}{\partial y}\right)\right]\right.$$
$$\left. + \mu\left(\frac{4}{3}\frac{\partial u}{\partial x} - \frac{2}{3}\frac{\partial v}{\partial y}\right)\frac{\partial u}{\partial x} + \mu\left(\frac{4}{3}\frac{\partial v}{\partial y} - \frac{2}{3}\frac{\partial u}{\partial x}\right)\frac{\partial v}{\partial y} + \mu\left(\frac{\partial u}{\partial y} + \frac{\partial v}{\partial x}\right)^2\right\}, \tag{3.3.2c}$$

$$\frac{\partial \rho}{\partial t} + \frac{\partial \rho u}{\partial x} + \frac{\partial \rho v}{\partial y} = 0, \tag{3.3.2d}$$

$$h = \frac{1}{(\gamma-1)M_\infty^2}\frac{1}{\rho} + \frac{\gamma}{\gamma-1}\frac{p}{\rho}. \tag{3.3.2e}$$

---

[13]For the dimensional form of the Navier–Stokes equations, the reader is referred to page 66 in Part 1 of this book series.

Here Pr is the Prandtl number, and γ is the specific heat ratio; for air $Pr \simeq 0.713$, $\gamma = 7/5$. The Reynolds number Re is defined as

$$Re = \frac{\rho_\infty V_\infty L}{\mu_\infty}.$$

In this study, we shall assume that Re is large, while the free-stream Mach number,

$$M_\infty = \frac{V_\infty}{a_\infty}, \tag{3.3.3}$$

remains finite. For subsonic flows considered here, $M_\infty < 1$. Remember that the speed of sound in (3.3.3) is calculated as[14]

$$a_\infty = \sqrt{\gamma \frac{p_\infty}{\rho_\infty}}.$$

3.3.2   Unperturbed flow

Our first task is to describe the steady unperturbed flow ahead of the roughness. At large values of the Reynolds number, the classical boundary-layer theory of Prandtl (1904) can be used for this purpose. According to this theory, the flow field should be divided into two regions: the inviscid region occupying the majority of the flow and the thin boundary layer that forms on the surface of the plate.

Inviscid flow

In the inviscid region, the flow remains unperturbed in the leading-order approximation with the fluid-dynamic functions preserving their values in the oncoming flow ahead of the plate:

$$u = 1, \quad v = 0, \quad p = 0, \quad \rho = 1, \quad h = \frac{1}{(\gamma - 1)M_\infty^2}. \tag{3.3.4}$$

Boundary layer

When dealing with the boundary layer we assume that

$$x = O(1), \quad Y = Re^{1/2}y = O(1), \quad Re \to \infty,$$

and represent the corresponding solution of the Navier–Stokes equations (3.3.2) in the form

$$\left.\begin{aligned}
&u(t, x, y; Re) = U_0(x, Y) + \cdots, && v(t, x, y; Re) = Re^{-1/2}V_0(x, Y) + \cdots, \\
&\rho(t, x, y; Re) = \rho_0(x, Y) + \cdots, && p(t, x, y; Re) = Re^{-1/2}P_0(x, Y) + \cdots, \\
&h(t, x, y; Re) = h_0(x, Y) + \cdots, && \mu(t, x, y; Re) = \mu_0(x, Y) + \cdots.
\end{aligned}\right\} \tag{3.3.5}$$

Substituting (3.3.5) into the Navier–Stokes equations (3.3.2) and setting $Re \to \infty$, we obtain in the leading-order approximation

---

[14]See Section 4.1.2 in Part 1 of this book series.

$$\rho_0 U_0 \frac{\partial U_0}{\partial x} + \rho_0 V_0 \frac{\partial U_0}{\partial Y} = \frac{\partial}{\partial Y}\left(\mu_0 \frac{\partial U_0}{\partial Y}\right), \tag{3.3.6a}$$

$$\rho_0 U_0 \frac{\partial h_0}{\partial x} + \rho_0 V_0 \frac{\partial h_0}{\partial Y} = \frac{1}{Pr}\frac{\partial}{\partial Y}\left(\mu_0 \frac{\partial h_0}{\partial Y}\right) + \mu_0 \left(\frac{\partial U_0}{\partial Y}\right)^2, \tag{3.3.6b}$$

$$\frac{\partial(\rho_0 U_0)}{\partial x} + \frac{\partial(\rho_0 V_0)}{\partial Y} = 0, \tag{3.3.6c}$$

$$h_0 = \frac{1}{(\gamma-1)M_\infty^2}\frac{1}{\rho_0}. \tag{3.3.6d}$$

This set of equations requires the following boundary conditions. At the leading edge of the plate, we have

$$U_0 = 1, \quad h_0 = \frac{1}{(\gamma-1)M_\infty^2} \quad \text{at} \quad x=0, \ Y \in [0,\infty). \tag{3.3.7}$$

The boundary conditions at the outer edge of the boundary layer

$$U_0 = 1, \quad h_0 = \frac{1}{(\gamma-1)M_\infty^2} \quad \text{at} \quad Y=\infty, \ x \in [0,1] \tag{3.3.8}$$

are obtained by matching with the inviscid solution (3.3.4). On the plate surface, the velocity components satisfy the no-slip conditions:

$$U_0 = V_0 = 0 \quad \text{at} \quad Y=0, \ x \in [0,1]. \tag{3.3.9}$$

They should be supplemented with a thermal condition for the enthalpy h_0. If, for example, the plate surface is thermally isolated, then

$$\frac{\partial h_0}{\partial Y} = 0 \quad \text{at} \quad Y=0, \ x \in [0,1]. \tag{3.3.10a}$$

Alternatively, one can assume that the wall temperature $\hat{T}_w$ is given, and then (3.3.10a) should be replaced by

$$h_0 = \frac{c_p \hat{T}_w}{V_\infty^2} \quad \text{at} \quad Y=0, \ x \in [0,1], \tag{3.3.10b}$$

where c_p is the specific heat at constant pressure.

The boundary-value problem (3.3.6)–(3.3.10) admits a self-similar solution for a thermally isolated wall, as well as in the case of constant wall temperature.[15] However, we do not need to restrict ourselves to these particular flow conditions. To proceed further, we only need to know that the solution to (3.3.6)–(3.3.10) remains smooth as

---

[15] A detailed discussion of such solutions may be found in Section 1.10 in Part 3 of this book series.

the boundary layer approaches the wall roughness, and the fluid-dynamic functions U_0, V_0, ρ_0, h_0, and μ_0 may be represented in the form of Taylor expansions:

$$\left.\begin{aligned}
U_0(x,Y) &= U_{00}(Y) + (x-1)U_{01}(Y) + \cdots, \\
V_0(x,Y) &= V_{00}(Y) + (x-1)V_{01}(Y) + \cdots, \\
\rho_0(x,Y) &= \rho_{00}(Y) + (x-1)\rho_{01}(Y) + \cdots, \\
h_0(x,Y) &= h_{00}(Y) + (x-1)h_{01}(Y) + \cdots, \\
\mu_0(x,Y) &= \mu_{00}(Y) + (x-1)\mu_{01}(Y) + \cdots
\end{aligned}\right\} \quad \text{as} \quad x \to 1-0. \tag{3.3.11}$$

Near the body surface, the leading-order terms in (3.3.11) exhibit the following behaviour:

$$\left.\begin{aligned}
U_{00}(Y) &= \lambda Y + \cdots, \\
V_{00}(Y) &= O(Y^2) + \cdots, \\
\rho_{00}(Y) &= \rho_w + \cdots, \\
h_{00}(Y) &= h_w + \cdots, \\
\mu_{00}(Y) &= \mu_w + \cdots
\end{aligned}\right\} \quad \text{as} \quad Y \to 0, \tag{3.3.12}$$

where λ, ρ_w, h_w, and μ_w are positive constants representing the dimensionless skin friction, density, enthalpy, and viscosity coefficient on the body surface immediately ahead of the roughness.

3.3.3 Acoustic noise

In practice, the acoustic noise interacting with the boundary layer, has a rather wide spectrum. Fortunately, if the amplitude of the acoustic waves is small, then each harmonic in this spectrum may be considered separate from the others. As mentioned before, our interest is in the harmonics that are in tune with the Tollmien–Schlichting waves, i.e. their frequency is an $O(Re^{1/4})$ quantity and, correspondingly, the wave length ℓ is estimated as $\ell \sim Re^{-1/4}$. To describe such a wave, we scale the time t and the longitudinal coordinate x as

$$t = Re^{-1/4}t_*, \qquad x = Re^{-1/4}x'. \tag{3.3.13}$$

Acoustic wave outside the boundary layer

We start with the inviscid flow region lying outside the boundary layer. Perturbing (3.3.4), we express the fluid-dynamic functions in this region in the form of the asymptotic expansions

$$\left.\begin{aligned}
u &= 1 + Re^{-1/8}u' + \cdots, \qquad p = Re^{-1/8}p' + \cdots, \\
\rho &= 1 + Re^{-1/8}\rho' + \cdots, \qquad h = \frac{1}{(\gamma-1)M_\infty^2} + Re^{-1/8}h' + \cdots.
\end{aligned}\right\} \tag{3.3.14}$$

For simplicity, we shall assume that the acoustic wave propagates in the direction parallel to the plate surface, with the wave front being perpendicular to the plate. In this case the perturbation functions u', p', ρ', and h' are functions of t_* and x' only. The wave amplitude, $Re^{-1/8}$, is chosen to make the pressure gradient

$$\frac{\partial p}{\partial x} = Re^{1/8}\frac{\partial p'}{\partial x'}$$

of the same order as that in the triple-deck region; the latter may be calculated using the asymptotic expansion (3.1.16c) for the pressure p and the scaling (3.1.15) for x in the triple-deck region. Taking into account that in such a wave, $v = 0$, we find upon substitution of (3.3.14) and (3.3.13) into the Navier–Stokes equations (3.3.2) that the perturbation field outside the boundary layer is described by the equations

$$\frac{\partial u'}{\partial t_*} + \frac{\partial u'}{\partial x'} = -\frac{\partial p'}{\partial x'}, \tag{3.3.15a}$$

$$\frac{\partial h'}{\partial t_*} + \frac{\partial h'}{\partial x'} = \frac{\partial p'}{\partial t_*} + \frac{\partial p'}{\partial x'}, \tag{3.3.15b}$$

$$\frac{\partial \rho'}{\partial t_*} + \frac{\partial \rho'}{\partial x'} + \frac{\partial u'}{\partial x'} = 0, \tag{3.3.15c}$$

$$h' = \frac{\gamma}{\gamma - 1}p' - \frac{1}{(\gamma - 1)M_\infty^2}\rho'. \tag{3.3.15d}$$

The above set of equations may be reduced to a single equation for p'. To perform this task, we start by substituting the state equation (3.3.15d) into the energy equation (3.3.15b). This leads to

$$\frac{\partial \rho'}{\partial t_*} + \frac{\partial \rho'}{\partial x'} = M_\infty^2\left(\frac{\partial p'}{\partial t_*} + \frac{\partial p'}{\partial x'}\right). \tag{3.3.16}$$

Now we can eliminate ρ' by combining (3.3.16) with the continuity equation (3.3.15c). We have

$$M_\infty^2\left(\frac{\partial p'}{\partial t_*} + \frac{\partial p'}{\partial x'}\right) + \frac{\partial u'}{\partial x'} = 0. \tag{3.3.17}$$

It remains to eliminate u' from (3.3.17) and (3.3.15a). Differentiation of (3.3.15a) with respect to x' yields

$$\frac{\partial^2 u'}{\partial t_* \partial x'} + \frac{\partial^2 u'}{\partial x'^2} = -\frac{\partial^2 p'}{\partial x'^2}. \tag{3.3.18}$$

The two terms on the left-hand side of (3.3.18) may be found by differentiating (3.3.17) with respect to t_* and with respect to x':

$$\frac{\partial^2 u'}{\partial t_* \partial x'} = -M_\infty^2\left(\frac{\partial^2 p'}{\partial t_*^2} + \frac{\partial^2 p'}{\partial t_* \partial x'}\right), \qquad \frac{\partial^2 u'}{\partial x'^2} = -M_\infty^2\left(\frac{\partial^2 p'}{\partial t_* \partial x'} + \frac{\partial^2 p'}{\partial x'^2}\right). \tag{3.3.19}$$

Substitution of (3.3.19) into (3.3.18) results in the following equation for p':

$$\frac{\partial^2 p'}{\partial t_*^2} + 2\frac{\partial^2 p'}{\partial t_* \partial x'} + \left(1 - \frac{1}{M_\infty^2}\right)\frac{\partial^2 p'}{\partial x'^2} = 0. \tag{3.3.20}$$

If we introduce new independent variables

$$\xi = t_* - \frac{M_\infty}{1 + M_\infty} x', \qquad \eta = t_* + \frac{M_\infty}{1 - M_\infty} x', \qquad (3.3.21)$$

then (3.3.20) assumes the form

$$\frac{\partial^2 p'}{\partial \xi \partial \eta} = 0. \qquad (3.3.22)$$

The general solution to (3.3.22) is written as

$$p' = f(\xi) + g(\eta),$$

where $f(\xi)$ and $g(\eta)$ are arbitrary functions of their respective arguments. It is easily seen from (3.3.21) that $f(\xi)$ represents an acoustic wave propagating downstream with the speed $1 + 1/M_\infty$, while $g(\eta)$ stands for an acoustic wave propagating upstream, since its speed $1 - 1/M_\infty$ is negative in subsonic flows. Either of these may be considered when analysing the receptivity of the boundary layer. In fact, the receptivity theory remains largely unchanged even if the boundary layer is irradiated with an oblique acoustic wave making an arbitrary angle with the plate surface. Still, to be definite, we shall assume that the acoustic field is represented by the downstream propagating monochromatic wave. Hence, we set $g(\eta)$ to zero, and take $f(\xi)$ to be a periodic function of ξ. We have

$$p' = a_* \sin(\omega_* \xi), \qquad (3.3.23)$$

where a_* and ω_* are the amplitude and the frequency of the acoustic wave.

Now, the rest of the fluid-dynamic functions may be found with the help of equations (3.3.15a)–(3.3.15c):

$$u' = a_* M_\infty \sin(\omega_* \xi), \qquad \rho' = a_* M_\infty^2 \sin(\omega_* \xi), \qquad h' = a_* \sin(\omega_* \xi). \qquad (3.3.24)$$

Main part of the boundary layer

Our next task is to learn how the acoustic wave penetrates the boundary layer. The unperturbed flow in the boundary layer is represented by the asymptotic expansions (3.3.5). The pressure oscillations in the acoustic wave (3.3.14) cause $O(Re^{-1/8})$ perturbations to (3.3.5). Therefore, we shall seek the solution in the boundary layer in the form:

$$\left.\begin{aligned}
u(t, x, y; Re) &= U_0(x, Y) + Re^{-1/8}\widetilde{U}'(x, Y)\sin(\omega_* \xi) + \cdots, \\
v(t, x, y; Re) &= Re^{-3/8}\widetilde{V}'(x, Y)\cos(\omega_* \xi) + \cdots, \\
p(t, x, y; Re) &= Re^{-1/8}\widetilde{P}'(x, Y)\sin(\omega_* \xi) + \cdots, \\
\rho(t, x, y; Re) &= \rho_0(x, Y) + Re^{-1/8}\widetilde{R}'(x, Y)\sin(\omega_* \xi) + \cdots, \\
h(t, x, y; Re) &= h_0(x, Y) + Re^{-1/8}\widetilde{H}'(x, Y)\sin(\omega_* \xi) + \cdots, \\
\mu(t, x, y; Re) &= \mu_0(x, Y) + \cdots.
\end{aligned}\right\} \qquad (3.3.25)$$

Here x and $Y = Re^{1/2}y$ are standard boundary-layer variables; ξ is defined by (3.3.21) and (3.3.13). The form of the asymptotic expansion for the lateral velocity component v

has been determined, as usual, by balancing the following terms in the continuity equation (3.3.2d):

$$\frac{\partial \rho u}{\partial x} \sim \frac{\partial \rho v}{\partial y}. \tag{3.3.26}$$

Substitution of (3.3.25) into the Navier–Stokes equations (3.3.2) yields

$$\rho_0 \left[\left(1 - \frac{M_\infty}{1 + M_\infty} U_0 \right) \widetilde{U}' + \frac{1}{\omega_*} \frac{\partial U_0}{\partial Y} \widetilde{V}' \right] = \frac{M_\infty}{1 + M_\infty} \widetilde{P}', \tag{3.3.27a}$$

$$\frac{\partial \widetilde{P}'}{\partial Y} = 0, \tag{3.3.27b}$$

$$\rho_0 \left[\left(1 - \frac{M_\infty}{1 + M_\infty} U_0 \right) \widetilde{H}' + \frac{1}{\omega_*} \frac{\partial h_0}{\partial Y} \widetilde{V}' \right] = \left(1 - \frac{M_\infty}{1 + M_\infty} U_0 \right) \widetilde{P}', \tag{3.3.27c}$$

$$\left(1 - \frac{M_\infty}{1 + M_\infty} U_0 \right) \widetilde{R}' - \frac{M_\infty}{1 + M_\infty} \rho_0 \widetilde{U}' + \frac{1}{\omega_*} \frac{\partial \rho_0 \widetilde{V}'}{\partial Y} = 0, \tag{3.3.27d}$$

$$\widetilde{H}' = \frac{\gamma}{\gamma - 1} \frac{\widetilde{P}'}{\rho_0} - \frac{1}{(\gamma - 1) M_\infty^2} \frac{\widetilde{R}'}{\rho_0^2}. \tag{3.3.27e}$$

Interestingly, equations (3.3.27) do not contain viscous terms. This means that the flow in the main part of the boundary layer may be treated as inviscid, and instead of the no-slip conditions, we have to use the impermeability condition on the plate surface:

$$\widetilde{V}' \Big|_{Y=0} = 0. \tag{3.3.28}$$

We start the analysis of the boundary-value problem with the lateral momentum equation (3.3.27b). It shows that the pressure perturbations induced by the acoustic wave (3.3.23) penetrate the boundary layer without change. Hence,

$$\widetilde{P}' = a_* \tag{3.3.29}$$

everywhere inside the boundary layer.

Turning to the longitudinal momentum equation (3.3.27a), one needs to remember that in the unperturbed flow, $U_0 = 0$ on the plate surface. Therefore, setting $Y = 0$ in (3.3.27a), and using (3.3.28) and (3.3.29), we find that

$$\widetilde{U}' \Big|_{Y=0} = \frac{M_\infty a_*}{(1 + M_\infty) \rho_0(x, 0)}. \tag{3.3.30}$$

Similarly, it can be deduced from the energy equation (3.3.27c) that

$$\widetilde{H}' \Big|_{Y=0} = \frac{a_*}{\rho_0(x, 0)}. \tag{3.3.31}$$

Finally, it follows from the state equation (3.3.27e) that

$$\widetilde{R}' \Big|_{Y=0} = M_\infty^2 a_* \rho_0(x, 0). \tag{3.3.32}$$

Clearly, the longitudinal velocity component (3.3.30) is not zero on the plate surface, which means that an additional viscous layer should be introduced in its proximity. It is termed the *Stokes layer*; see Figure 3.17.

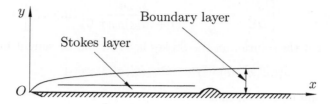

Fig. 3.17: Stokes layer formation in the boundary layer.

Stokes layer

The flow in the viscous layer is characterized by the same time and longitudinal coordinate scales (3.3.13) as the main part of the boundary layer and the flow outside the boundary layer. The thickness of the viscous layer may be determined by balancing the local acceleration term on the left-hand side of the momentum equation (3.3.2a) with the dominant viscous term on the right-hand side:

$$\rho\frac{\partial u}{\partial t} \sim \frac{1}{Re}\frac{\partial}{\partial y}\left(\mu\frac{\partial u}{\partial y}\right). \tag{3.3.33}$$

Representing the derivatives in (3.3.33) by finite differences, we have

$$\rho\frac{u}{t} \sim \frac{\mu}{Re}\frac{u}{y^2}.$$

It remains to take into account that the density ρ and viscosity coefficient μ remain finite near the plate surface, and we can conclude that

$$y \sim \sqrt{Re^{-1}t} \sim Re^{-5/8}.$$

Thus, the characteristic lateral coordinate Y_* should be introduced through the scaling

$$y = Re^{-5/8}Y_*. \tag{3.3.34}$$

To determine the form of the asymptotic expansions of the fluid-dynamic functions in the Stokes layer, we shall use the matching with the solution (3.3.25) in the main part of the boundary layer. Let us consider, for example, the longitudinal velocity component:

$$u = U_0(x, Y) + Re^{-1/8}\widetilde{U}'(x, Y)\sin(\omega_*\xi) + \cdots. \tag{3.3.35}$$

The first term on the right-hand side of (3.3.35) represents the solution for the steady boundary layer, which satisfies the no-slip condition: $U_0(x, 0) = 0$. Therefore, the Taylor expansion of $U_0(x, Y)$ for small values of Y is written as

$$U_0(x, Y) = \frac{\partial U_0}{\partial Y}(x, 0)Y + \cdots. \tag{3.3.36}$$

Substituting (3.3.36) and (3.3.30) into (3.3.35), and using the fact that $Y = Re^{-1/8}Y_*$, we find that the 'inner expansion of the outer solution' has the form

$$u = Re^{-1/8}\left[\frac{\partial U_0}{\partial Y}(x,0)Y_* + \frac{M_\infty a_*}{(1+M_\infty)\rho_0(x,0)}\sin(\omega_*\xi)\right] + \cdots.$$

This suggests that the solution in the Stokes layer should be sought in the form

$$u(t,x,y;Re) = Re^{-1/8}U'_*(t_*,x',Y_*;x) + \cdots, \tag{3.3.37}$$

where the function U'_* is such that

$$U'_* = \frac{\partial U_0}{\partial Y}(x,0)Y_* + \frac{M_\infty a_*}{(1+M_\infty)\rho_0(x,0)}\sin(\omega_*\xi) \quad \text{as} \quad Y_* \to \infty. \tag{3.3.38}$$

In a similar way it may be inferred that the asymptotic expansions of the enthalpy h, density ρ and viscosity coefficient μ are written as

$$\left.\begin{aligned}
h(t,x,y;Re) &= h_0(x,0) + Re^{-1/8}h'_*(t_*,x',Y_*;x) + \cdots, \\
\rho(t,x,y;Re) &= \rho_0(x,0) + Re^{-1/8}\rho'_*(t_*,x',Y_*;x) + \cdots, \\
\mu(t,x,y;Re) &= \mu_0(x,0) + Re^{-1/8}\mu'_*(t_*,x',Y_*;x) + \cdots.
\end{aligned}\right\} \tag{3.3.39}$$

As far as the pressure p is concerned, we know that it does not change across the boundary layer, which means that the asymptotic expansion of p in the Stokes layer should have the same form as that in the main part of the boundary layer:

$$p(t,x,y;Re) = Re^{-1/8}P'_*(t_*,x',Y_*;x) + \cdots. \tag{3.3.40}$$

Finally, the form of the asymptotic expansion of the lateral velocity component

$$v(t,x,y;Re) = Re^{-1/2}V'_*(t_*,x',Y_*;x) + \cdots \tag{3.3.41}$$

is obtained by means of the usual balancing (3.3.26) in the continuity equation (3.3.2d).

By substituting (3.3.37), (3.3.39)–(3.3.41) together with (3.3.13) and (3.3.34) into the longitudinal momentum equation (3.3.2a), we find that the longitudinal velocity oscillations U'_* in the Stokes layer are governed by the equation

$$\rho_0(x,0)\frac{\partial U'_*}{\partial t_*} = -\frac{\partial P'_*}{\partial x'} + \mu_0(x,0)\frac{\partial^2 U'_*}{\partial Y_*^2}. \tag{3.3.42}$$

It further follows from the lateral momentum equation (3.3.2b) that

$$\frac{\partial P'_*}{\partial Y'_*} = 0,$$

which means that the pressure in the Stokes layer coincides with the pressure in the main part of the boundary layer, that is,

$$P'_* = a_*\sin(\omega_*\xi). \tag{3.3.43}$$

Equation (3.3.42) should be solved subject to the no-slip condition on the plate surface

$$U'_* = 0 \quad \text{at} \quad Y_* = 0, \tag{3.3.44}$$

and the condition of matching (3.3.38) with the solution in the main part of the boundary layer.

Guided by (3.3.38) we seek the solution to (3.3.42) in the form

$$U'_* = \frac{\partial U_0}{\partial Y}(x,0)Y_* + Q(Y_*;x)e^{i\omega_*\xi} + \overline{Q}(Y_*;x)e^{-i\omega_*\xi}. \tag{3.3.45}$$

Here the function $Q(Y_*;x)$ depends on x as a parameter; $\overline{Q}$ stands for the complex conjugate of Q, which ensures that U'_* remains real. Substitution of (3.3.45) and (3.3.43) into (3.3.42) yields the following equation for $Q(Y_*;x)$:

$$\mu_0(x,0)\frac{d^2Q}{dY_*^2} - i\omega_*\rho_0(x,0)Q = -\frac{a_*M_\infty\omega_*}{2(1+M_\infty)}. \tag{3.3.46}$$

The boundary conditions for this equation are obtained by substituting (3.3.45) into (3.3.38) and (3.3.44). We see that

$$Q\Big|_{Y_*=\infty} = \frac{M_\infty a_*}{2i(1+M_\infty)\rho_0(x,0)}, \qquad Q\Big|_{Y_*=0} = 0. \tag{3.3.47}$$

Equation (3.3.46) is an ordinary differential equation with constant coefficients. Its general solution is composed of a particular integral

$$Q_p = \frac{M_\infty a_*}{2i(1+M_\infty)\rho_0(x,0)}$$

and two complementary solutions of the homogeneous part of (3.3.46). The latter are sought in the form $Q = Ce^{\lambda Y_*}$, with λ satisfying the characteristic equation

$$\mu_0(x,0)\lambda^2 - i\omega_*\rho_0(x,0) = 0. \tag{3.3.48}$$

The two roots of (3.3.48) are

$$\lambda_{1,2} = \pm(1+i)\sqrt{\frac{\omega_*\rho_0(x,0)}{2\mu_0(x,0)}},$$

which allows us to conclude that the general solution of (3.3.46) is written as

$$Q = \frac{d}{2i} + C_1 e^{(1+i)\varkappa Y_*} + C_2 e^{-(1+i)\varkappa Y_*},$$

with constants d and $\varkappa$ given by

$$d = \frac{M_\infty a_*}{(1+M_\infty)\rho_0(x,0)}, \qquad \varkappa = \sqrt{\frac{\omega_*\rho_0(x,0)}{2\mu_0(x,0)}}. \tag{3.3.49}$$

To find the constants C_1 and C_2, we need to use boundary conditions (3.3.47). The first condition shows that $Q(Y_*;x)$ remains finite as $Y_* \to \infty$, which is only possible if $C_1 = 0$. Using the second condition in (3.3.47), we find that $C_2 = -d/2i$. Thus,

$$Q = \frac{d}{2i}\Big[1 - e^{-(1+i)\varkappa Y_*}\Big]. \tag{3.3.50}$$

It remains to substitute (3.3.50) into (3.3.45), and we can conclude that in the Stokes layer

$$U'_* = \frac{\partial U_0}{\partial Y}(x, 0)Y_* + d\sin(\omega_*\xi) - de^{-\varkappa Y_*}\sin(\omega_*\xi - \varkappa Y_*). \tag{3.3.51}$$

The solution for the other fluid-dynamic functions is constructed in the same way; see Problem 2 in Exercises 12.

3.3.4   Triple-deck region

Now, we shall consider the flow in the vicinity of the roughness; see Figure 3.18. Unlike in Terent'ev's problem (Section 3.1), we shall assume here that the shape of the roughness does not change with time, and is given by the equation

$$y = Re^{-5/8}f\left(\frac{x-1}{Re^{-3/8}}\right). \tag{3.3.52}$$

The perturbations induced in the boundary layer by such a roughness are described by triple-deck theory. A detailed discussion of this theory applied to steady subsonic flow past a wall irregularity may be found in Section 4.2 in Part 3 of this book series. Here we shall show how the theory can be adapted to include the unsteadiness caused by an acoustic wave.

Upper tier (region 1)

We start with the upper tier shown in Figure 3.18 as region 1. In this region, order one independent variables are

$$t_* = \frac{t}{Re^{-1/4}}, \qquad x_* = \frac{x-1}{Re^{-3/8}}, \qquad y_* = \frac{y}{Re^{-3/8}}. \tag{3.3.53}$$

If there were no acoustic wave, then the perturbations in the triple-deck region would be produced solely by the wall roughness, and the asymptotic expansions of the velocity components and the pressure would be given by (3.1.48). For compressible flow, considered here, these should be supplemented with the following asymptotic expansions for the density and enthalpy[16]

$$\rho = 1 + Re^{-1/4}\rho_* + \cdots, \qquad h = \frac{1}{(\gamma - 1)M_\infty^2} + Re^{-1/4}h_* + \cdots. \tag{3.3.54}$$

We shall now see how (3.1.48), (3.3.54) should be modified in the presence of the acoustic wave.

Substituting (3.3.23), (3.3.24) into (3.3.14), we have the solution for the acoustic wave in the form

$$\left.\begin{aligned}
u &= 1 + Re^{-1/8}a_* M_\infty \sin(\omega_*\xi) + \cdots, \\
p &= Re^{-1/8}a_* \sin(\omega_*\xi) + \cdots, \\
\rho &= 1 + Re^{-1/8}a_* M_\infty^2 \sin(\omega_*\xi) + \cdots, \\
h &= \frac{1}{(\gamma - 1)M_\infty^2} + Re^{-1/8}a_* \sin(\omega_*\xi) + \cdots.
\end{aligned}\right\} \tag{3.3.55}$$

---

[16]See equations (4.2.79) on page 295 in Part 3 of this book series.

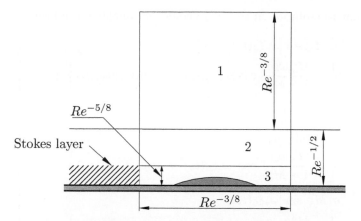

Fig. 3.18: The flow in the vicinity of the wall roughness.

This solution is valid in the inviscid part of the flow outside the boundary layer, and thus, may be matched with the solution in the upper tier (region 1) of the triple-deck structure. To perform the matching, we need to re-expand (3.3.55) in terms of the variables of region 1. It follows from (3.3.13) and (3.3.53) that

$$x' = x_0' + Re^{-1/8}x_*. \tag{3.3.56}$$

Here x_0' denotes the value of x' at the location of the roughness. Using (3.3.56) in the equation for ξ in (3.3.21), we have

$$\xi = \xi_0(t_*) - Re^{-1/8}\frac{M_\infty}{1 + M_\infty}x_* \quad \text{with} \quad \xi_0(t_*) = t_* - \frac{M_\infty x_0'}{1 + M_\infty}. \tag{3.3.57}$$

Taking into account that $Re^{-1/8}$ in (3.3.57) is small, we apply the Taylor expansion to $\sin(\omega_*\xi)$:

$$\sin(\omega_*\xi) = \sin\left[\omega_*\xi_0(t_*)\right] - Re^{-1/8}\frac{M_\infty\omega_*}{1 + M_\infty}\cos\left[\omega_*\xi_0(t_*)\right]x_* + \cdots . \tag{3.3.58}$$

It remains to substitute (3.3.58) into (3.3.55), and we will have the 'inner expansion of the outer solution':

$$u = 1 + Re^{-1/8}a_*M_\infty \sin\left[\omega_*\xi_0(t_*)\right] - Re^{-1/4}\frac{M_\infty^2\omega_*a_*}{1 + M_\infty}\cos\left[\omega_*\xi_0(t_*)\right]x_* + \cdots ,$$

$$p = Re^{-1/8}a_* \sin\left[\omega_*\xi_0(t_*)\right] - Re^{-1/4}\frac{M_\infty\omega_*a_*}{1 + M_\infty}\cos\left[\omega_*\xi_0(t_*)\right]x_* + \cdots ,$$

$$\rho = 1 + Re^{-1/8}a_*M_\infty^2 \sin\left[\omega_*\xi_0(t_*)\right] - Re^{-1/4}\frac{M_\infty^3\omega_*a_*}{1 + M_\infty}\cos\left[\omega_*\xi_0(t_*)\right]x_* + \cdots ,$$

$$h = \frac{1}{(\gamma - 1)M_\infty^2} + Re^{-1/8}a_* \sin\left[\omega_*\xi_0(t_*)\right]$$

$$- Re^{-1/4}\frac{M_\infty\omega_*a_*}{1 + M_\infty}\cos\left[\omega_*\xi_0(t_*)\right]x_* + \cdots .$$

This suggests that the solution in region 1 should be sought in the form

$$
\left.
\begin{aligned}
u &= 1 + Re^{-1/8} a_* M_\infty \sin\left[\omega_* \xi_0(t_*)\right] \\
&\quad + Re^{-1/4}\left\{-\frac{M_\infty^2 \omega_* a_*}{1 + M_\infty}\cos\left[\omega_* \xi_0(t_*)\right]x_* + u_*(t_*, x_*, y_*)\right\} + \cdots, \\
v &= Re^{-1/4} v_*(t_*, x_*, y_*) + \cdots, \\
p &= Re^{-1/8} a_* \sin\left[\omega_* \xi_0(t_*)\right] \\
&\quad + Re^{-1/4}\left\{-\frac{M_\infty \omega_* a_*}{1 + M_\infty}\cos\left[\omega_* \xi_0(t_*)\right]x_* + p_*(t_*, x_*, y_*)\right\} + \cdots, \\
\rho &= 1 + Re^{-1/8} a_* M_\infty^2 \sin\left[\omega_* \xi_0(t_*)\right] \\
&\quad + Re^{-1/4}\left\{-\frac{M_\infty^3 \omega_* a_*}{1 + M_\infty}\cos\left[\omega_* \xi_0(t_*)\right]x_* + \rho_*(t_*, x_*, y_*)\right\} + \cdots, \\
h &= \frac{1}{(\gamma - 1)M_\infty^2} + Re^{-1/8} a_* \sin\left[\omega_* \xi_0(t_*)\right] \\
&\quad + Re^{-1/4}\left\{-\frac{M_\infty \omega_* a_*}{1 + M_\infty}\cos\left[\omega_* \xi_0(t_*)\right]x_* + h_*(t_*, x_*, y_*)\right\} + \cdots.
\end{aligned}
\right\} \tag{3.3.59}
$$

Substitution of (3.3.59) into the Navier–Stokes equations (3.3.2) results in the following set of equations:

$$
\frac{\partial u_*}{\partial x_*} = -\frac{\partial p_*}{\partial x_*}, \tag{3.3.60a}
$$

$$
\frac{\partial v_*}{\partial x_*} = -\frac{\partial p_*}{\partial y_*}, \tag{3.3.60b}
$$

$$
\frac{\partial h_*}{\partial x_*} = \frac{\partial p_*}{\partial x_*}, \tag{3.3.60c}
$$

$$
\frac{\partial \rho_*}{\partial x_*} + \frac{\partial u_*}{\partial x_*} + \frac{\partial v_*}{\partial y_*} = 0, \tag{3.3.60d}
$$

$$
h_* = \frac{\gamma}{\gamma - 1}p_* - \frac{1}{(\gamma - 1)M_\infty^2}\rho_* - a_*^2 M_\infty^2 \sin^2\left[\omega_* \xi_0(t_*)\right]. \tag{3.3.60e}
$$

It may be reduced to a single equation for the pressure p_*. Indeed, substitution of the state equation (3.3.60e) into the energy equation (3.3.60c) results in

$$
\frac{\partial \rho_*}{\partial x_*} = M_\infty^2 \frac{\partial p_*}{\partial x_*}. \tag{3.3.61}
$$

Now, we substitute (3.3.61) together with the x-momentum equation (3.3.60a) into the continuity equation (3.3.60d). We find that

$$
(M_\infty^2 - 1)\frac{\partial p_*}{\partial x_*} + \frac{\partial v_*}{\partial y_*} = 0. \tag{3.3.62}
$$

It remains to eliminate v_*, which is achieved by cross-differentiating (3.3.62) with the y-momentum equation (3.3.60b). This leads to the following equation for p_*:

$$(1 - M_\infty^2)\frac{\partial^2 p_*}{\partial x_*^2} + \frac{\partial^2 p_*}{\partial y_*^2} = 0, \qquad (3.3.63)$$

known as the equation of thin aerofoil theory.[17]

We shall formulate the boundary conditions for this equation after analysing the flow in the viscous sublayer (region 3) and the main part of the boundary layer (region 2).

Viscous sublayer (region 3)

As usual, the asymptotic analysis of the Navier–Stokes equations in region 3 is conducted based on the limit procedure where

$$t_* = \frac{t}{Re^{-1/8}}, \qquad x_* = \frac{x-1}{Re^{-3/8}}, \qquad Y_* = \frac{y}{Re^{-5/8}} \qquad (3.3.64)$$

remain finite as $Re \to \infty$. Interestingly enough, the characteristic thickness, $y \sim Re^{-5/8}$, of region 1 is the same as that of the Stokes layer; see Figure 3.18.

The asymptotic expansions of the fluid-dynamic functions are represented in this region in the form

$$\left.\begin{aligned}
&u = Re^{-1/8}U_*(t_*, x_*, Y_*) + \cdots, \qquad v = Re^{-3/8}V_*(t_*, x_*, Y_*) + \cdots, \\
&p = Re^{-1/8}a_* \sin\left[\omega_*\xi_0(t_*)\right] \\
&\quad + Re^{-1/4}\left\{-\frac{M_\infty\omega_*a_*}{1+M_\infty}\cos\left[\omega_*\xi_0(t_*)\right]x_* + P_*(t_*, x_*, Y_*)\right\} + \cdots, \\
&\rho = \rho_w + O(Re^{-1/8}), \qquad \mu = \mu_w + O(Re^{-1/8}).
\end{aligned}\right\} \quad (3.3.65)$$

Here it is taken into account that in region 3 the fluid velocity is small, and therefore, the flow may be treated as incompressible, with the density and viscosity coefficient assuming their values (3.3.12) in the unperturbed boundary layer at the position of the roughness.[18]

Substitution of (3.3.65), (3.3.64) into the Navier–Stokes equations (3.3.2) yields

$$\rho_w\left(\frac{\partial U_*}{\partial t_*} + U_*\frac{\partial U_*}{\partial x_*} + V_*\frac{\partial U_*}{\partial Y_*}\right) = \frac{M_\infty\omega_*a_*}{1+M_\infty}\cos\left[\omega_*\xi_0(t_*)\right]$$

$$-\frac{\partial P_*}{\partial x_*} + \mu_w\frac{\partial^2 U_*}{\partial Y_*^2}, \qquad (3.3.66a)$$

$$\frac{\partial P_*}{\partial Y_*} = 0, \qquad (3.3.66b)$$

$$\frac{\partial U_*}{\partial x_*} + \frac{\partial V_*}{\partial Y_*} = 0. \qquad (3.3.66c)$$

---

[17]See Section 2.2 in Part 2 of this book series.

[18]A detailed discussion of this issue is presented in Section 2.2.4 in Part 3 of this book series.

Equations (3.3.66) should be solved with the no-slip conditions on the surface of the roughness (3.3.52):

$$U_* = V_* = 0 \quad \text{at} \quad Y_* = f(x_*),\tag{3.3.67}$$

and the condition of matching with the solution (3.3.51) in the Stokes layer:

$$U_* = \lambda Y_* + d_0 \sin\big[\omega_* \xi_0(t_*)\big]$$
$$- d_0 e^{-\varkappa_0 Y_*} \sin\big[\omega_* \xi_0(t_*) - \varkappa_0 Y_*\big] \quad \text{at} \quad x_* = -\infty.\tag{3.3.68}$$

Here λ is the skin friction immediately upstream of the triple-deck region, defined by (3.3.12); d_0 and $\varkappa_0$ are the values of d and $\varkappa$ which are obtained by setting $x = 1$ in (3.3.49), i.e.

$$d_0 = \frac{M_\infty a_*}{(1 + M_\infty)\rho_0(1,0)}, \qquad \varkappa_0 = \sqrt{\frac{\omega_* \rho_0(1,0)}{2\mu_0(1,0)}}.$$

The analysis of the behaviour of the solution to (3.3.66) at large values of Y_* is conducted in the same way as it was done in Terent'ev's problem. In fact, the arguments expressed by equations (3.1.26)–(3.1.31) remain valid for the flow considered here, except equation (3.1.32) should now be written as

$$A_*\Big|_{x_* = -\infty} = \begin{cases} d_0 \sin\big[\omega_* \xi_0(t_*)\big] & \text{if} \quad \alpha = 1, \\ 0 & \text{if} \quad \alpha \neq 1. \end{cases}$$

Keeping this in mind, we shall redefine A_* as $A_* + d_0 \sin\big[\omega_* \xi_0(t_*)\big]$. Equation (3.1.30) is then modified to

$$\Psi_*(t_*, x_*, Y_*) = \frac{1}{2}\lambda Y_*^2 + \Big\{ A_*(t_*, x_*) + d_0 \sin\big[\omega_* \xi_0(t_*)\big] \Big\} Y_* + \cdots,\tag{3.3.69}$$

where $A_*(t_*, x_*)$ is such that

$$A_*(t_*, x_*) = 0 \quad \text{at} \quad x_* = -\infty.$$

It remains to substitute (3.3.69) into (3.1.26), and we can conclude that at the outer edge of region 3

$$\left.\begin{aligned} U_* &= \lambda Y_* + A_*(t_*, x_*) + d_0 \sin\big[\omega_* \xi_0(t_*)\big] + \cdots, \\ V_* &= -\frac{\partial A_*}{\partial x_*} Y_* + \cdots \end{aligned}\right\} \quad \text{as} \quad Y_* \to \infty.\tag{3.3.70}$$

Main part of the boundary layer (region 2)

Region 2, the middle tier of the triple-deck structure (see Figure 3.18), represents a continuation of the standard boundary layer, that forms on the plate surface ahead of the roughness. The thickness of region 2 is estimated as $y \sim Re^{-1/2}$. Of course, the longitudinal extent of region 2 coincides with that of the interaction region as a whole, being estimated as $|x - 1| = O(Re^{-3/8})$. Consequently, the asymptotic analysis of the

Navier–Stokes equations (3.3.2) in region 2 has to be performed based on the limit with

$$t_* = \frac{t}{Re^{-1/4}}, \quad x_* = \frac{x-1}{Re^{-3/8}}, \quad Y = \frac{y}{Re^{-1/2}} = O(1) \tag{3.3.71}$$

remaining finite as $Re \to \infty$.

The form of the asymptotic expansions of the fluid-dynamic functions in region 2 may be predicted using the matching with the solution in region 3. At the outer edge of region 3, the velocity components are given by the asymptotic expansions (3.3.70). If we substitute (3.3.70) into the asymptotic expansions of u and v in (3.3.65) and express the result in term of variables (3.3.71) of region 2, using the fact that $Y = Re^{-1/8}Y_*$, then we will see that at the 'bottom' of region 2

$$\left. \begin{aligned} u &= \lambda Y + Re^{-1/8}\Big\{A_*(t_*, x_*) + d_0 \sin\big[\omega_* \xi_0(t_*)\big]\Big\} + \cdots, \\ v &= Re^{-1/4}\Big(-\frac{\partial A_*}{\partial x_*}Y\Big) + \cdots. \end{aligned} \right\} \tag{3.3.72}$$

This suggests that the solution in region 2 should be sought in the form

$$\left. \begin{aligned} u &= U_{00}(Y) + Re^{-1/8}\widetilde{U}_1(t_*, x_*, Y) + \cdots, \\ v &= Re^{-1/4}\widetilde{V}_1(t_*, x_*, Y) + \cdots. \end{aligned} \right\} \tag{3.3.73}$$

The leading-order term $U_{00}(Y)$ in the expansion for u coincides with the velocity profile (3.3.11) in the boundary layer immediately ahead of the interaction region. According to (3.3.12)

$$U_{00} = \lambda Y + \cdots \quad \text{as} \quad Y \to 0. \tag{3.3.74}$$

It further follows from (3.3.72) that the perturbation terms $\widetilde{U}_1(x_*, Y)$ and $\widetilde{V}_1(x_*, Y)$ in (3.3.73) should exhibit the following behaviour at the 'bottom' of region 2:

$$\left. \begin{aligned} \widetilde{U}_1 &= A_*(t_*, x_*) + d_0 \sin\big[\omega_* \xi_0(t_*)\big] + \cdots, \\ \widetilde{V}_1 &= -\frac{\partial A_*}{\partial x_*}Y + \cdots \end{aligned} \right\} \quad \text{as} \quad Y \to 0. \tag{3.3.75}$$

By analogy with the longitudinal velocity component u in (3.3.73), we shall seek the enthalpy h, the density ρ, and the viscosity μ in region 2 in the form of the asymptotic expansions

$$\left. \begin{aligned} h &= h_{00}(Y) + Re^{-1/8}\widetilde{h}_1(t_*, x_*, Y) + \cdots, \\ \rho &= \rho_{00}(Y) + Re^{-1/8}\widetilde{\rho}_1(t_*, x_*, Y) + \cdots, \\ \mu &= \mu_{00}(Y) + Re^{-1/8}\widetilde{\mu}_1(t_*, x_*, Y) + \cdots. \end{aligned} \right\} \tag{3.3.76}$$

Finally, we expect the pressure p to remain unchanged across the boundary layer. Consequently, the asymptotic representation of p in region 2 should have the same form

$$\begin{aligned} p = Re^{-1/8}a_* \sin\big[\omega_*\xi_0(t_*)\big] \\ + Re^{-1/4}\Big\{-\frac{M_\infty \omega_* a_*}{1+M_\infty}\cos\big[\omega_*\xi_0(t_*)\big]x_* + \widetilde{P}_1(t_*, x_*, Y_*)\Big\} + \cdots \end{aligned} \tag{3.3.77}$$

as that in region 3; see the equation for p in (3.3.65).

The substitution of (3.3.73), (3.3.76), and (3.3.77) into the Navier–Stokes equations (3.3.2) results in

$$U_{00}(Y)\frac{\partial \tilde{U}_1}{\partial x_*} + \tilde{V}_1\frac{dU_{00}}{dY} = 0, \tag{3.3.78a}$$

$$\frac{\partial \tilde{P}_1}{\partial Y} = 0, \tag{3.3.78b}$$

$$U_{00}(Y)\frac{\partial \tilde{h}_1}{\partial x_*} + \tilde{V}_1\frac{dh_{00}}{dY} = 0, \tag{3.3.78c}$$

$$\rho_{00}(Y)\frac{\partial \tilde{U}_1}{\partial x_*} + U_{00}(Y)\frac{\partial \tilde{\rho}_1}{\partial x_*} + \rho_{00}(Y)\frac{\partial \tilde{V}_1}{\partial Y} + \tilde{V}_1\frac{d\rho_{00}}{dY} = 0, \tag{3.3.78d}$$

$$h_{00} = \frac{1}{(\gamma-1)M_\infty^2}\frac{1}{\rho_{00}}, \qquad \tilde{h}_1 = -\frac{1}{(\gamma-1)M_\infty^2}\frac{\tilde{\rho}_1}{\rho_{00}^2}. \tag{3.3.78e}$$

It is interesting to note that the above equations do not involve viscous terms, which means that the flow in region 2 may be treated as inviscid in the leading-order approximation.

Equations (3.3.78) are easily solved using the following elimination procedure. We first substitute (3.3.78e) into the energy equation (3.3.78c). This leads to

$$U_{00}(Y)\frac{\partial \tilde{\rho}_1}{\partial x_*} + \tilde{V}_1\frac{d\rho_{00}}{dY} = 0,$$

which means that the continuity equation (3.3.78d) may be written as

$$\frac{\partial \tilde{U}_1}{\partial x_*} + \frac{\partial \tilde{V}_1}{\partial Y} = 0. \tag{3.3.79}$$

Now, using (3.3.79), we can eliminate $\partial \tilde{U}_1/\partial x_*$ from the longitudinal momentum equation (3.3.78a). This results in

$$U_{00}(Y)\frac{\partial \tilde{V}_1}{\partial Y} - \tilde{V}_1\frac{dU_{00}}{dY} = 0. \tag{3.3.80}$$

Dividing both terms in (3.3.80) by U_{00}^2, we have

$$\frac{1}{U_{00}(Y)}\frac{\partial \tilde{V}_1}{\partial Y} - \frac{\tilde{V}_1}{U_{00}^2}\frac{dU_{00}}{dY} = 0,$$

or, equivalently,

$$\frac{\partial}{\partial Y}\left(\frac{\tilde{V}_1}{U_{00}}\right) = 0.$$

We see that the ratio $\tilde{V}_1/U_{00}$ is a function of t_* and x_* only, say $G(t_*, x_*)$:

$$\frac{\tilde{V}_1}{U_{00}} = G(t_*, x_*). \tag{3.3.81}$$

This function may be found by making use of equations (3.3.74), (3.3.75) describing the behaviour of U_{00} and $\tilde{V}_1$ at the bottom of region 2. We see that

$$G(t_*, x_*) = -\frac{1}{\lambda}\frac{\partial A_*}{\partial x_*}. \tag{3.3.82}$$

Substituting (3.3.82) back into (3.3.81) we can conclude that

$$\frac{\tilde{V}_1}{U_{00}} = -\frac{1}{\lambda}\frac{\partial A_*}{\partial x_*}. \tag{3.3.83}$$

Now let us return to the asymptotic expansions (3.3.73) of the velocity components, and calculate the streamline slope angle in region 2:

$$\vartheta = \arctan\frac{v}{u} = Re^{-1/4}\frac{\tilde{V}_1}{U_{00}} + \cdots = Re^{-1/4}\left(-\frac{1}{\lambda}\frac{\partial A_*}{\partial x_*}\right) + \cdots. \tag{3.3.84}$$

We see that ϑ does not depend on Y, which allows us to use (3.3.84) for matching with the solution in the upper tier (region 1).

3.3.5 Viscous-inviscid interaction problem

We are ready now to formulate the viscous-inviscid interaction problem. To describe the flow in the vicinity of the roughness, we need to solve the boundary-layer equations (3.3.66) for the viscous sublayer, subject to the boundary conditions (3.3.67), (3.3.68), and (3.3.70). Unlike in classical boundary-layer theory, the pressure gradient $\partial P_*/\partial x_*$ in (3.3.66a) is not known in advance, and has to be found by solving equation (3.3.63) for the upper tier. The boundary conditions for (3.3.63) are formulated as follows.

First, using the asymptotic expansions (3.3.59) for the velocity components (u, v) we calculate the streamline slope angle in region 1:

$$\vartheta = \arctan\frac{v}{u} = Re^{-1/4}v_*(t_*, x_*, y_*) + \cdots. \tag{3.3.85}$$

Comparing (3.3.85) with (3.3.84) one can easily see that the sought matching condition can be written as

$$v_*\Big|_{y_*=0} = -\frac{1}{\lambda}\frac{\partial A_*}{\partial x_*}. \tag{3.3.86}$$

To convert (3.3.86) into a condition on the pressure p_* we set $y_* = 0$ in the y-momentum equation (3.3.60b), and using (3.3.86) on the left-hand side of (3.3.60b), we arrive at the conclusion that

$$\frac{\partial p_*}{\partial y_*}\Big|_{y_*=0} = \frac{1}{\lambda}\frac{\partial^2 A_*}{\partial x_*^2}. \tag{3.3.87}$$

The second boundary condition, that we will be using when solving equation (3.3.63), is the condition of attenuation of the perturbations at large distance from the roughness:

$$p_* \to 0 \quad \text{as} \quad x_*^2 + y_*^2 \to \infty. \tag{3.3.88}$$

Strictly speaking, the pressure p_* can tend to a non-zero 'background' $\mathcal{P}$ that is a function of time t_* only. Since the viscous sublayer equations (3.3.66) do not involve the pressure but only the pressure gradient, such a background pressure will not affect the velocity field.

To complete the formulation of the viscous-inviscid interaction problem, we take into account that the pressure does not change across the middle and lower tiers, and therefore, we can find the pressure in the lower tier through the equation

$$P_* = p_*\big|_{y_*=0}. \tag{3.3.89}$$

The viscous-inviscid interaction problem involves four parameters: μ_w, ρ_w, λ, and $\beta = \sqrt{1 - M_\infty^2}$. These may be excluded from the problem formulation by performing the following scaling of the variables

$$
\left.
\begin{aligned}
t_* &= \frac{\mu_w^{-1/2}}{\lambda^{3/2}\beta^{1/2}}\,\bar{t} + \frac{M_\infty x_0'}{1 + M_\infty}, & x_* &= \frac{\mu_w^{-1/4}\rho_w^{-1/2}}{\lambda^{5/4}\beta^{3/4}}\,\bar{x}, \\[2mm]
Y_* &= \frac{\mu_w^{1/4}\rho_w^{-1/2}}{\lambda^{3/4}\beta^{1/4}}\,\bar{Y} + f(x_*), & f &= \frac{\mu_w^{1/4}\rho_w^{-1/2}}{\lambda^{3/4}\beta^{1/4}}\,\bar{f}, \\[2mm]
U_* &= \frac{\mu_w^{1/4}\rho_w^{-1/2}}{\lambda^{-1/4}\beta^{1/4}}\,\bar{U}, & V_* &= \frac{\mu_w^{3/4}\rho_w^{-1/2}}{\lambda^{-3/4}\beta^{-1/4}}\,\bar{V} + U_*\frac{df}{dx_*}, \\[2mm]
P_* &= \frac{\mu_w^{1/2}}{\lambda^{-1/2}\beta^{1/2}}\,\bar{P}, & A_* &= \frac{\mu_w^{1/4}\rho_w^{-1/2}}{\lambda^{-1/4}\beta^{1/4}}\,\bar{A} - \lambda f(x_*)
\end{aligned}
\right\} \tag{3.3.90}
$$

that combine standard affine transformations of the triple-deck theory with Prandtl's transposition.[19] The latter introduces the body-fitted coordinates $(\bar{x}, \bar{Y})$ with $\bar{x}$ measured along the body surface and $\bar{Y}$ in the normal direction. With (3.3.90), the boundary-value problem (3.3.66)–(3.3.68), (3.3.70) for the viscous sublayer assumes the following canonical form:

$$
\left.
\begin{aligned}
\frac{\partial \bar{U}}{\partial \bar{t}} + \bar{U}\frac{\partial \bar{U}}{\partial \bar{x}} + \bar{V}\frac{\partial \bar{U}}{\partial \bar{Y}} &= a\cos(\omega\bar{t}) - \frac{d\bar{P}}{d\bar{x}} + \frac{\partial^2 \bar{U}}{\partial \bar{Y}^2}, \\[2mm]
\frac{\partial \bar{U}}{\partial \bar{x}} + \frac{\partial \bar{V}}{\partial \bar{Y}} &= 0, \\[2mm]
\bar{U} = \bar{V} = 0 \qquad &\text{at}\quad \bar{Y} = 0, \\[2mm]
\bar{U} = \bar{Y} + \frac{a}{\omega}\sin(\omega\bar{t}) & \\[1mm]
-\frac{a}{\omega}e^{-\sqrt{\omega/2}\,\bar{Y}}\sin\left(\omega\bar{t} - \sqrt{\frac{\omega}{2}}\,\bar{Y}\right) \quad &\text{at}\quad \bar{x} = -\infty, \\[2mm]
\bar{U} = \bar{Y} + \bar{A}(\bar{t}, \bar{x}) + \frac{a}{\omega}\sin(\omega\bar{t}) + \cdots \quad &\text{as}\quad \bar{Y} \to \infty.
\end{aligned}
\right\} \tag{3.3.91}
$$

Here, the acoustic wave amplitude and frequency are transformed as

$$
a = a_* \frac{M_\infty \omega_*}{1 + M_\infty} \frac{\mu_w^{-3/4}\rho_w^{-1/2}}{\lambda^{7/4}\beta^{1/4}}, \qquad \omega = \omega_* \frac{\mu_w^{-1/2}}{\lambda^{3/2}\beta^{1/2}}.
$$

---

[19]See Section 1.4.3 in Part 3 of this book series.

Corresponding to (3.3.90), we transform the variables in the upper tier as

$$y_* = \frac{\mu_w^{-1/4}\,\rho_w^{-1/2}}{\lambda^{5/4}\beta^{7/4}}\,\bar{y}, \qquad p_* = \frac{\mu_w^{1/2}}{\lambda^{-1/2}\beta^{1/2}}\,\bar{p}.$$

This turns the upper tier problem (3.3.63), (3.3.87), (3.3.88) into

$$\left.\begin{aligned}
\frac{\partial^2 \bar{p}}{\partial \bar{x}^2} + \frac{\partial^2 \bar{p}}{\partial \bar{y}^2} &= 0, \\
\frac{\partial \bar{p}}{\partial \bar{y}} = \frac{\partial^2 \bar{A}}{\partial \bar{x}^2} - \frac{d^2 \bar{f}}{d\bar{x}^2} \quad &\text{at} \quad \bar{y} = 0, \\
\bar{p} \to 0 \quad \text{as} \quad \bar{x}^2 + \bar{y}^2 &\to \infty.
\end{aligned}\right\}
\tag{3.3.92}$$

The solutions in regions 1 and 3 are related to one another through equation (3.3.89) which is written in the new variables as

$$\bar{P} = \bar{p}\Big|_{\bar{y}=0}. \tag{3.3.93}$$

3.3.6 Linear receptivity

We shall now assume that the amplitude of the acoustic wave, a, is small. We shall also assume that the roughness is 'shallow', namely,

$$\bar{f}(\bar{x}) = \varepsilon F(\bar{x}), \tag{3.3.94}$$

where ε is a small parameter. Then the solution to the lower tier problem (3.3.91) may be sought in the form

$$\bar{U} = \bar{Y} + a U_s(\bar{t}, \bar{Y}) + \varepsilon U_r(\bar{x}, \bar{Y}) + a\varepsilon U'(\bar{t}, \bar{x}, \bar{Y}) + \cdots, \tag{3.3.95a}$$

$$\bar{V} = \varepsilon V_r(\bar{x}, \bar{Y}) + a\varepsilon V'(\bar{t}, \bar{x}, \bar{Y}) + \cdots, \tag{3.3.95b}$$

$$\bar{P} = \varepsilon P_r(\bar{x}) + a\varepsilon P'(\bar{t}, \bar{x}) + \cdots, \tag{3.3.95c}$$

$$\bar{A} = \varepsilon A_r(\bar{x}) + a\varepsilon A'(\bar{t}, \bar{x}) + \cdots. \tag{3.3.95d}$$

In the asymptotic expansion (3.3.95a) for $\bar{U}$, the leading-order term, $\bar{U} = \bar{Y}$, represents the unperturbed steady boundary layer, and would be the only term if neither the acoustic wave nor the roughness were present. The next term, $a U_s(\bar{t}, \bar{Y})$ with

$$U_s(\bar{t}, \bar{Y}) = \frac{1}{\omega}\left[\sin(\omega \bar{t}) - e^{-\sqrt{\omega/2}\,\bar{Y}}\sin\left(\omega \bar{t} - \sqrt{\frac{\omega}{2}}\,\bar{Y}\right)\right], \tag{3.3.96}$$

represents the perturbations produced by the Stokes layer. The third term, $\varepsilon U_r(\bar{x}, \bar{Y})$, stands for the steady perturbations produced by the wall roughness. Finally, the fourth term, $a\varepsilon U'(\bar{t}, \bar{x}, \bar{Y})$, represents the perturbations forming in the boundary layer due to the interaction of the Stokes layer with the steady perturbation field around the roughness.

Corresponding to (3.3.95), the solution in the upper tier is represented in the form

$$\bar{p} = \varepsilon p_r(\bar{x}, \bar{y}) + a\varepsilon p'(\bar{t}, \bar{x}, \bar{y}) + \cdots .\tag{3.3.97}$$

We start our analysis with the steady perturbations produced by the roughness.

Steady problem

Substituting (3.3.97), (3.3.95d), and (3.3.94) into (3.3.92), and working with $O(\varepsilon)$ terms, we find that, in the upper tier, the steady perturbations are described by the equation

$$\frac{\partial^2 p_r}{\partial \bar{x}^2} + \frac{\partial^2 p_r}{\partial \bar{y}^2} = 0,\tag{3.3.98a}$$

together with the boundary conditions

$$\frac{\partial p_r}{\partial \bar{y}} = \frac{d^2 A_r}{d\bar{x}^2} - \frac{d^2 F}{d\bar{x}^2} \quad \text{at} \quad \bar{y} = 0,\tag{3.3.98b}$$

$$p_r \to 0 \quad \text{as} \quad \bar{x}^2 + \bar{y}^2 \to \infty.\tag{3.3.98c}$$

The solution of the boundary-value problem (3.3.98) is easily found with the help of the method of Fourier transforms. We define the Fourier transform $\breve{p}_r(\alpha; \bar{y})$ of $p_r(\bar{x}, \bar{y})$ as

$$\breve{p}_r(\alpha; \bar{y}) = \int_{-\infty}^{\infty} p_r(\bar{x}, \bar{y})e^{-i\alpha\bar{x}}\, d\bar{x},$$

and then (3.3.98) is converted to

$$-\alpha^2 \breve{p}_r + \frac{d^2 \breve{p}_r}{d\bar{y}^2} = 0,\tag{3.3.99a}$$

$$\frac{d\breve{p}_r}{d\bar{y}} = -\alpha^2(\breve{A}_r - \breve{F}) \quad \text{at} \quad \bar{y} = 0,\tag{3.3.99b}$$

$$\breve{p}_r \to 0 \quad \text{as} \quad \bar{y} \to \infty.\tag{3.3.99c}$$

The general solution of equation (3.3.99a) is written as

$$\breve{p}_r = C_1 e^{\alpha\bar{y}} + C_2 e^{-\alpha\bar{y}}.$$

Here α is a positive or negative real quantity. Applying boundary condition (3.3.99c) we have to write

$$\breve{p}_r = C e^{-|\alpha|\bar{y}},$$

with the constant C found from condition (3.3.99b) to be

$$C = |\alpha|(\breve{A}_r - \breve{F}).$$

We therefore conclude that in the upper tier

$$\breve{p}_r = |\alpha|(\breve{A}_r - \breve{F})e^{-|\alpha|\bar{y}}.\tag{3.3.100}$$

To study the perturbations produced by the roughness in the lower tier, we substitute (3.3.95) into (3.3.91) and select the $O(\varepsilon)$ terms. We have

$$
\left.\begin{aligned}
&\bar{Y}\frac{\partial U_r}{\partial \bar{x}} + V_r = -\frac{dP_r}{d\bar{x}} + \frac{\partial^2 U_r}{\partial \bar{Y}^2}, \\
&\frac{\partial U_r}{\partial \bar{x}} + \frac{\partial V_r}{\partial \bar{Y}} = 0, \\
&U_r = V_r = 0 \quad \text{at} \quad \bar{Y} = 0, \\
&U_r = 0 \qquad \text{at} \quad \bar{x} = -\infty, \\
&U_r = A_r \qquad \text{at} \quad \bar{Y} = \infty.
\end{aligned}\right\}
\tag{3.3.101}
$$

The boundary-value problem (3.3.101) is written in terms of Fourier transforms as

$$
i\alpha\bar{Y}\breve{U}_r + \breve{V}_r = -i\alpha\breve{P}_r + \frac{d^2\breve{U}_r}{d\bar{Y}^2},
\tag{3.3.102a}
$$

$$
i\alpha\breve{U}_r + \frac{d\breve{V}_r}{d\bar{Y}} = 0,
\tag{3.3.102b}
$$

$$
\breve{U}_r = \breve{V}_r = 0 \quad \text{at} \quad \bar{Y} = 0,
\tag{3.3.102c}
$$

$$
\breve{U}_r = \breve{A}_r \qquad \text{at} \quad \bar{Y} = \infty.
\tag{3.3.102d}
$$

Here the Fourier transform of the pressure is found by setting $\bar{y} = 0$ in (3.3.100):

$$
\breve{P}_r = |\alpha|(\breve{A}_r - \breve{F}).
\tag{3.3.103}
$$

The boundary-value problem (3.3.102) is solved in the same way as problem (3.1.78). We start with differentiation of the momentum equation (3.3.102a) with respect to $\bar{Y}$, which yields

$$
i\alpha\breve{U}_r + i\alpha\bar{Y}\frac{d\breve{U}_r}{d\bar{Y}} + \frac{d\breve{V}_r}{d\bar{Y}} = \frac{d^3\breve{U}_r}{d\bar{Y}^3}.
\tag{3.3.104}
$$

Now we can use the continuity equation (3.3.102b) to eliminate $d\breve{V}_r/d\bar{Y}$ from (3.3.104). This results in the following equation for $\breve{U}_r$:

$$
i\alpha\bar{Y}\frac{d\breve{U}_r}{d\bar{Y}} = \frac{d^3\breve{U}_r}{d\bar{Y}^3}.
\tag{3.3.105a}
$$

Two boundary conditions for (3.3.105a) follow from (3.3.102c), (3.3.102d) as

$$
\breve{U}_r = 0 \quad \text{at} \quad \bar{Y} = 0,
\tag{3.3.105b}
$$

$$
\breve{U}_r = \breve{A}_r \quad \text{at} \quad \bar{Y} = \infty.
\tag{3.3.105c}
$$

The third condition

$$
\left.\frac{d^2\breve{U}_r}{d\bar{Y}^2}\right|_{\bar{Y}=0} = i\alpha|\alpha|(\breve{A}_r - \breve{F})
\tag{3.3.105d}
$$

is obtained by setting $\bar{Y} = 0$ in (3.3.102a) and using equation (3.3.103) for $\breve{P}_r$.

The substitution of the independent variable

$$\zeta = (i\alpha)^{1/3}\bar{Y} \qquad (3.3.106)$$

allows us to transform equation (3.3.105a) into the Airy equation

$$\frac{d^3\breve{U}_r}{d\zeta^3} - \zeta\frac{d\breve{U}_r}{d\zeta} = 0 \qquad (3.3.107a)$$

for the derivative $d\breve{U}_r/d\zeta$. In terms of the new variable ζ, the boundary conditions (3.3.105b), (3.3.105c), and (3.3.105d) are written as

$$\breve{U}_r = 0 \qquad\qquad \text{at}\quad \zeta = 0, \qquad (3.3.107b)$$

$$\frac{d^2\breve{U}_r}{d\zeta^2} = (i\alpha)^{1/3}|\alpha|\big(\breve{A}_r - \breve{F}\big) \quad \text{at}\quad \zeta = 0, \qquad (3.3.107c)$$

$$\breve{U}_r = \breve{A}_r \qquad\qquad \text{at}\quad \zeta = \infty. \qquad (3.3.107d)$$

The general solution of the equation (3.3.107a) is known to be

$$\frac{d\breve{U}_r}{d\zeta} = C_1 Ai(\zeta) + C_2 Bi(\zeta). \qquad (3.3.108)$$

Here $Ai(\zeta)$ and $Bi(\zeta)$ are two complementary solutions of the Airy equation (see, for example, Abramowitz and Stegun, 1965). Our task now will be to find the constants C_1 and C_2. We notice, first of all, that according to (3.3.107d), the function $\breve{U}_r$ should remain finite at large ζ, which is only possible if $d\breve{U}_r/d\zeta$ tends to zero as $\zeta \to \infty$. Remember that at large values of ζ

$$Ai(\zeta) = \frac{\zeta^{-1/4}}{2\sqrt{\pi}}e^{-\frac{2}{3}\zeta^{3/2}} + \cdots, \qquad Bi(\zeta) = \frac{\zeta^{-1/4}}{\sqrt{\pi}}e^{\frac{2}{3}\zeta^{3/2}} + \cdots.$$

Therefore, if we make a branch cut in the complex α-plane along the positive imaginary semi-axis (see Figure 3.19) and, for $\alpha = |\alpha|e^{i\phi}$, define $(i\alpha)^{1/3}$ as

$$(i\alpha)^{1/3} = \big(e^{i\pi/2}|\alpha|e^{i\phi}\big)^{1/3} = |\alpha|^{1/3}e^{i(\pi/6+\phi/3)},$$

with $\phi \in \left(-\frac{3}{2}\pi, \frac{1}{2}\pi\right)$, then for any real positive $\bar{Y}$ the argument of $\zeta = (i\alpha)^{1/3}\bar{Y}$ will lie within the interval $\arg\zeta \in \left(-\frac{1}{3}\pi, \frac{1}{3}\pi\right)$, such that $Ai(\zeta)$ decays exponentially as $\zeta \to \infty$, while $Bi(\zeta)$ exhibits exponential growth for all α in the complex α-plane. Therefore, we have to set $C_2 = 0$, which reduces (3.3.108) to

$$\frac{d\breve{U}_r}{d\zeta} = C_1 Ai(\zeta). \qquad (3.3.109)$$

Substitution of (3.3.109) into (3.3.107c) yields

$$C_1 Ai'(0) = (i\alpha)^{1/3}|\alpha|\big(\breve{A}_r - \breve{F}\big). \qquad (3.3.110)$$

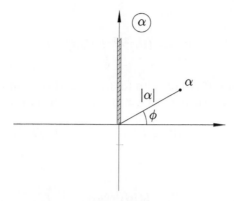

Fig. 3.19: Complex α-plane.

In order to deduce the second equation relating the two unknown constants, C_1 and $\breve{A}_r$, we integrate (3.3.109) with the initial condition (3.3.107b), to obtain

$$\breve{U}_r = C_1 \int_0^\zeta Ai(s)ds, \tag{3.3.111}$$

and, after setting $\zeta = \infty$ in (3.3.111), use the boundary condition (3.3.107d). This gives[20]

$$\breve{A}_r = \frac{1}{3}C_1. \tag{3.3.112}$$

It remains to solve the pair of equations (3.3.110) and (3.3.112) for C_1 and $\breve{A}_r$. We find that

$$C_1 = \frac{3(i\alpha)^{1/3}|\alpha|}{(i\alpha)^{1/3}|\alpha| - 3Ai'(0)}\breve{F}(\alpha), \tag{3.3.113}$$

$$\breve{A}_r = \frac{(i\alpha)^{1/3}|\alpha|}{(i\alpha)^{1/3}|\alpha| - 3Ai'(0)}\breve{F}(\alpha).$$

With C_1 known, the Fourier transform of the longitudinal velocity, $\breve{U}_r$, can be calculated with the help of equation (3.3.111). In order to find the Fourier transform of the lateral velocity, $\breve{V}_r$, one needs to integrate the continuity equation (3.3.102b). This equation is written in terms of the new variable (3.3.106) as

$$\frac{d\breve{V}_r}{d\zeta} = -(i\alpha)^{2/3}\breve{U}_r. \tag{3.3.114}$$

According to (3.3.102c), equation (3.3.114) should be integrated with the initial condition

$$\breve{V}_r = 0 \quad \text{at} \quad \zeta = 0,$$

---

[20]Remember that

$$\int_0^\infty Ai(s)\,ds = \frac{1}{3}.$$

which leads to

$$\check{V}_r = -(i\alpha)^{2/3} \int\limits_0^\zeta \check{U}_r(\zeta')\, d\zeta'. \tag{3.3.115}$$

Finally, we substitute (3.3.113) into (3.3.111) and then into (3.3.115). We see that the solution for steady perturbations can be expressed in the form

$$\check{U}_r = \check{F}(\alpha)\Phi(\zeta;\alpha), \qquad \check{V}_r = \check{F}(\alpha)\Psi(\zeta;\alpha), \tag{3.3.116}$$

where

$$\Phi(\zeta;\alpha) = \frac{3(i\alpha)^{1/3}|\alpha|}{(i\alpha)^{1/3}|\alpha| - 3Ai'(0)} \int\limits_0^\zeta Ai(s)\, ds, \tag{3.3.117}$$

$$\Psi(\zeta;\alpha) = -(i\alpha)^{2/3} \int\limits_0^\zeta \Phi(\zeta';\alpha)\, d\zeta'. \tag{3.3.118}$$

Unsteady problem

We are ready now to consider the $O(a\varepsilon)$ terms in (3.3.95) and (3.3.97). These represent the perturbations produced in the flow through the interaction of the Stokes layer with the wall roughness. To study these perturbations we have to solve the following equations in the viscous sublayer:

$$\left.\begin{aligned} \frac{\partial U'}{\partial \bar{t}} + \bar{Y}\frac{\partial U'}{\partial \bar{x}} + V' + U_s\frac{\partial U_r}{\partial \bar{x}} + V_r\frac{\partial U_s}{\partial \bar{Y}} &= -\frac{\partial P'}{\partial \bar{x}} + \frac{\partial^2 U'}{\partial \bar{Y}^2}, \\ \frac{\partial U'}{\partial \bar{x}} + \frac{\partial V'}{\partial \bar{Y}} &= 0. \end{aligned}\right\} \tag{3.3.119}$$

The boundary conditions for (3.3.119) are

$$\left.\begin{aligned} U' = V' = 0 \quad &\text{at} \quad \bar{Y} = 0, \\ U' = A' \quad &\text{at} \quad \bar{Y} = \infty, \\ U' = 0 \quad &\text{at} \quad \bar{x} = -\infty. \end{aligned}\right\} \tag{3.3.120}$$

In the upper tier, the $O(a\varepsilon)$ pressure perturbations are described by the equation

$$\frac{\partial^2 p'}{\partial \bar{x}^2} + \frac{\partial^2 p'}{\partial \bar{y}^2} = 0, \tag{3.3.121}$$

which has to be solved subject to the boundary conditions

$$\left.\begin{aligned} \frac{\partial p'}{\partial \bar{y}} = \frac{\partial^2 A'}{\partial \bar{x}^2} \quad &\text{at} \quad \bar{y} = 0, \\ p' \to 0 \quad &\text{as} \quad \bar{x}^2 + \bar{y}^2 \to \infty. \end{aligned}\right\} \tag{3.3.122}$$

Once the solution to (3.3.121), (3.3.122) is found, we can obtain the pressure in the viscous sublayer by setting

$$P' = p'\big|_{\bar{y}=0}.\tag{3.3.123}$$

The Stokes velocity profile (3.3.96) is periodic in time. It may be expressed in the form

$$U_s(\bar{t}, \bar{Y}) = e^{i\omega\bar{t}}U_s^\star(\bar{Y}) + (c.c.),\tag{3.3.124}$$

where

$$U_s^\star(\bar{Y}) = \frac{1}{2i\omega}\left[1 - e^{-(1+i)\sqrt{\omega/2}\,\bar{Y}}\right].\tag{3.3.125}$$

In view of the nature of the time dependence in (3.3.124) we shall seek the solution of the viscous sublayer problem (3.3.119), (3.3.120) in the form

$$\left.\begin{aligned}
U'(\bar{t}, \bar{x}, \bar{Y}) &= e^{i\omega\bar{t}}U^\star(\bar{x}, \bar{Y}) + (c.c.),\\
V'(\bar{t}, \bar{x}, \bar{Y}) &= e^{i\omega\bar{t}}V^\star(\bar{x}, \bar{Y}) + (c.c.),\\
P'(\bar{t}, \bar{x}) &= e^{i\omega\bar{t}}P^\star(\bar{x}) + (c.c.),\\
A'(\bar{t}, \bar{x}) &= e^{i\omega\bar{t}}A^\star(\bar{x}) + (c.c.).
\end{aligned}\right\}\tag{3.3.126}$$

Substitution of (3.3.126), (3.3.124) into (3.3.119) and (3.3.120) results in

$$\left.\begin{aligned}
i\omega U^\star + \bar{Y}\frac{\partial U^\star}{\partial \bar{x}} + V^\star + U_s^\star\frac{\partial U_r}{\partial \bar{x}} + V_r\frac{dU_s^\star}{d\bar{Y}} &= -\frac{\partial P^\star}{\partial \bar{x}} + \frac{\partial^2 U^\star}{\partial \bar{Y}^2},\\
\frac{\partial U^\star}{\partial \bar{x}} + \frac{\partial V^\star}{\partial \bar{Y}} &= 0,\\
U^\star = V^\star = 0 \quad \text{at} \quad \bar{Y} &= 0,\\
U^\star = A^\star \quad \text{at} \quad \bar{Y} &= \infty,\\
U^\star = 0 \quad \text{at} \quad \bar{x} &= -\infty.
\end{aligned}\right\}\tag{3.3.127}$$

Similarly, we represent the pressure in the upper tier in the form

$$p'(\bar{t}, \bar{x}, \bar{Y}) = e^{i\omega\bar{t}}p^\star(\bar{x}, \bar{Y}) + (c.c.).\tag{3.3.128}$$

Substitution of (3.3.128) into (3.3.121), (3.3.122) results in the following boundary-value problem for $p^\star$:

$$\frac{\partial^2 p^\star}{\partial \bar{x}^2} + \frac{\partial^2 p^\star}{\partial \bar{y}^2} = 0,$$

$$\frac{\partial p^\star}{\partial \bar{y}} = \frac{\partial^2 A^\star}{\partial \bar{x}^2} \quad \text{at} \quad \bar{y} = 0,$$

$$p^\star \to 0 \quad \text{as} \quad \bar{x}^2 + \bar{y}^2 \to \infty.$$

This problem is written in terms of Fourier transforms as

$$\left.\begin{aligned} -\alpha^2 \breve{p} + \frac{d^2 \breve{p}}{d\bar{y}^2} &= 0, \\ \frac{d\breve{p}}{d\bar{y}} &= -\alpha^2 \breve{A} \quad \text{at} \quad \bar{y} = 0, \\ \breve{p} &= 0 \qquad \text{at} \quad \bar{y} = \infty. \end{aligned}\right\} \tag{3.3.129}$$

Here $\breve{p}$, $\breve{A}$ are the Fourier transforms of $p^\star$ and $A^\star$, respectively.

The solution to (3.3.129) is constructed in the usual way, and is found to be

$$\breve{p} = |\alpha| \breve{A} e^{-|\alpha|\bar{y}}. \tag{3.3.130}$$

It remains to set $\bar{y} = 0$, as equation (3.3.123) suggests, and we see that the Fourier transform $\breve{P}$ of the pressure $P^\star$ in the viscous sublayer proves to be

$$\breve{P} = |\alpha| \breve{A}. \tag{3.3.131}$$

Now we turn our attention to the boundary-value problem (3.3.127) that describes the flow in the viscous sublayer. It is written in terms of Fourier transforms as

$$i(\omega + \alpha \bar{Y})\breve{U} + \breve{V} + i\alpha U_s^\star \breve{U}_r + \breve{V}_r \frac{dU_s^\star}{d\bar{Y}} = -i\alpha\breve{P} + \frac{d^2 \breve{U}}{d\bar{Y}^2}, \tag{3.3.132a}$$

$$i\alpha\breve{U} + \frac{d\breve{V}}{d\bar{Y}} = 0, \tag{3.3.132b}$$

$$\breve{U} = \breve{V} = 0 \quad \text{at} \quad \bar{Y} = 0, \tag{3.3.132c}$$

$$\breve{U} = \breve{A} \qquad \text{at} \quad \bar{Y} = \infty. \tag{3.3.132d}$$

Here $\breve{U}$, $\breve{V}$ denote the Fourier transforms of $U^\star$ and $V^\star$. As usual, we eliminate $\breve{V}$ from (3.3.132a) and (3.3.132b) by differentiating (3.3.132a) with respect to $\bar{Y}$ and using (3.3.102b) and (3.3.132b). We find that

$$i(\omega + \alpha \bar{Y})\frac{d\breve{U}}{d\bar{Y}} + i\alpha U_s^\star \frac{d\breve{U}_r}{d\bar{Y}} + \breve{V}_r \frac{d^2 U_s^\star}{d\bar{Y}^2} = \frac{d^3 \breve{U}}{d\bar{Y}^3}. \tag{3.3.133}$$

Since equation (3.3.133) is a third order differential equation, it requires an additional boundary condition. The latter may be obtained by setting $\bar{Y} = 0$ in (3.3.132a), and using the no-slip conditions for the steady and unsteady perturbations; these are given by (3.3.102c) and (3.3.132c), respectively. We find

$$\left.\frac{d^2 \breve{U}}{d\bar{Y}^2}\right|_{\bar{Y}=0} = i\alpha\breve{P} = i\alpha|\alpha|\breve{A}. \tag{3.3.134}$$

By introducing a new independent variable

$$z = z_0 + (i\alpha)^{1/3}\bar{Y}, \qquad z_0 = \frac{i\omega}{(i\alpha)^{2/3}}, \tag{3.3.135}$$

we can transform equation (3.3.133) into the inhomogeneous Airy equation

$$\frac{d^3 \breve{U}}{dz^3} - z\frac{d\breve{U}}{dz} = \breve{F}(\alpha)H(z), \tag{3.3.136a}$$

where the function $H(z)$ on the right-hand side of (3.3.136a) is calculated with the help of (3.3.116) and (3.3.125), and is found to be

$$H(z) = (i\alpha)^{1/3}U_s^\star\frac{d\Phi}{d\zeta} + (i\alpha)^{-1}\Psi\frac{d^2 U_s^\star}{d\breve{Y}^2}. \tag{3.3.136b}$$

The boundary conditions (3.3.132c), (3.3.132d), and (3.3.134) can be written in terms of the new variable z as

$$\breve{U} = 0 \qquad \text{at} \quad z = z_0, \tag{3.3.136c}$$

$$\frac{d^2 \breve{U}}{dz^2} = (i\alpha)^{1/3}|\alpha|\breve{A} \quad \text{at} \quad z = z_0, \tag{3.3.136d}$$

$$\breve{U} = \breve{A} \qquad \text{at} \quad z = \infty. \tag{3.3.136e}$$

The general solution of (3.3.136a) for $d\breve{U}/dz$ is a composition of the two complementary solutions of the Airy equation, $Ai(z)$ and $Bi(z)$, and a particular integral, $\breve{F}(\alpha)\varphi(z)$:

$$\frac{d\breve{U}}{dz} = C_1 Ai(z) + C_2 Bi(z) + \breve{F}(\alpha)\varphi(z). \tag{3.3.137}$$

Here we choose $\varphi(z)$ to be the solution to the following boundary-value problem:

$$\frac{d^2\varphi}{dz^2} - z\varphi = H(z), \tag{3.3.138a}$$

$$\frac{d\varphi}{dz} = 0 \quad \text{at} \quad z = z_0, \tag{3.3.138b}$$

$$\varphi = 0 \quad \text{at} \quad z = \infty. \tag{3.3.138c}$$

To avoid exponential growth of $d\breve{U}/dz$ as $z \to \infty$, we set $C_2 = 0$ in (3.3.137). Thus, we have

$$\frac{d\breve{U}}{dz} = C_1 Ai(z) + \breve{F}(\alpha)\varphi(z). \tag{3.3.139}$$

Then, applying boundary condition (3.3.136d), we see that

$$C_1 Ai'(z_0) = (i\alpha)^{1/3}|\alpha|\breve{A} \tag{3.3.140}$$

This is the first relation between C_1 and $\breve{A}$. To deduce a second relation, we integrate (3.3.139) with the initial condition (3.3.136c) to obtain

$$\breve{U} = C_1 \int_{z_0}^{z} Ai(s)\, ds + \breve{F}(\alpha)\int_{z_0}^{z} \varphi(s)\, ds,$$

and apply condition (3.3.136e), which yields the second relation between C_1 and $\breve{A}$:

$$C_1 \int\limits_{z_0}^{\infty} Ai(s)\,ds + \breve{F}(\alpha) \int\limits_{z_0}^{\infty} \varphi(s)\,ds = \breve{A}. \qquad (3.3.141)$$

It remains to eliminate C_1 from (3.3.140) and (3.3.141), and we can conclude that

$$\breve{A} = \frac{Ai'(z_0)\breve{F}(\alpha) \int\limits_{z_0}^{\infty} \varphi(z)\,dz}{Ai'(z_0) - (i\alpha)^{1/3}|\alpha| \int\limits_{z_0}^{\infty} Ai(z)\,dz}. \qquad (3.3.142)$$

Finally, we substitute (3.3.142) into (3.3.131). We find that the Fourier transform of the pressure in the lower tier (region 3 in Figure 3.18) is given by

$$\breve{P} = \frac{Ai'(z_0)|\alpha|\breve{F}(\alpha) \int\limits_{z_0}^{\infty} \varphi(z)\,dz}{Ai'(z_0) - (i\alpha)^{1/3}|\alpha| \int\limits_{z_0}^{\infty} Ai(z)\,dz}. \qquad (3.3.143)$$

To return to physical variables, one needs to apply the inverse Fourier transform to (3.3.143) and substitute the result into the equation for $P'(\bar{t}, \bar{x})$ in (3.3.126). This yields

$$P'(\bar{t}, \bar{x}) = \Re\left\{ \frac{e^{i\omega\bar{t}}}{\pi} \int\limits_{-\infty}^{\infty} \frac{Ai'(z_0)|\alpha| \int\limits_{z_0}^{\infty} \varphi(z)\,dz}{Ai'(z_0) - (i\alpha)^{1/3}|\alpha| \int\limits_{z_0}^{\infty} Ai(z)\,dz} \breve{F}(\alpha) e^{i\alpha\bar{x}}\,d\alpha \right\}, \qquad (3.3.144)$$

where $\Re$ stands for the real part of the expression in the curly brackets.

3.3.7   Receptivity coefficient

Equation (3.3.144) allows us to determine the distribution of the pressure perturbations $P'(\bar{t}, \bar{x})$ in the entire vicinity of the roughness. However, our main interest is in the behaviour of the perturbations behind the roughness, where a Tollmien–Schlichting wave is expected to form. To study this process, we shall repeat the procedure that was used in Section 3.1.6. We split the integral in (3.3.144) into two, one part along the negative real semi-axis C_-, and the other along the positive real semi-axis C_+, as shown in Figure 3.5; for convenience, we reproduce this figure on the next page and refer to it now as Figure 3.20. We have

$$P'(\bar{t}, \bar{x}) = \Re\left\{ \frac{e^{i\omega\bar{t}}}{\pi} \big[\mathcal{N}(\bar{x}; \omega) + \mathcal{M}(\bar{x}; \omega) \big] \right\}, \qquad (3.3.145)$$

where

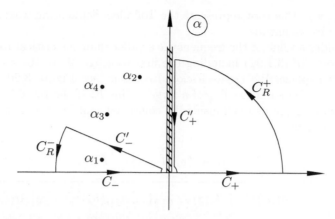

Fig. 3.20: Deformation of the contour of integration in (3.3.146) and (3.3.147).

$$\mathcal{N}(\bar{x};\omega) = -\int_{-\infty}^{0} \frac{\alpha Ai'(z_0) \int_{z_0}^{\infty} \varphi(z)\,dz}{Ai'(z_0) + (i\alpha)^{1/3}\alpha \int_{z_0}^{\infty} Ai(z)\,dz} \check{F}(\alpha)e^{i\alpha\bar{x}}\,d\alpha, \qquad (3.3.146)$$

and

$$\mathcal{M}(\bar{x};\omega) = \int_{0}^{\infty} \frac{\alpha Ai'(z_0) \int_{z_0}^{\infty} \varphi(z)\,dz}{Ai'(z_0) - (i\alpha)^{1/3}\alpha \int_{z_0}^{\infty} Ai(z)\,dz} \check{F}(\alpha)e^{i\alpha\bar{x}}\,d\alpha. \qquad (3.3.147)$$

We first examine the function $\mathcal{N}(\bar{x};\omega)$. We wish to determine its asymptotic behaviour at large values of $\bar{x}$. To perform this task, we consider an analytic extension of the integrand in (3.3.146) into the second quadrant of the complex α-plane, with the intention of deforming the integration path. Of course, when doing this one needs to be mindful of the poles of the integrand. Notice that these are defined by the same dispersion equation (3.1.92) as in Terent'ev's problem:

$$(i\alpha)^{1/3}|\alpha| \int_{z_0}^{\infty} Ai(z)\,dz - Ai'(z_0) = 0, \qquad z_0 = \frac{i\omega}{(i\alpha)^{2/3}}. \qquad (3.3.148)$$

Remember that this dispersion equation has an infinite (countable) number of roots. These are represented by the points α_1, α_2, ... in Figure 3.20. The position of each root depends on the frequency ω. The trajectories of the first five roots, as ω varies from zero to infinity, are shown in Figure 2.3. All the roots originate at $\omega = 0$ from the coordinate origin, and all of them, except the first one, remain in the second quadrant for all $\omega \in (0,\infty)$. The behaviour of the first root is different. It stays in the second quadrant until the frequency reaches its critical value, $\omega_* \simeq 2.29797$, and then it crosses the real axis at the point $\alpha_* \simeq -1.00049$ and remains in the third quadrant for all $\omega \subset (\omega_*, \infty)$. This root represents the Tollmien–Schlichting wave, and our task is to determine its amplitude.

Let us consider a value of the frequency ω smaller than the critical frequency, ω_*.[21] Then all the roots of (3.1.92) including the first root, α_1, lie in the second quadrant of the complex α-plane. We create a closed contour (see Figure 3.20) by adding to C_- a ray C'_- and a circular arc C_R^- of a large radius R. If the ray C'_- is chosen such that only one root, α_1, finds itself inside the combined contour, then, using the residue theorem, we will have

$$\mathcal{N}(\bar{x}; \omega) = -2\pi i \frac{\left[\alpha Ai'(z_0) \int\limits_{z_0}^{\infty} \varphi(z)\, dz\right] \check{F}(\alpha) e^{i\alpha\bar{x}}}{\frac{4}{3}(i\alpha)^{1/3} \int\limits_{z_0}^{\infty} Ai(z)\, dz - \frac{2}{3}(z_0/\alpha)\left[i(i\alpha)^{4/3} + z_0\right] Ai(z_0)}\Bigg|_{\alpha=\alpha_1}$$

$$+ \int\limits_{C'_-} \frac{\alpha Ai'(z_0) \int\limits_{z_0}^{\infty} \varphi(z)\, dz}{Ai'(z_0) + (i\alpha)^{1/3}\alpha \int\limits_{z_0}^{\infty} Ai(z)\, dz} \check{F}(\alpha) e^{i\alpha\bar{x}}\, d\alpha. \qquad (3.3.149)$$

Here it is taken into account that, according to Jordan's lemma, the integral along C_R^- tends to zero as $R \to \infty$. The asymptotic behaviour of the integral along C'_- may be determined with the help of Watson's lemma.[22] To apply this lemma, we need to know the behaviour of the integrand in (3.3.149) in the limit

$$\omega = O(1), \qquad \alpha \to 0. \qquad (3.3.150)$$

Let us consider the integral

$$\int\limits_{z_0}^{\infty} \varphi(z)\, dz. \qquad (3.3.151)$$

Remember that the function $\varphi(z)$ is defined by the boundary-value problem (3.3.138). The asymptotic solution of this problem may be found with the help of the method of matched asymptotic expansions.[23] It is easily seen that in the limit (3.3.150), two regions should be considered. The first one (we shall call it the inner region) occupies a small vicinity of the lower limit z_0 of the integration in (3.3.151). The size Δz of this region is obtained by balancing the two terms on the left-hand side of (3.3.138a). We have

$$\frac{\varphi}{(\Delta z)^2} \sim z\varphi. \qquad (3.3.152)$$

Here, the factor z in the second term may be substituted by z_0. Indeed, we know that

$$z_0 = \frac{i\omega}{(i\alpha)^{2/3}}, \qquad (3.3.153)$$

[21] This restriction may be lifted in the same way as in Terent'ev's problem; see Section 3.2.

[22] See Section 1.2.2 in Part 2 of this book series.

[23] See Section 1.4 in Part 2 of this book series.

which is large in the limit (3.3.150). Meanwhile Δz proves to be small, as is easily shown by substituting (3.3.153) into (3.3.152) and solving the resulting equation for Δz. We find that

$$\Delta z = z - z_0 \sim \alpha^{1/3}.$$

It then follows from the first equation in (3.3.135) that $\bar{Y}$ is an order one quantity in the inner region. It further follows from (3.3.106) that $\zeta \sim \alpha^{1/3}$. Keeping this in mind and using (3.3.117), (3.3.118), and (3.3.125) in (3.3.136b), we find that H, the right-hand side of (3.3.138a), scales with α as

$$H \sim \alpha^{5/3}.$$

An estimate for the function φ may then be obtained by balancing either term on the left-hand side of equation (3.3.138a) with the function H on its right-hand side:

$$\varphi \sim \frac{H}{z_0} \sim \alpha^{7/3}.$$

Hence, the contribution of the inner region into the integral (3.3.151) is estimated as

$$\int \varphi(z)\, dz \sim \varphi \Delta z = O(\alpha^{8/3}). \tag{3.3.154}$$

The size of the second (outer) region,

$$|z - z_0| \sim O(1),$$

is dictated by the form of the function H. In this region, $\bar{Y} = (z - z_0)/(i\alpha)^{1/3}$ is large, and it follows from (3.3.125) that U_s^* in the first term in (3.3.136b) may be substituted by $1/(2i\omega)$ while the second term is transcendentally small. We can also ignore $(i\alpha)^{1/3}|\alpha|$ in the denominator in (3.3.117). This reduces (3.3.136b) to

$$H = -\frac{(i\alpha)^{2/3}|\alpha|}{2i\omega\, Ai'(0)} Ai(\zeta).$$

In the outer region, the first term on the left-hand side of equation (3.3.138a) may be disregarded and the factor z in the second term may be substituted by z_0. This allows us to write the leading-order term of the asymptotic solution to (3.3.138a) as

$$\varphi = -\frac{H}{z_0} = -\frac{(i\alpha)^{4/3}|\alpha|}{2\omega^2\, Ai'(0)} Ai(\zeta). \tag{3.3.155}$$

It remains to substitute (3.3.155) into (3.3.151), and we can conclude that[24]

$$\int_{z_0}^{\infty} \varphi(z)\, dz = \frac{(i\alpha)^{4/3}|\alpha|}{6\omega^2\, Ai'(0)} \quad | \cdots \quad \text{as} \quad \alpha \to 0. \tag{3.3.156}$$

[24] Here we take into account that $z = z_0 + \zeta$, and therefore,

$$\int_{z_0}^{\infty} Ai(\zeta)\, dz = \int_{0}^{\infty} Ai(\zeta)\, d\zeta = \frac{1}{3}.$$

Notice that, in the leading-order approximation, the contribution (3.3.154) of the inner region may be disregarded.

Let us now return to the integral along C'_- in (3.3.149). According to Watson's lemma, at large values of $\bar{x}$ the 'region of dominant contribution' for this integral is represented by a small section of C'_- lying in a vicinity of the point $\alpha = 0$; see Figure 3.20. In this vicinity the integrand in (3.3.149) can be simplified with the help of inequality (3.1.97). We have

$$\frac{\alpha Ai'(z_0) \int\limits_{z_0}^{\infty} \varphi(z)\,dz}{Ai'(z_0) + (i\alpha)^{1/3}\alpha \int\limits_{z_0}^{\infty} Ai(z)\,dz} \breve{F}(\alpha)e^{i\alpha\bar{x}} = \breve{F}(0)\alpha e^{i\alpha\bar{x}}\int\limits_{z_0}^{\infty}\varphi(z)\,dz. \qquad (3.3.157)$$

Since we are dealing with an analytic extension of the integrand from the real negative semi-axis, we have to set $|\alpha| = -\alpha$ in (3.3.156), which renders the right-hand side of (3.3.157) in the form

$$\breve{F}(0)\frac{(i\alpha)^{4/3}\alpha^2}{6\omega^2 Ai'(0)}e^{i\alpha\bar{x}}. \qquad (3.3.158)$$

Approximating the integrand in (3.3.149) with (3.3.158), we find that (see Problem 3 in Exercises 12)

$$\int\limits_{C'_-}\frac{\alpha Ai'(z_0)\int\limits_{z_0}^{\infty}\varphi(z)\,dz}{Ai'(z_0)+(i\alpha)^{1/3}\alpha\int\limits_{z_0}^{\infty}Ai(z)\,dz}\breve{F}(\alpha)e^{i\alpha\bar{x}}\,d\alpha$$

$$=\frac{\breve{F}(0)}{6\omega^2 Ai'(0)}\int\limits_{C'_-}(i\alpha)^{4/3}\alpha^2 e^{i\alpha\bar{x}}\,d\alpha = \frac{\breve{F}(0)e^{i\pi/6}}{6\omega^2 Ai'(0)}\frac{\Gamma(13/3)}{\bar{x}^{13/3}}+\cdots \qquad (3.3.159)$$

as $\bar{x}\to\infty$, where Γ stands for the gamma function.

This completes the analysis of the first integral (3.3.146) in (3.3.145). The second integral (3.3.147) is analysed in the same way, except there are no roots of the dispersion equation (3.3.148) in the first quadrant of the complex α-plane. Therefore, the integration along the real positive semi-axis C_+ can be simply replaced by the integration along the right-hand side edge C'_+ of the branch cut in Figure 3.20. The integral along C'_+ is a Laplace type integral. Its asymptotic behaviour at large values of $\bar{x}$ may be determined by once again using Watson's lemma. We find that

$$\mathcal{M}(\bar{x};\omega) = \frac{\breve{F}(0)e^{-i\pi/6}}{6\omega^2 Ai'(0)}\frac{\Gamma(13/3)}{\bar{x}^{13/3}}+\cdots \quad \text{as} \quad \bar{x}\to\infty. \qquad (3.3.160)$$

It remains to substitute (3.3.159) into (3.3.149) and then, together with (3.3.160), into (3.3.145). We see that downstream of the roughness

$$P'(\bar{t},\bar{x}) = \Re\left\{\mathcal{K}(\omega)\breve{F}(\alpha_1)e^{i(\omega\bar{t}+\alpha_1\bar{x})} + \frac{\Gamma(13/3)}{2\sqrt{3}\,\omega^2 Ai'(0)}\frac{\breve{F}(0)e^{i\omega\bar{t}}}{\pi\bar{x}^{13/3}}+\cdots\right\}, \qquad (3.3.161)$$

where

$$\mathcal{K}(\omega) = -\frac{2i\alpha Ai'(z_0)\int\limits_{z_0}^{\infty}\varphi(z)\,dz}{\frac{4}{3}(i\alpha)^{1/3}\int\limits_{z_0}^{\infty}Ai(z)\,dz - \frac{2}{3}(z_0/\alpha)\big[i(i\alpha)^{4/3}+z_0\big]Ai(z_0)}\bigg|_{\alpha=\alpha_1} \qquad (3.3.162)$$

is the *receptivity coefficient*. It should be noted that $\mathcal{K}(\omega)$ is a function of ω only, being independent of the wall roughness shape.

The numerical calculations of $\mathcal{K}(\omega)$ are performed in the following way. First, for a given real positive value of ω, the first root α_1 of the dispersion equation (3.3.148) is found as described in Section 2.3.3. Then we turn to the boundary-value problem (3.3.138). To find the function $\varphi(z)$, we consider an interval $\bar{Y} \in [0, \bar{Y}_{\max}]$ with a sufficiently large value of $\bar{Y}_{\max}$. This interval is partitioned using the mesh $\{\bar{Y}_j\}$ with uniform spacing $\Delta\bar{Y}$, such that

$$\bar{Y}_j = j\Delta\bar{Y}, \qquad j = 0, 1, 2\ldots N.$$

The corresponding points in the complex ζ-plane are

$$\zeta_j = (i\alpha)^{1/3}\bar{Y}_j.$$

At each ζ_j, we calculate $d\Phi/d\zeta$ and Ψ using (3.3.117) and (3.3.118), respectively. The results of these calculations are used, together with (3.3.125), to determine the values H_j of the function (3.3.136b) at the mesh points. Then equation (3.3.138a) is written in finite-difference form as

$$\frac{\varphi_{j+1} - 2\varphi_j + \varphi_{j-1}}{(\Delta z)^2} - z_j\varphi_j = H_j, \qquad (3.3.163)$$

where $\Delta z = (i\alpha)^{1/3}\Delta Y$ and $z_j = z_0 + \zeta_j$. The solution to (3.3.163), satisfying boundary conditions (3.3.138b) and (3.3.138c), is easily found with the help of the Thomas

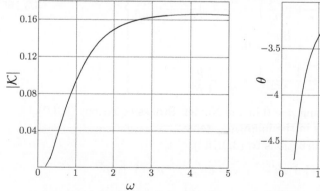
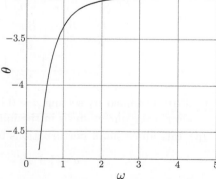

(a) Modulus of the receptivity coefficient.   (b) Argument of the receptivity coefficient.

Fig. 3.21: The receptivity coefficient $\mathcal{K} = |\mathcal{K}|e^{i\theta}$ as a function of the frequency ω.

algorithm.[25] Once this is done, the receptivity coefficient $\mathcal{K}(\omega)$ is calculated directly from (3.3.162). The results of the numerical calculations of (3.3.162) are displayed in Figure 3.21, where the modulus and the argument of $\mathcal{K}(\omega)$ are shown.

Equation (3.3.161) is interpreted in the same way as equation (3.1.100) in Terent'ev's problem. It is easily seen that the second term on the right-hand side of (3.3.161) decays with $\bar{x}$ uniformly with respect to ω, provided that $\omega \geq M$, where M is a positive constant. The first term in (3.3.161) represents the generated Tollmien–Schlichting wave. It decays for subcritical values of the frequency ($\omega < \omega_*$), becomes neutral when ω reaches its critical value ω_* and grows when $\omega > \omega_*$. The amplitude of the Tollmien–Schlichting wave, $\mathcal{K}(\omega)\breve{F}(\alpha_1)$, is proportional to the Fourier transform $\breve{F}(\alpha_1)$ of the wall roughness shape function $F(\bar{x})$ taken at the first root, $\alpha_1(\omega)$, of the dispersion equation (3.3.148).

Exercises 12

1. Consider uniform flow of a perfect gas with the velocity components (u, v), density ρ, pressure p, and enthalpy h being

$$u = 1, \qquad v = 0, \qquad \rho = 1, \qquad p = 0, \qquad h = \frac{1}{(\gamma - 1)M_\infty^2}. \qquad (3.3.164)$$

Here we use Cartesian coordinates with the x-axis directed parallel to the fluid flow.

Your task is to demonstrate that any small perturbation to (3.3.164) may be represented as a superposition of *acoustic waves*, *vorticity waves*, and *entropy waves*. For simplicity you may assume that the perturbations are two-dimensional and described by the Euler equations:

$$\left.\begin{aligned}
\rho\left(\frac{\partial u}{\partial t} + u\frac{\partial u}{\partial x} + v\frac{\partial u}{\partial y}\right) &= -\frac{\partial p}{\partial x}, \\
\rho\left(\frac{\partial v}{\partial t} + u\frac{\partial v}{\partial x} + v\frac{\partial v}{\partial y}\right) &= -\frac{\partial p}{\partial y}, \\
\rho\left(\frac{\partial h}{\partial t} + u\frac{\partial h}{\partial x} + v\frac{\partial h}{\partial y}\right) &= \frac{\partial p}{\partial t} + u\frac{\partial p}{\partial x} + v\frac{\partial p}{\partial y}, \\
\frac{\partial \rho}{\partial t} + \frac{\partial \rho u}{\partial x} + \frac{\partial \rho v}{\partial y} &= 0, \\
h = \frac{1}{(\gamma - 1)M_\infty^2}\frac{1}{\rho} &+ \frac{\gamma}{\gamma - 1}\frac{p}{\rho},
\end{aligned}\right\} \qquad (3.3.165)$$

that are obtained by setting $\mu = 0$ in the Navier–Stokes equations (3.3.2).

Perform your analysis in the following steps:

(a) Introduce small perturbations to (3.3.164)

$$u = 1 + \varepsilon u', \qquad v = \varepsilon v', \qquad \rho = 1 + \varepsilon \rho',$$

$$p = \varepsilon p', \qquad h = \frac{1}{(\gamma - 1)M_\infty^2} + \varepsilon h',$$

---

[25] See Section 1.9.1 in Part 3 of this book series.

and show that for $\varepsilon \ll 1$ the Euler equations (3.3.165) become

$$\frac{\partial u'}{\partial t} + \frac{\partial u'}{\partial x} = -\frac{\partial p'}{\partial x}, \tag{3.3.166a}$$

$$\frac{\partial v'}{\partial t} + \frac{\partial v'}{\partial x} = -\frac{\partial p'}{\partial y}, \tag{3.3.166b}$$

$$\frac{\partial h'}{\partial t} + \frac{\partial h'}{\partial x} = \frac{\partial p'}{\partial t} + \frac{\partial p'}{\partial x}, \tag{3.3.166c}$$

$$\frac{\partial \rho'}{\partial t} + \frac{\partial \rho'}{\partial x} + \frac{\partial u'}{\partial x} + \frac{\partial v'}{\partial y} = 0, \tag{3.3.166d}$$

$$h' = \frac{\gamma}{\gamma - 1} p' - \frac{1}{(\gamma - 1)M_\infty^2} \rho'. \tag{3.3.166e}$$

(b) It is convenient to eliminate h' from (3.3.166) which is achieved by substituting the state equation (3.3.166e) into the energy equation (3.3.166c). Show that this results in

$$\frac{\partial \rho'}{\partial t} + \frac{\partial \rho'}{\partial x} = M_\infty^2 \left(\frac{\partial p'}{\partial t} + \frac{\partial p'}{\partial x} \right). \tag{3.3.167}$$

(c) We seek the perturbations in the simple wave form:

$$\left. \begin{array}{ll} u' = u_a e^{i(\omega t + \alpha x + \beta y)} + (c.c.), & v' = v_a e^{i(\omega t + \alpha x + \beta y)} + (c.c.), \\ p' = p_a e^{i(\omega t + \alpha x + \beta y)} + (c.c.), & \rho' = \rho_a e^{i(\omega t + \alpha x + \beta y)} + (c.c.), \end{array} \right\} \tag{3.3.168}$$

where u_a, v_a, p_a, and ρ_a are constants representing the wave amplitude. The frequency ω and the wave numbers (α, β) are assumed real.

By substituting (3.3.168) into (3.3.166a), (3.3.166b), (3.3.167), and (3.3.166d) deduce the following set of linear homogenous equations for u_a, v_a, p_a, ρ_a:

$$\left. \begin{array}{l} (\omega + \alpha)u_a + \alpha p_a = 0, \\ (\omega + \alpha)v_a + \beta p_a = 0, \\ M_\infty^2(\omega + \alpha)p_a - (\omega + \alpha)\rho_a = 0, \\ (\omega + \alpha)\rho_a + \alpha u_a + \beta v_a = 0. \end{array} \right\} \tag{3.3.169}$$

(d) Calculate the determinant

$$\begin{vmatrix} \omega + \alpha & 0 & \alpha & 0 \\ 0 & \omega + \alpha & \beta & 0 \\ 0 & 0 & M_\infty^2(\omega + \alpha) & -(\omega + \alpha) \\ \alpha & \beta & 0 & \omega + \alpha \end{vmatrix}$$

of the above set of equations, and deduce that a non-trivial solution to (3.3.169) exists only if

$$(\omega + \alpha)^2 \left[M_\infty^2(\omega + \alpha)^2 - (\alpha^2 + \beta^2) \right] = 0. \tag{3.3.170}$$

Confirm that the four roots of (3.3.170) are

$$\omega_{1,2} = -\alpha, \qquad \omega_{3,4} = -\alpha \pm \frac{\sqrt{\alpha^2 + \beta^2}}{M_\infty},$$

with the corresponding eigen-vectors

$$\begin{pmatrix} 1 \\ -\frac{\alpha}{\beta} \\ 0 \\ 0 \end{pmatrix} u_a, \qquad \begin{pmatrix} 0 \\ 0 \\ 0 \\ 1 \end{pmatrix} \rho_a, \qquad \begin{pmatrix} \frac{\alpha M_\infty}{\sqrt{\alpha^2+\beta^2}} \\ \frac{\beta M_\infty}{\sqrt{\alpha^2+\beta^2}} \\ 1 \\ M_\infty^2 \end{pmatrix} p_a, \qquad \begin{pmatrix} -\frac{\alpha M_\infty}{\sqrt{\alpha^2+\beta^2}} \\ -\frac{\beta M_\infty}{\sqrt{\alpha^2+\beta^2}} \\ 1 \\ M_\infty^2 \end{pmatrix} p_a.$$

These are, in turn, the vorticity wave, the entropy wave, and the two acoustic waves, one propagating downstream with respect to the moving gas, and another propagating upstream.

(e) Calculate the vorticity $\omega = \partial v/\partial x - \partial u/\partial y$ for the first of these modes.

2. Return to the flow in the Stokes layer that lies on the plate surface upstream of the roughness; see Figure 3.22.

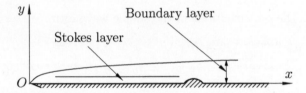

Fig. 3.22: Stokes layer formation in the boundary layer.

You may use without proof the fact that in this layer the fluid-dynamic functions are represented by the asymptotic expansions (3.3.37), (3.3.39)–(3.3.41), namely,

$$\left.\begin{aligned}
u(t,x,y;Re) &= Re^{-1/8}U'_*(t_*,x',Y_*;x) + \cdots, \\
v(t,x,y;Re) &= Re^{-1/2}V'_*(t_*,x',Y_*;x) + \cdots, \\
p(t,x,y;Re) &= Re^{-1/8}P'_*(t_*,x',Y_*;x) + \cdots, \\
h(t,x,y;Re) &= h_0(x,0) + Re^{-1/8}h'_*(t_*,x',Y_*;x) + \cdots, \\
\rho(t,x,y;Re) &= \rho_0(x,0) + Re^{-1/8}\rho'_*(t_*,x',Y_*;x) + \cdots, \\
\mu(t,x,y;Re) &= \mu_0(x,0) + Re^{-1/8}\mu'_*(t_*,x',Y_*;x) + \cdots,
\end{aligned}\right\} \qquad (3.3.171)$$

where the independent variables t_*, x', and Y_* are introduced through the scalings

$$t = Re^{-1/4}t_*, \qquad x = Re^{-1/4}x', \qquad y = Re^{-5/8}Y_*. \qquad (3.3.172)$$

You can also use without proof equation (3.3.43) for the pressure in the Stokes layer:

$$P'_* = a_* \sin(\omega_* \xi), \qquad \xi = t_* - \frac{M_\infty}{1 + M_\infty} x'. \tag{3.3.173}$$

Your task is to determine the enthalpy perturbations in the Stokes layer. You may perform this task in the following steps:

(a) By substituting (3.3.171)–(3.3.173) into the energy equation

$$\rho \left(\frac{\partial h}{\partial t} + u \frac{\partial h}{\partial x} + v \frac{\partial h}{\partial y} \right) = \frac{\partial p}{\partial t} + u \frac{\partial p}{\partial x} + v \frac{\partial p}{\partial y}$$

$$+ \frac{1}{Re} \left\{ \frac{1}{Pr} \left[\frac{\partial}{\partial x} \left(\mu \frac{\partial h}{\partial x} \right) + \frac{\partial}{\partial y} \left(\mu \frac{\partial h}{\partial y} \right) \right] \right.$$

$$\left. + \mu \left(\frac{4}{3} \frac{\partial u}{\partial x} - \frac{2}{3} \frac{\partial v}{\partial y} \right) \frac{\partial u}{\partial x} + \mu \left(\frac{4}{3} \frac{\partial v}{\partial y} - \frac{2}{3} \frac{\partial u}{\partial x} \right) \frac{\partial v}{\partial y} + \mu \left(\frac{\partial u}{\partial y} + \frac{\partial v}{\partial x} \right)^2 \right\}$$

show that the function h'_* satisfies the equation

$$\rho_0(x,0) \frac{\partial h'_*}{\partial t_*} = a_* \omega_* \cos(\omega_* \xi) + \frac{\mu_0(x,0)}{Pr} \frac{\partial^2 h'_*}{\partial Y_*^2}. \tag{3.3.174}$$

(b) Assume that h'_* is periodic in ξ, and seek the solution to (3.3.174) in the form

$$h'_*(t_*, x', Y_*; x) = e^{i\omega_* \xi} H(Y_*; x) + (c.c.).$$

Show that the function $H(Y_*; x)$, where x serves the role of a parameter, satisfies the following ordinary differential equation:

$$i\omega_* \rho_0(x,0) H = \frac{1}{2} a_* \omega_* + \frac{\mu_0(x,0)}{Pr} \frac{d^2 H}{dY_*^2}. \tag{3.3.175}$$

(c) Find the solution to (3.3.175) for (i) a thermally isolated wall, and (ii) in the case when the wall temperature is kept unperturbed, that is $H = 0$ at $Y_* = 0$.

    Hint: Keep in mind that the matching with the solution in the main part of the boundary layer is only possible if H does not grow exponentially when $Y_* \to \infty$.

3. Let C'_- be a semi-infinite ray that originates from the coordinate origin and lies in the second quadrant of the complex α-plane; see Figure 3.23. Your task is to calculate the integral

$$\int_{C'_-} (i\alpha)^{4/3} \alpha^2 e^{i\alpha \bar{x}} \, d\alpha. \tag{3.3.176}$$

You may use without proof the following formula:

$$\int_0^\infty t^b e^{-at} \, dt = \frac{\Gamma(b+1)}{a^{b+1}}. \tag{3.3.177}$$

In (3.3.177), the integration variable t is real, as are constants a and b. Of course, for the integral on the left-hand side of (3.3.177) to converge, the constant a should be positive.

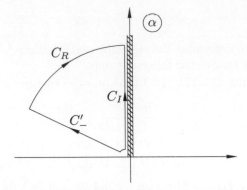

Fig. 3.23: Deformation of the contour of integration in (3.3.176).

Suggestion: Create a closed contour by adding to C'_- a ray C_I drawn along the left-hand side of the branch cut and a circular arc C_R connecting C'_- and C_I. Argue that for any $\bar{x} > 0$, the integral along C_R satisfies Jordan's lemma, and may be disregarded when the arc's radius R tends to infinity. When dealing with the integral along C_I, you have to set $\alpha = e^{-i3\pi/2}t$.

4. In this chapter, we have been dealing with the stability of subsonic flows. In fact, in Sections 3.1 and 3.2 we simply assumed the flow incompressible which corresponds to $M_\infty = 0$. Then, in Section 3.3, the theory was extended to compressible flows with $M_\infty < 1$, and the asymptotic analysis of the Navier–Stokes equations was performed assuming the parameter $\beta = \sqrt{1 - M_\infty^2}$ to be an order one quantity. If one is interested in the transonic flow regime, then the asymptotic procedure should be re-examined to see when the governing asymptotic equations start to change as $\beta \to 0$. According to Timoshin (1990) and Bowles and Smith (1993), this first happens in the upper tier of the triple-deck structure where the continuity equation

$$\frac{\partial \rho}{\partial t} + \frac{\partial \rho u}{\partial x} + \frac{\partial \rho v}{\partial y} = 0 \qquad (3.3.178)$$

can no longer be treated as quasi-steady if

$$M_\infty = 1 + Re^{-1/9}M_1, \qquad (3.3.179)$$

with M_1 being an order one parameter, positive or negative.

Your task is to confirm this result. You may use without proof the fact that in subsonic flow, the fluid-dynamic functions are represented in the upper tier as

$$\left.\begin{array}{ll} u = 1 + Re^{-1/4}\dfrac{\mu_w^{1/2}}{\lambda^{-1/2}\beta^{1/2}}\bar{u}, & v = Re^{-1/4}\dfrac{\mu_w^{1/2}}{\lambda^{-1/2}\beta^{-1/2}}\bar{v}, \\[3mm] p = Re^{-1/4}\dfrac{\mu_w^{1/2}}{\lambda^{-1/2}\beta^{1/2}}\bar{p}, & \rho = 1 + Re^{-1/4}\dfrac{\mu_w^{1/2}}{\lambda^{-1/2}\beta^{1/2}}\bar{\rho}, \\[3mm] \multicolumn{2}{c}{h = \dfrac{1}{(\gamma - 1)M_\infty^2} + Re^{-1/4}\dfrac{\mu_w^{1/2}}{\lambda^{-1/2}\beta^{1/2}}\bar{h},} \end{array}\right\} \qquad (3.3.180)$$

with the independent variables being

$$t = Re^{-1/4} \frac{\mu_w^{-1/2}}{\lambda^{3/2}\beta^{1/2}}\bar{t}, \qquad x = 1 + Re^{-3/8} \frac{\mu_w^{-1/4}\rho_w^{-1/2}}{\lambda^{5/4}\beta^{3/4}}\bar{x},$$
$$y = Re^{-3/8} \frac{\mu_w^{-1/4}\rho_w^{-1/2}}{\lambda^{5/4}\beta^{7/4}}\bar{y}. \tag{3.3.181}$$

We suggest you perform your analysis in the following steps:

(a) Start with the longitudinal momentum equation (3.3.2a). Using (3.3.180) and (3.3.181) show that for all $\beta \gg Re^{-1/2}$

$$\frac{\partial u}{\partial t} \ll u\frac{\partial u}{\partial x}.$$

Hence, conclude that the longitudinal momentum equation stays quasi-steady, and assumes the form

$$\frac{\partial \bar{u}}{\partial \bar{x}} = -\frac{\partial \bar{p}}{\partial \bar{x}}. \tag{3.3.182}$$

(b) Similarly, confirm that substitution of (3.3.180) and (3.3.181) into the lateral momentum equation (3.3.2b), the energy equation (3.3.2c), and the state equation (3.3.2d) results in

$$\frac{\partial \bar{v}}{\partial \bar{x}} = -\frac{\partial \bar{p}}{\partial \bar{y}}, \tag{3.3.183}$$

$$\frac{\partial \bar{h}}{\partial \bar{x}} = \frac{\partial \bar{p}}{\partial \bar{x}}, \tag{3.3.184}$$

$$\bar{h} = \frac{\gamma}{\gamma - 1}\bar{p} - \frac{1}{(\gamma - 1)M_\infty^2}\bar{\rho}. \tag{3.3.185}$$

(c) Substitute (3.3.185) into (3.3.184), and deduce that

$$\frac{\partial \bar{\rho}}{\partial \bar{x}} = M_\infty^2 \frac{\partial \bar{p}}{\partial \bar{x}}. \tag{3.3.186}$$

(d) Now, consider the continuity equation (3.3.178). Using (3.3.180), (3.3.181), (3.3.182), and (3.3.186), show that the three terms in this equation are

$$\frac{\partial \rho}{\partial t} = \mu_w \lambda^2 \frac{\partial \bar{\rho}}{\partial \bar{t}},$$

$$\frac{\partial \rho u}{\partial x} = -Re^{1/8} \frac{\mu_w^{3/4}\rho_w^{1/2}}{\lambda^{-7/4}\beta^{-9/4}} \frac{\partial \bar{p}}{\partial \bar{x}},$$

$$\frac{\partial \rho v}{\partial y} = Re^{1/8} \frac{\mu_w^{3/4}\rho_w^{1/2}}{\lambda^{-7/4}\beta^{-9/4}} \frac{\partial \bar{v}}{\partial \bar{y}}.$$

Compare these with each other, to confirm Timoshin's (1990) conclusion that the flow in the upper tier becomes unsteady when

$$\beta = O(Re^{-1/18}), \tag{3.3.187}$$

which alternatively may be expressed by equation (3.3.179).

(e) By substituting (3.3.187) into (3.3.180) and (3.3.180) obtain the following estimate for the fluid-dynamic functions and independent variables in the upper tier:

$$\left.\begin{aligned}
u - 1 &\sim Re^{-2/9}, & v &\sim Re^{-5/18}, & p &\sim Re^{-2/9}, \\
\rho - 1 &\sim Re^{-2/9}, & h - \frac{1}{(\gamma-1)M_\infty^2} &\sim Re^{-2/9}, \\
t &\sim Re^{-2/9}, & x - 1 &\sim Re^{-1/3}, & y &\sim Re^{-5/18}.
\end{aligned}\right\} \quad (3.3.188)$$

(f) Guided by (3.3.188), represent the solution in the upper tier of the triple-deck region in the form

$$\left.\begin{aligned}
u &= 1 + Re^{-2/9} u_1^*(t_*, x_*, y_*) + Re^{-3/9} u_2^*(t_*, x_*, y_*) + \cdots, \\
v &= Re^{-5/18} v_1^*(t_*, x_*, y_*) + Re^{-7/18} v_2^*(t_*, x_*, y_*) + \cdots, \\
p &= Re^{-2/9} p_1^*(t_*, x_*, y_*) + Re^{-3/9} p_2^*(t_*, x_*, y_*) + \cdots, \\
\rho &= 1 + Re^{-2/9} \rho_1^*(x_*, y_*) + Re^{-3/9} \rho_2^*(t_*, x_*, y_*) + \cdots, \\
h &= \frac{1}{\gamma-1} - Re^{-1/9} \frac{2M_1}{\gamma-1} \\
&\quad + Re^{-2/9} h_1^*(t_*, x_*, y_*) + Re^{-3/9} h_2^*(t_*, x_*, y_*) + \cdots,
\end{aligned}\right\} \quad (3.3.189)$$

with the independent variables defined as

$$t = Re^{-2/9} t_*, \qquad x = 1 + Re^{-1/3} x_*, \qquad y = Re^{-5/18} y_*. \quad (3.3.190)$$

(g) Substitute (3.3.189), (3.3.190) together with (3.3.179) into the Navier–Stokes equations (3.3.2), and working with the leading-order perturbation terms, deduce that

$$\frac{\partial u_1^*}{\partial x_*} + \frac{\partial p_1^*}{\partial x_*} = 0, \quad (3.3.191a)$$

$$\frac{\partial v_1^*}{\partial x_*} + \frac{\partial p_1^*}{\partial y_*} = 0, \quad (3.3.191b)$$

$$\frac{\partial h_1^*}{\partial x_*} - \frac{\partial p_1^*}{\partial x_*} = 0, \quad (3.3.191c)$$

$$\frac{\partial u_1^*}{\partial x_*} + \frac{\partial \rho_1^*}{\partial x_*} = 0, \quad (3.3.191d)$$

$$h_1^* + \frac{\rho_1^*}{\gamma-1} - \frac{\gamma}{\gamma-1} p_1^* = \frac{3M_1^2}{\gamma-1}. \quad (3.3.191e)$$

Eliminate h_1^* from the above set of equations by substituting (3.3.191e) into (3.3.191c). Confirm that it leads to the following equation relating ρ_1^* and p_1^*:

$$\frac{\partial \rho_1^*}{\partial x_*} = \frac{\partial p_1^*}{\partial x_*}. \quad (3.3.192)$$

One would expect that the second equation for ρ_1^* and p_1^* could be obtained by eliminating $\partial u_1^*/\partial x_*$ from (3.3.191a) and (3.3.191d). However, this leads again to equation (3.3.192), which means that the leading-order set of equations (3.3.191) is degenerate. It does not allow us to determine the pressure p_1^* in the upper tier.

(h) Hence, consider the equations for the next-order terms:

$$\frac{\partial u_2^*}{\partial x_*} + \frac{\partial p_2^*}{\partial x_*} = -\frac{\partial u_1^*}{\partial t_*},$$

$$\frac{\partial v_2^*}{\partial x_*} + \frac{\partial p_2^*}{\partial y_*} = -\frac{\partial v_1^*}{\partial t_*},$$

$$\frac{\partial h_2^*}{\partial x_*} - \frac{\partial p_2^*}{\partial x_*} = \frac{\partial p_1^*}{\partial t_*} - \frac{\partial h_1^*}{\partial t_*},$$

$$\frac{\partial u_2^*}{\partial x_*} + \frac{\partial \rho_2^*}{\partial x_*} = -\frac{\partial \rho_1^*}{\partial t_*} - \frac{\partial v_1^*}{\partial y_*},$$

$$h_2^* + \frac{\rho_2^*}{\gamma-1} - \frac{\gamma}{\gamma-1}p_2^* = \frac{2M_1}{\gamma-1}\rho_1^* - \frac{4M_1^3}{\gamma-1}.$$

By manipulating these equations show that the sought equation for p_1^* is

$$2M_1\frac{\partial^2 p_1^*}{\partial x_*^2} + 2\frac{\partial^2 p_1^*}{\partial t_*\partial x_*} - \frac{\partial^2 p_1^*}{\partial y_*^2} = 0.$$

3.4 Further Advances in Receptivity Theory

In this chapter, we studied the receptivity of the boundary layer through the asymptotic analysis of the Navier–Stokes equations at large values of the Reynolds number. An alternative approach was developed by Zhigulev and Tumin (1987), Choudhari (1994), Ng and Crouch (1999), and Schrader *et al.* (2009), among others. In this approach, the Reynolds number is assumed finite, and the solution of the Navier–Stokes equations is represented as a superposition of solutions of the Orr–Sommerfeld equation with forcing terms representing the external perturbations. Of course, one more possibility is the direct numerical simulation of the receptivity process by means of numerical solution of the full Navier–Stokes equations.

The asymptotic theory has obvious advantages compared to finite-Reynolds-number numerical methods. Firstly, it allows us to deduce an explicit formula for the amplitude of the generated Tollmien–Schlichting wave that may be used, for example, if the receptivity process is to be suppressed through passive or active flow control, such as in Brennan *et al.* (2021). Secondly, the asymptotic theory represents an ideal tool for 'uncovering' the physical processes leading to the generation of the instability modes in the boundary layer. Thirdly, the asymptotic theory has proved to be instrumental in identifying possible mechanisms of boundary-layer receptivity. These include the generation of Görtler vortices by wall roughness (see Denier *et al.*, 1991), and the receptivity of the boundary layer to vorticity waves. The latter were shown by Duck *et al.* (1996) to produce Tollmien–Schlichting waves when encountering wall roughness. Later this theory was extended to the boundary-layer receptivity to entropy waves; see Ruban *et al.* (2021). The possibility of obtaining efficient receptivity in the absence of a wall roughness was demonstrated by Wu (1999) who showed that the double-resonance conditions may be satisfied when an acoustic wave interacts with free-stream turbulence. He also extended the theory of Ruban (1984) and Goldstein (1985) to the case of distributed wall roughness; see Wu (2001). The generation of Tollmien–Schlichting waves by acoustic waves propagating through the wing structure was studied by Ruban *et al.* (2013). Asymptotic receptivity theory is easily adjusted to different flow speed regimes. In particular, Ruban *et al.* (2016) used it to study the receptivity of the boundary layer to acoustic waves in transonic flows while Dong *et al.* (2020) performed the corresponding analysis for supersonic flows.

As is always the case with an asymptotic theory, the receptivity theory presented in this chapter, involves a degree of approximation when applied to finite-Reynolds-number fluid flows. The accuracy of such a theory may be established by comparing with experiments and numerical solutions of the full Navier–Stokes equations. Wu (2001) was the first to demonstrate that the predictions of asymptotic receptivity theory are in a close agreement with the experimental observations. A comparison with the Navier–Stokes simulations was performed by De Tullio and Ruban (2015) for the boundary-layer receptivity to acoustic waves (see Section 3.3). A surprisingly good agreement was observed.

4
Weakly Nonlinear Stability Theory

4.1 Landau's Concept of Laminar-Turbulent Transition

The equation of weakly nonlinear stability theory

$$\frac{d|A|^2}{dt} = 2\sigma_r|A|^2 - l_r|A|^4 \tag{4.1.1}$$

was first published by Landau (1944). In this equation, A denotes the amplitude of perturbations, t is time, and σ_r and l_r are real constants. The latter is referred to as the *Landau constant*. Landau did not suggest a mathematical procedure to derive this equation. Instead he offered the following physical explanation. When A is small, the second term on the right-hand side of (4.1.1) can be disregarded, and the evolution of the perturbations is described by linear stability theory (see Chapter 1). The corresponding solution of (4.1.1) is written as

$$|A| = Ce^{\sigma_r t}. \tag{4.1.2}$$

If we consider, as an example, plane Poiseuille flow, then we can compare (4.1.2) with (1.2.32), and we see that the amplification rate σ_r is calculated as

$$\sigma_r = \alpha c_i,$$

which is positive everywhere inside the neutral curve; see Figure 1.8(a). As $|A|$ increases, the second term on the right-hand side of (4.1.1) comes into play. It accounts for nonlinear effects in the perturbation field. Assuming l_r positive, Landau argued that the nonlinear term in (4.1.1) acts to suppress the growth of the perturbations. When the amplitude reaches the value

$$|A| = \sqrt{2\sigma_r/l_r}, \tag{4.1.3}$$

the right-hand side in (4.1.1) becomes zero and the perturbations stop growing.

In his paper, Landau (1944) also put forward the following scenario of laminar-turbulent transition in boundary layers. If, for example, we consider the boundary layer on the surface of a flat plate (see Figure 1.10), then near the leading edge of the plate, where the Reynolds number Re_* is smaller than its critical value (1.3.24), the flow remains laminar and is described by the Blasius solution (1.3.2), (1.3.3). As soon as Re_* exceeds the critical value (we shall denote it Re_{*1}) the flow becomes unstable, and the perturbations grow until the limit amplitude (4.1.3) is reached. This is referred to as *bifurcation* of the flow. As a result, instead of Blasius flow, a new flow state is achieved, where the Blasius flow is superimposed with an instability wave. This new flow state becomes unstable at a higher Reynolds number, Re_{*2}. It bifurcates again leading to a flow where the Blasius solution is superimposed with two waves; we shall denote their frequencies as ω_1 and ω_2, respectively. This process continues, and Landau conjectured that the set of Reynolds numbers Re_{*i} has a limiting value Re_{*t} that is achieved at a position $\hat{x}_t$ on the body surface. By this stage, the flow pulsations will involve an infinite number of frequencies ω_i, which is why Landau suggested that at $\hat{x}_t$ the flow becomes turbulent.

Landau's concept of laminar-turbulent transition attracted significant interest in the research community but, unfortunately, did not find experimental confirmation. As far as equation (4.1.1) is concerned, it still proved to be useful in understanding the initial stages of the transition process. A formal derivation of this equation was given by Stuart (1960) and Watson (1960). Since then this equation is referred to as the *Landau–Stuart equation*.

4.2 Landau–Stuart Equation

We shall now present the mathematical procedure developed by Stuart (1960) and Watson (1960).

4.2.1 Problem formulation

While weakly nonlinear stability theory is applicable to a wide variety of fluid flows, for definiteness, we shall consider here plane Poiseuille flow. This is the flow of an incompressible fluid in the channel formed by two parallel infinite plates; see Figure 4.1.

The linear stability of this flow was discussed in Section 1.2. As for any other parallel flow, plane Poiseuille flow obeys Squire's theorem, which allows us to restrict our attention to two-dimensional perturbations. Of course, for nonlinear perturbations, Squire's theorem might not hold. Nevertheless a two-dimensional flow assumption is a good starting point for the nonlinear stability theory to be developed.

For two-dimensional flows, the Navier–Stokes equations are written as

$$\frac{\partial u}{\partial t} + u\frac{\partial u}{\partial x} + v\frac{\partial u}{\partial y} = -\frac{\partial p}{\partial x} + \frac{1}{Re}\left(\frac{\partial^2 u}{\partial x^2} + \frac{\partial^2 u}{\partial y^2}\right), \tag{4.2.1a}$$

$$\frac{\partial v}{\partial t} + u\frac{\partial v}{\partial x} + v\frac{\partial v}{\partial y} = -\frac{\partial p}{\partial y} + \frac{1}{Re}\left(\frac{\partial^2 v}{\partial x^2} + \frac{\partial^2 v}{\partial y^2}\right), \tag{4.2.1b}$$

$$\frac{\partial u}{\partial x} + \frac{\partial v}{\partial y} = 0. \tag{4.2.1c}$$

Here we use the non-dimensional variables defined by equations (1.2.10), namely, the coordinates x and y are referred to the half-gap h between the plates, the velocity components (u, v) are made dimensionless by referring them to the maximum velocity $\hat{u}_{\max}$, as it is defined by the laminar flow solution (1.2.9). Also, the time t is referred to $h/\hat{u}_{\max}$, and the pressure p to $\rho\hat{u}_{\max}^2$. The Reynolds number in (4.2.1) is defined by

$$Re = \frac{\rho\hat{u}_{\max}h}{\mu},$$

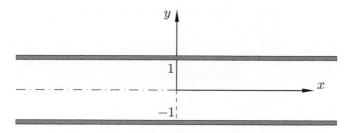

Fig. 4.1: Geometrical configuration for plane Poiseuille flow.

with the density ρ and dynamic viscosity coefficient μ assumed constant over the entire flow field.

Choosing the x-coordinate to be along the middle line of the channel, and y in the perpendicular direction, we can write the no-slip conditions on the two plates as

$$u = v = 0 \quad \text{at} \quad y = -1 \text{ and } y = 1. \tag{4.2.2}$$

In Section 1.2 we found that the basic laminar solution of (4.2.1), (4.2.2) is given by (1.2.11):

$$U = 1 - y^2, \qquad V = 0, \qquad P = -\frac{2}{Re}x. \tag{4.2.3}$$

We then introduced small perturbations

$$u = U(y) + \delta u'(t, x, y), \qquad v = \delta v'(t, x, y), \qquad p = P(x) + \delta p'(t, x, y),$$

and sought functions u', v', and p' in the normal-mode form

$$\left. \begin{aligned} u' = e^{i\alpha(x-ct)}\breve{u}(y) + (c.c.), \qquad v' = e^{i\alpha(x-ct)}\breve{v}(y) + (c.c.), \\ p' = e^{i\alpha(x-ct)}\breve{p}(y) + (c.c.), \end{aligned} \right\} \tag{4.2.4}$$

which led to the Orr–Sommerfeld equation (1.2.33a). The results of the solution of this equation are shown in Figure 1.8, where the neutral curves are displayed in the (Re, α)- and (Re, c_r)-planes.

Let us consider Figure 1.8(a). Remember that for each point in the (Re, α)-plane, the solution of the Orr–Sommerfeld problem (1.2.33) gives a complex value of the phase velocity

$$c = c_r + ic_i. \tag{4.2.5}$$

With (4.2.5), the exponential function in (4.2.4) is written as

$$e^{i\alpha(x-ct)} = e^{\alpha c_i t}e^{i\alpha(x-c_r t)}, \tag{4.2.6}$$

which shows that the perturbations with positive c_i are growing, and those with negative c_i are decaying. They are separated by the neutral curve where $c_i = 0$; see Figure 4.2. In what follows, we will study the behaviour of the perturbations in a small vicinity of the neutral curve. For this purpose, we consider a point on the neutral curve, and denote it as point B; see Figure 4.2. We further denote the values of the

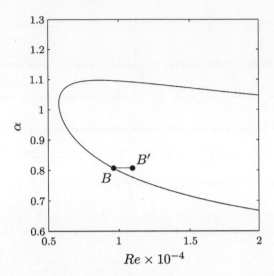

Fig. 4.2: Neutral curve for plane Poiseuille flow showing a neutral point B and a neighbouring point B' at which the flow is unstable.

Reynolds number and the wave number at this point as Re_0 and α_0. To study growing perturbations, one needs to 'move' inside the neutral curve. This may be done in various ways. We choose to keep the wave number α unchanged and increase the Reynolds number slightly. As a result, point B moves to point B', as shown in Figure 4.2, with the Reynolds number at B' being

$$Re = Re_0 + \delta r. \qquad (4.2.7)$$

Here δ is small parameter, and r is an order one constant. For the situation shown in Figure 4.2, r is positive, but if we want point B' to lie outside the neutral curve, then r should be taken negative.

The solution of the eigen-value problem (1.2.33) defines $c = c_r + ic_i$ as a function of Re and α. Since this function is smooth, it is clear that at point B'

$$c_r = c_{r0} + \delta c_{r1}, \qquad c_i = \delta c_{i1}. \qquad (4.2.8)$$

With (4.2.8), the right-hand side of equation (4.2.6) assumes the form

$$e^{\alpha_0(c_{i1} - ic_{r1})\delta t} e^{i(\alpha_0 x - \omega_0 t)}, \qquad (4.2.9)$$

where α_0 and $\omega_0 = \alpha_0 c_{r0}$ are the wave number and the frequency at point B. It is easily seen that there are two time scales involved in (4.2.9): the first one, with $t = O(1)$, corresponds to periodic oscillations of the perturbations, and the second, with $\delta t = O(1)$, describes the slow growth of the amplitude of the perturbations. This suggests that the problem at hand should be dealt with using the method of multiple scales.[1]

---

[1]See Section 1.5 in Part 2 of this book series.

Before applying the corresponding asymptotic procedure to the Navier–Stokes equations (4.2.1), it is convenient to introduce the stream function $\psi(t, x, y)$. It is related to the velocity components by means of the equations

$$u = \frac{\partial \psi}{\partial y}, \qquad v = -\frac{\partial \psi}{\partial x}. \tag{4.2.10}$$

We shall also perform cross-differentiation of (4.2.1a) and (4.2.1b), which allows us to eliminate the pressure p, and leads to the following equation

$$\frac{\partial \omega}{\partial t} + u \frac{\partial \omega}{\partial x} + v \frac{\partial \omega}{\partial y} = \frac{1}{Re} \left(\frac{\partial^2 \omega}{\partial x^2} + \frac{\partial^2 \omega}{\partial y^2} \right) \tag{4.2.11}$$

for the vorticity

$$\omega = \frac{\partial v}{\partial x} - \frac{\partial u}{\partial y}. \tag{4.2.12}$$

4.2.2   Asymptotic procedure

Now we turn to the asymptotic procedure. We assume that the Reynolds number is given by (4.2.7), and represent the stream function in the form of the asymptotic expansion

$$\psi(t, x, y) = \Psi(y) + \varepsilon \psi_1(\tilde{t}, \xi, y) + \varepsilon^2 \psi_2(\tilde{t}, \xi, y) + \varepsilon^3 \psi_3(\tilde{t}, \xi, y) + \cdots. \tag{4.2.13}$$

Here $\Psi(y)$ stands for the basic flow solution; it is related to the longitudinal velocity $U(y)$ in (4.2.3) by the equation $\Psi'(y) = U(y)$. The amplitude of the perturbations, ε, is assumed small, while functions ψ_1, ψ_2, and ψ_3 are order one quantities. The arguments of these functions are the 'slow time'

$$\tilde{t} = \delta t, \tag{4.2.14}$$

the phase variable

$$\xi = \alpha_0 x - \omega_0 t, \tag{4.2.15}$$

and the lateral coordinate y.

Substituting (4.2.13) together with (4.2.15) into (4.2.10) we find that

$$u = U(y) + \varepsilon \frac{\partial \psi_1}{\partial y} + \varepsilon^2 \frac{\partial \psi_2}{\partial y} + \varepsilon^3 \frac{\partial \psi_3}{\partial y} + \cdots, \tag{4.2.16a}$$

$$v = -\varepsilon \alpha_0 \frac{\partial \psi_1}{\partial \xi} - \varepsilon^2 \alpha_0 \frac{\partial \psi_2}{\partial \xi} - \varepsilon^3 \alpha_0 \frac{\partial \psi_3}{\partial \xi} + \cdots. \tag{4.2.16b}$$

Next, we substitute (4.2.16) into (4.2.12), which yields

$$\omega = -U'(y) - \varepsilon \left(\frac{\partial^2 \psi_1}{\partial y^2} + \alpha_0^2 \frac{\partial^2 \psi_1}{\partial \xi^2} \right)$$
$$- \varepsilon^2 \left(\frac{\partial^2 \psi_2}{\partial y^2} + \alpha_0^2 \frac{\partial^2 \psi_2}{\partial \xi^2} \right) - \varepsilon^3 \left(\frac{\partial^2 \psi_3}{\partial y^2} + \alpha_0^2 \frac{\partial^2 \psi_3}{\partial \xi^2} \right) + \cdots. \tag{4.2.17}$$

It remains to substitute (4.2.17) together with (4.2.16), (4.2.15), (4.2.14), and (4.2.7) into the vorticity equation (4.2.11), and we will have the following equations for functions ψ_1, ψ_2, and ψ_3:

$$
\frac{1}{Re_0}\left(\alpha_0^4\frac{\partial^4\psi_i}{\partial\xi^4} + 2\alpha_0^2\frac{\partial^4\psi_i}{\partial\xi^2\partial y^2} + \frac{\partial^4\psi_i}{\partial y^4}\right)
$$

$$
+ (\omega_0 - U\alpha_0)\left(\frac{\partial^3\psi_i}{\partial\xi\partial y^2} + \alpha_0^2\frac{\partial^3\psi_i}{\partial\xi^3}\right) + \alpha_0\frac{d^2U}{dy^2}\frac{\partial\psi_i}{\partial\xi} = Q_i, \quad i = 1, 2, 3, \quad (4.2.18)
$$

where

$$
Q_1 = 0, \tag{4.2.19}
$$

$$
Q_2 = \alpha_0\frac{\partial\psi_1}{\partial y}\left(\frac{\partial^3\psi_1}{\partial\xi\partial y^2} + \alpha_0^2\frac{\partial^3\psi_1}{\partial\xi^3}\right) - \alpha_0\frac{\partial\psi_1}{\partial\xi}\left(\frac{\partial^3\psi_1}{\partial y^3} + \alpha_0^2\frac{\partial^3\psi_1}{\partial\xi^2\partial y}\right), \tag{4.2.20}
$$

$$
Q_3 = \alpha_0\frac{\partial\psi_1}{\partial y}\left(\frac{\partial^3\psi_2}{\partial\xi\partial y^2} + \alpha_0^2\frac{\partial^3\psi_2}{\partial\xi^3}\right) + \alpha_0\frac{\partial\psi_2}{\partial y}\left(\frac{\partial^3\psi_1}{\partial\xi\partial y^2} + \alpha_0^2\frac{\partial^3\psi_1}{\partial\xi^3}\right)
$$

$$
- \alpha_0\frac{\partial\psi_2}{\partial\xi}\left(\frac{\partial^3\psi_1}{\partial y^3} + \alpha_0^2\frac{\partial^3\psi_1}{\partial\xi^2\partial y}\right) - \alpha_0\frac{\partial\psi_1}{\partial\xi}\left(\frac{\partial^3\psi_2}{\partial y^3} + \alpha_0^2\frac{\partial^3\psi_2}{\partial\xi^2\partial y}\right)
$$

$$
+ \frac{\delta}{\varepsilon^2}\left[\frac{\partial^3\psi_1}{\partial\tilde{t}\partial y^2} + \alpha_0^2\frac{\partial^3\psi_1}{\partial\tilde{t}\partial\xi^2} + \frac{r}{Re_0^2}\left(\alpha_0^4\frac{\partial^4\psi_1}{\partial\xi^4} + 2\alpha_0^2\frac{\partial^4\psi_1}{\partial\xi^2\partial y^2} + \frac{\partial^4\psi_1}{\partial y^4}\right)\right]. \tag{4.2.21}
$$

To formulate the boundary conditions for ψ_1, ψ_2, and ψ_3, we substitute (4.2.16) into (4.2.2). We see that

$$
\frac{\partial\psi_i}{\partial\xi} = \frac{\partial\psi_i}{\partial y} = 0 \quad \text{at} \quad y = -1 \text{ and } y = 1, \quad i = 1, 2, 3. \tag{4.2.22}
$$

4.2.3   Linear perturbations

We start with the function ψ_1 that represents the linear perturbations in (4.2.13). We know that ψ_1 should be periodic with respect to the phase variable ξ, while the amplitude of the oscillations is allowed to depend on the slow time $\tilde{t}$. Hence, we write

$$
\psi_1(\tilde{t}, \xi, y) = A(\tilde{t})\, e^{i\xi}\phi_1(y) + \overline{A}(\tilde{t})\, e^{-i\xi}\,\overline{\phi}_1(y). \tag{4.2.23}
$$

Here $\overline{A}$ and $\overline{\phi}_1$ are complex conjugates of A and ϕ_1, respectively. The second term is added in (4.2.23) to ensure that, with complex A and ϕ_1, the stream function remains real.

Substituting (4.2.23) into equation (4.2.18) with (4.2.19), and working with the terms proportional to $e^{i\xi}$, we arrive at the Orr–Sommerfeld equation[2]

$$
\frac{1}{i\alpha_0 Re_0}\left(\frac{d^4\phi_1}{dy^4} - 2\alpha_0^2\frac{d^2\phi_1}{dy^2} + \alpha_0^4\phi_1\right)
$$

$$
+ (\omega_0/\alpha_0 - U)\left(\frac{d^2\phi_1}{dy^2} - \alpha_0^2\phi_1\right) + \frac{d^2U}{dy^2}\phi_1 = 0. \tag{4.2.24a}
$$

---

[2] The terms proportional to $e^{-i\xi}$ produce an equation that may be obtained by taking complex conjugates of all the terms in (4.2.24a).

Substitution of (4.2.23) into (4.2.22) shows that the boundary conditions for ϕ_1 are

$$\phi_1 = \frac{d\phi_1}{dy} = 0 \quad \text{at} \quad y = -1 \text{ and } y = 1. \tag{4.2.24b}$$

The eigen-value problem (4.2.24) defines $\phi_1(y)$ to within an arbitrary constant. Hence, to have the amplitude A in (4.2.23) properly defined, we need to normalize $\phi_1(y)$. Since our interest is in the unstable (sinuous) mode,[3] we shall achieve this by setting

$$\phi_1 = 1 \quad \text{at} \quad y = 0.$$

4.2.4 Quadratic approximation

Substitution of (4.2.23) into (4.2.20) yields

$$Q_2 = A\overline{A}\mathcal{H}(y) + A^2 e^{2i\xi} \mathcal{F}(y) + \overline{A}^2 e^{-2i\xi} \overline{\mathcal{F}}(y), \tag{4.2.25}$$

where

$$\mathcal{H} = i\alpha_0 \frac{d}{dy}\left(\frac{d^2\phi_1}{dy^2}\overline{\phi}_1 - \frac{d^2\overline{\phi}_1}{dy^2}\phi_1\right), \tag{4.2.26}$$

$$\mathcal{F} = i\alpha_0 \left(\frac{d^2\phi_1}{dy^2}\frac{d\phi_1}{dy} - \frac{d^3\phi_1}{dy^3}\phi_1\right).$$

By taking the complex conjugate of (4.2.26) it is easily shown that $\overline{\mathcal{H}} = \mathcal{H}$, which proves that the function $\mathcal{H}(y)$ is real.

Now we can return to equation (4.2.18); this time we set $i = 2$. Guided by (4.2.25), we seek ψ_2 in the form

$$\psi_2(\tilde{t}, \xi, y) = A\overline{A}\, h(y) + A^2 e^{i2\xi} \phi_2(y) + \overline{A}^2 e^{-i2\xi}\, \overline{\phi}_2(y). \tag{4.2.27}$$

We start with the function $\phi_2(y)$. The equation

$$\frac{1}{i\alpha_0 Re_0}\left(\frac{d^4\phi_2}{dy^4} - 8\alpha_0^2\frac{d^2\phi_2}{dy^2} + 16\alpha_0^4\phi_2\right)$$

$$+ 2\left(\omega_0/\alpha_0 - U\right)\left(\frac{d^2\phi_2}{dy^2} - 4\alpha_0^2\phi_2\right) + 2\frac{d^2U}{dy^2}\phi_2 = -\frac{i}{\alpha_0}\mathcal{F}(y), \tag{4.2.28}$$

for $\phi_2(y)$ is obtained, as usual, by substituting (4.2.27) and (4.2.25) into (4.2.18) and working with the $e^{i2\xi}$ terms. Substitution of (4.2.27) into (4.2.22) yields the following boundary conditions for (4.2.28):

$$\phi_2 = \frac{d\phi_2}{dy} = 0 \quad \text{at} \quad y = -1 \text{ and } y = 1. \tag{4.2.29}$$

The situation with the function $h(y)$ appears to be more intricate. Substitution of (4.2.27) into (4.2.18) yields the following fourth order differential equation for $h(y)$:

---

[3]See Section 1.2.2.

$$\frac{d^4 h}{dy^4} = Re_0 \mathcal{H}(y).$$ (4.2.30)

However, the substitution of (4.2.27) into (4.2.22) yields just two boundary conditions:

$$\frac{dh}{dy} = 0 \quad \text{at} \quad y = -1 \text{ and } y = 1.$$ (4.2.31)

Of course, we know that the stream function is only defined to within an arbitrary constant. Therefore, without loss of generality, we can choose h to be zero on the lower plate:

$$h = 0 \quad \text{at} \quad y = -1.$$ (4.2.32)

Still, one more condition is required.

To clarify the situation, we return to the asymptotic expansion (4.2.16a) of the longitudinal velocity component. Disregarding the $O(\varepsilon^3)$ term for now and using (4.2.23) and (4.2.27) for the linear and quadratic terms, we have

$$u = U(y) + \varepsilon\left[A(\tilde{t})e^{i\xi}\frac{d\phi_1}{dy} + \overline{A}(\tilde{t})e^{-i\xi}\frac{d\overline{\phi}_1}{dy} \right]$$
$$+ \varepsilon^2\left[A\overline{A}\frac{dh}{dy} + A^2 e^{i2\xi}\frac{d\phi_2}{dy} + \overline{A}^2 e^{-i2\xi}\frac{d\overline{\phi}_2}{dy} \right] + \cdots .$$ (4.2.33a)

$$v = -\varepsilon\left[i\alpha_0 A(\tilde{t})e^{i\xi}\phi_1 - i\alpha_0 \overline{A}(\tilde{t})e^{-i\xi}\overline{\phi}_1 \right]$$
$$- \varepsilon^2\left[2i\alpha_0 A^2 e^{2i\xi}\phi_2 - 2i\alpha_0 \overline{A}^2 e^{-i2\xi}\overline{\phi}_2 \right] + \cdots .$$ (4.2.33b)

If we perform averaging in (4.2.33a) with respect to the fast time t or, equivalently, with respect to the phase variable ξ, then we will see that the mean velocity $u_m(\tilde{t}, y)$ is given by

$$u_m(\tilde{t}, y) = U(y) + \varepsilon^2 A\overline{A}\frac{dh}{dy} + \cdots .$$ (4.2.34)

The leading-order term in (4.2.34) represents the basic velocity profile. In the dimensional variables, it is expressed by equation (1.2.8), from which it follows that the velocity of fluid motion through the channel is proportional to the pressure gradient $d\hat{p}/d\hat{x}$. The latter is given by (1.2.7), where the pressure difference between the channel ends, $\Delta\hat{p}$, may be treated as a free parameter controlled by an 'experimentalist'. Similarly, the second term on the right-hand side of (4.2.34) may be influenced by introducing $O(\epsilon^2)$ perturbations to $\Delta\hat{p}$. Keeping this in mind, and guided by (4.2.33), we write the asymptotic expansion of the pressure in the form

$$p = -\frac{2}{Re}x + \varepsilon\left[A(\tilde{t})e^{i\xi}p_1(y) + \overline{A}(\tilde{t})e^{-i\xi}\overline{p}_1(y) \right]$$
$$+ \varepsilon^2\left[A\overline{A}\lambda x + A^2 e^{i2\xi}p_2(y) + \overline{A}^2 e^{-i2\xi}\overline{p}_2(y) \right] + \cdots .$$ (4.2.35)

Here λ is the parameter controlling the $O(\varepsilon^2)$ perturbations of the imposed pressure gradient. Substituting (4.2.33) and (4.2.35) into the lateral momentum equation (4.2.1a), we arrive at the following equation for $h(y)$:

$$\frac{d^3 h}{dy^3} = i\alpha_0 Re_0 \left(\frac{d^2\phi_1}{dy^2}\overline{\phi}_1 - \frac{d^2\overline{\phi}_1}{dy^2}\phi_1 \right) + Re_0\lambda. \tag{4.2.36}$$

Traditionally, two formulations are considered in weakly nonlinear theory. In the first one, it is assumed that the pressure gradient remains unperturbed ($\lambda = 0$) and equation (4.2.36) is solved with the boundary conditions (4.2.31), (4.2.32). In the second formulation, the mean fluid flux through the channel is kept unchanged. In this case, we can use the fact that the fluid flux Q through a line connecting two points, M and M', in the flow field is calculated as[4]

$$Q = \psi(M') - \psi(M).$$

Placing point M on the lower boundary of the channel and M' on the upper boundary, we see that, in addition to (4.2.32), we have to pose the condition

$$h = 0 \quad \text{at} \quad y = 1. \tag{4.2.37}$$

This closes the boundary-value problem (4.2.30), (4.2.31), (4.2.32), and (4.2.37), making its solution unique.

As an aside it should be noted that for boundary-layer flow only the first of these scenarios is applicable. To prove this statement, we use equation (4.2.30) which is equally applicable to plane Poiseuille flow and boundary-layer flow. We substitute (4.2.26) into (4.2.30) and integrate the resulting equation with respect to y. We find that

$$\frac{d^3 h}{dy^3} = i\alpha_0 Re_0 \left(\frac{d^2\phi_1}{dy^2}\overline{\phi}_1 - \frac{d^2\overline{\phi}_1}{dy^2}\phi_1 \right) + C, \tag{4.2.38}$$

where C is the integration constant. Both $\phi_1(y)$ and $h(y)$ have to satisfy the perturbations attenuation condition at the outer edge of the boundary layer. Therefore, setting $y \to \infty$ in (4.2.38) we can see that $C = 0$, which is equivalent to setting $\lambda = 0$ in (4.2.36).

4.2.5 Cubic approximation

Finally, we need to consider the cubic term $\varepsilon^3\psi_3$ in (4.2.13). Remember that the function ψ_3 satisfies equation (4.2.18) with the forcing term Q_3 given by (4.2.21), where the small parameter δ is the scaling coefficient in the definition (4.2.14) of the slow time $\tilde{t}$. Up to now, its relationship with the amplitude of the perturbations, ε, was left undetermined. To progress further we choose

$$\delta = \varepsilon^2. \tag{4.2.39}$$

---

[4]See equation (3.3.13) on page 156 in Part 1 of this book series.

This choice is dictated by the principle of least degeneration.[5] According to this principle, in order to widen the applicability of an asymptotic theory, one has to retain in the governing equation as many terms as possible.

Substituting (4.2.23) and (4.2.27) into (4.2.21), and using (4.2.39), we see that Q_3 may be written in the form

$$Q_3 = A^3 e^{i3\xi} \mathcal{G}(y) + e^{i\xi} \left[\mathcal{R}(y) \frac{dA}{d\tilde{t}} + r\mathcal{S}(y)A + \mathcal{T}(y)A^2\overline{A} \right] + (c.c.), \qquad (4.2.40)$$

where

$$\mathcal{G}(y) = 2i\alpha_0 \frac{d\phi_1}{dy}\left(\frac{d^2\phi_2}{dy^2} - 4\alpha_0^2\phi_2 \right) + i\alpha_0 \frac{d\phi_2}{dy}\left(\frac{d^2\phi_1}{dy^2} - \alpha_0^2\phi_1 \right)$$

$$- 2i\alpha_0\phi_2\left(\frac{d^3\phi_1}{dy^3} - \alpha_0^2\frac{d\phi_1}{dy} \right) - i\alpha_0\phi_1\left(\frac{d^3\phi_2}{dy^3} - 4\alpha_0^2\frac{d\phi_2}{dy} \right),$$

$$\mathcal{R}(y) = \frac{d^2\phi_1}{dy^2} - \alpha_0^2\phi_1, \qquad (4.2.41)$$

$$\mathcal{S}(y) = \frac{1}{Re_0^2}\left(\frac{d^4\phi_1}{dy^4} - 2\alpha_0^2\frac{d^2\phi_1}{dy^2} + \alpha_0^4\phi_1 \right), \qquad (4.2.42)$$

$$\mathcal{T}(y) = 2i\alpha_0 \frac{d\overline{\phi}_1}{dy}\left(\frac{d^2\phi_2}{dy^2} - 4\alpha_0^2\phi_2 \right) + i\alpha_0 \frac{dh}{dy}\left(\frac{d^2\phi_1}{dy^2} - \alpha_0^2\phi_1 \right)$$

$$- i\alpha_0 \frac{d\phi_2}{dy}\left(\frac{d^2\overline{\phi}_1}{dy^2} - \alpha_0^2\overline{\phi}_1 \right) - 2i\alpha_0\phi_2\left(\frac{d^3\overline{\phi}_1}{dy^3} - \alpha_0^2\frac{d\overline{\phi}_1}{dy} \right)$$

$$+ i\alpha_0\overline{\phi}_1\left(\frac{d^3\phi_2}{dy^3} - 4\alpha_0^2\frac{d\phi_2}{dy} \right) - i\alpha_0\phi_1\frac{d^3h}{dy^3}. \qquad (4.2.43)$$

As before, we use the (c.c.) symbol to denote the complex conjugate of the preceding terms in (4.2.40).

Guided by (4.2.40) we seek ψ_3 in the form

$$\psi_3 = A^3 e^{i3\xi}\phi_3(y) + e^{i\xi}\Phi(\tilde{t}, y) + (c.c.). \qquad (4.2.44)$$

The method of multiple scales (see Section 1.5 in Part 2 of this book series) requires us to concentrate on the second term in (4.2.44) which has the same frequency as the leading-order perturbation (4.2.23). Substituting (4.2.44) together with (4.2.40) into (4.2.18), we find that the function $\Phi(\tilde{t}, y)$ satisfies the following equation:

$$\frac{1}{i\alpha_0 Re_0}\left(\frac{\partial^4\Phi}{\partial y^4} - 2\alpha_0^2\frac{\partial^2\Phi}{\partial y^2} + \alpha_0^4\Phi \right)$$

$$+ (\omega_0/\alpha_0 - U)\left(\frac{\partial^2\Phi}{\partial y^2} - \alpha_0^2\Phi \right) + \frac{d^2U}{dy^2}\Phi$$

$$= -\frac{i}{\alpha_0}\left[\mathcal{R}(y)\frac{dA}{d\tilde{t}} + r\mathcal{S}(y)A + \mathcal{T}(y)A^2\overline{A} \right]. \qquad (4.2.45)$$

---

[5]For a discussion of the principle of least degeneration the reader is referred to Section 1.4.2 in Part 2 of this book series.

The boundary conditions for (4.2.45) are

$$\Phi = \frac{\partial \Phi}{\partial y} = 0 \quad \text{at} \quad y = -1 \quad \text{and} \quad y = 1. \tag{4.2.46}$$

For a solution to the boundary-value problem (4.2.45), (4.2.46) to exist, the right-hand side of equation (4.2.45) should satisfy the following solvability condition:[6]

$$\int_{-1}^{1} \left[\mathcal{R}(y)\frac{dA}{d\tilde{t}} + r\mathcal{S}(y)A + \mathcal{T}(y)A^2\overline{A} \right] \Phi^*(y)\, dy = 0. \tag{4.2.47}$$

Here $\Phi^*(y)$ is a solution to the adjoint equation

$$\frac{d^4\Phi^*}{dy^4} - \left(2\alpha_0^2 + i\alpha_0 Re_0 U - i\omega_0 Re_0\right)\frac{d^2\Phi^*}{dy^2}$$
$$- 2i\alpha_0 Re_0 \frac{dU}{dy}\frac{d\Phi^*}{dy} + \left(\alpha_0^4 + i\alpha_0^3 Re_0 U - i\omega_0\alpha_0^2 Re_0\right)\Phi^* = 0,$$

subject to the boundary conditions

$$\Phi^* = \frac{d\Phi^*}{dy} = 0 \quad \text{at} \quad y = -1 \text{ and } y = 1.$$

It follows from (4.2.47) that the amplitude of perturbations $A(\tilde{t})$ must satisfy the equation

$$\frac{dA}{d\tilde{t}} = r\kappa A - \frac{l}{2}A^2\overline{A}, \tag{4.2.48}$$

where the complex constants κ and l are given by

$$\kappa = -\frac{\int_{-1}^{1} \mathcal{S}(y)\Phi^*(y)\, dy}{\int_{-1}^{1} \mathcal{R}(y)\Phi^*(y)\, dy}, \qquad l = 2\frac{\int_{-1}^{1} \mathcal{T}(y)\Phi^*(y)\, dy}{\int_{-1}^{1} \mathcal{R}(y)\Phi^*(y)\, dy}.$$

Equation (4.2.48) is known as the *Landau–Stuart equation.*

4.2.6 Properties of the Landau–Stuart equation

As we know now how the coefficients κ and l in the Landau–Stuart equation (4.2.48) can be calculated, we shall turn to the task of deriving the Landau equation (4.1.1). We start by taking complex conjugates on both sides of (4.2.48):

$$\frac{d\overline{A}}{d\tilde{t}} = r\overline{\kappa}\overline{A} - \frac{\overline{l}}{2}\overline{A}^2 A. \tag{4.2.49}$$

[6]See Problem 2 in Exercises 13.

We then multiply equation (4.2.48) by $\overline{A}$ and equation (4.2.49) by A. Adding the results together yields

$$\overline{A}\frac{dA}{d\tilde{t}} + A\frac{d\overline{A}}{d\tilde{t}} = r(\kappa + \overline{\kappa})A\overline{A} - \frac{l+\overline{l}}{2}A^2\overline{A}^2,$$

or, equivalently,

$$\frac{d}{d\tilde{t}}\left(A\overline{A}\right) = 2r\kappa_r A\overline{A} - l_r A^2\overline{A}^2,$$

where κ_r and l_r are the real parts of κ and l. It remains to note that $A\overline{A} = |A|^2$, and we have the Landau equation in the form

$$\frac{d|A|^2}{d\tilde{t}} = 2r\kappa_r|A|^2 - l_r|A|^4. \tag{4.2.50}$$

In the notations used here, the amplification rate of the linear perturbations is given by

$$\sigma_r = r\kappa_r.$$

We recall that r is the Reynolds number deviation parameter introduced by equation (4.2.7) with $\delta = \varepsilon^2$. For the situation shown in Figure 4.2, the parameter r is positive, and we know that inside the neutral curve $\sigma_r > 0$. If, instead we place the point B' to the left of the neutral curve, then both r and σ_r will be negative. This allows us to conclude that κ_r is always positive. As far as the Landau parameter l_r is concerned, it may be positive or negative depending on the flow considered or even on the position of point B on the neutral curve; see Figure 4.3.

Armed with this information, we shall now see how the position of the neutral curve in the (Re, α)-plane is affected by the amplitude $|A|$ of the perturbations. For

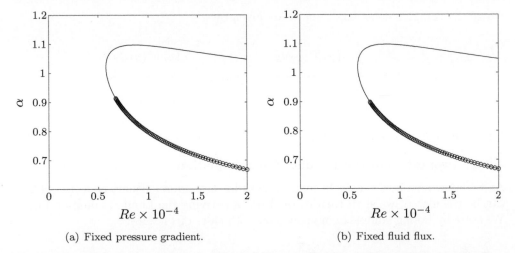

(a) Fixed pressure gradient.   (b) Fixed fluid flux.

Fig. 4.3: Results of computation of the Landau coefficient, l, for plane Poiseuille flow; circles mark the points on the neutral curve where the flow is supercritically unstable ($l_r > 0$); on the rest of the curve, the flow displays subcritical instability ($l_r < 0$).

neutral oscillations, $|A|$ is independent of time $\tilde{t}$. Hence, disregarding the left-hand side in (4.2.50) and solving the resulting equation for r, we have

$$r = \frac{l_r}{2\kappa_r} A_*^2. \tag{4.2.51}$$

Here, and in what follows, we denote the amplitude of the neutral perturbations as A_*. Equation (4.2.51) shows that if l_r is positive, then r is also positive. In this case, if we consider point B on the neutral curve (see Figure 4.2) and start increasing the amplitude of the perturbations, then the neutral value of the Reynolds number will increase from Re_0 to $Re_0 + \varepsilon^2 r$, and the neutral curve will move to the right. If, on the other hand, $l_r < 0$, then the neutral Reynolds number decreases. and the neutral curve moves to the left. Correspondingly, flows with $l_r > 0$ are said to be *supercritically unstable*, and those with $l_r < 0$ are referred to as *subcritically unstable*.

In Figure 4.3 we show the results of numerical calculations of the Landau constant l_r for plane Poiseuille flow. The calculations were performed for both formulations outlined above, with fixed pressure gradient (Figure 4.3a), and with fixed fluid flux (Figure 4.3b). The solid lines represent the neutral curve obtained from linear stability analysis. The circles mark the part of the curve where the flow is supercritically unstable. On the rest of the neutral curve, the flow displays subcritical instability. It should be noted there is not much difference between the two formulations. The corresponding results for the Blasius boundary layer are displayed in Figure 4.4. Of course, our main interest is in the vicinity of the critical point, where the Reynolds number assumes its minimal value Re_c. We see that the nonlinearity increases the critical Reynolds number for Blasius boundary layer, and decreases it for plane Poiseuille flow.

In addition to neutral oscillations, the Landau–Stuart theory also allows us to study the evolution of perturbations starting from an arbitrary initial state

$$A = A_0 \quad \text{at} \quad \tilde{t} = 0. \tag{4.2.52}$$

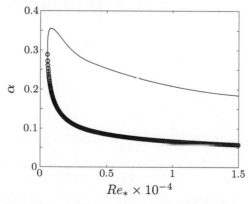

Fig. 4.4: Blasius boundary layer. Circles again mark the point where the flow shows supercritical instability. On the rest of the neutral curve the flow is subcritically unstable.

The general solution of equation (4.2.50) is written as (see Problem 3 in Exercises 13).

$$|A|^2 = \frac{2r\kappa_r C}{Cl_r - e^{-2r\kappa_r \tilde{t}}},$$ (4.2.53)

where the constant C is found from the initial condition (4.2.52) to be

$$C = \frac{|A_0|^2}{l_r|A_0|^2 - 2r\kappa_r}.$$ (4.2.54)

We shall now use (4.2.53), (4.2.54) to analyse the behaviour of the perturbations in a number of typical situations.

Supercritical case.   Let us first consider three possible scenarios for flows with supercritical instability ($l_r > 0$) when the neutral Reynolds number increases with the amplitude of perturbations as shown by the solid line in Figure 4.5(a). Point 1 in this figure represents an initial state (4.2.52) for a flow with Re less than Re_0. Since in this case $r < 0$, it follows from (4.2.54) that $Cl_r \in (0, 1)$. Meanwhile, $e^{-2r\kappa_r \tilde{t}}$ is larger than one for all $\tilde{t} > 0$, which precludes the denominator in (4.2.53) from becoming zero. In fact, $|Cl_r - e^{-2r\kappa_r \tilde{t}}|$ grows monotonically with time, making $|A|^2$ vanish in the limit as $\tilde{t} \to \infty$.

Let us now assume that $Re > Re_0$, and consider two points, 2 and 3, that lie above and below the neutral solution. For either point, $r > 0$, and we can use equation (4.2.51) to find the amplitude of the neutral oscillations to be

$$A_*^2 = \frac{2r\kappa_r}{l_r}.$$ (4.2.55)

With the help of (4.2.55) and (4.2.54), the solution (4.2.53) of the Landau equation may be expressed in the form

$$|A|^2 = \frac{A_*^2}{1 - \left(1 - \frac{A_*^2}{|A_0|^2}\right)e^{-2r\kappa_r \tilde{t}}}.$$ (4.2.56)

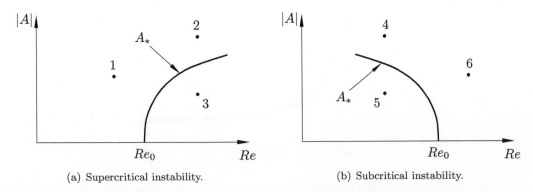

(a) Supercritical instability.   (b) Subcritical instability.

Fig. 4.5: Six possible scenarios of the development of perturbations according to the weakly nonlinear stability theory.

It is now easy to see that if we start from point 2, where $A_*/|A_0| < 1$, then the denominator in (4.2.56) will remain smaller than unity for all $\tilde{t} > 0$, but it monotonically increases towards unity as $\tilde{t} \to \infty$. As a result, the solution approaches the neutral state 'from above'.

The behaviour of the solution originating from point 3 (see Figure 4.5a) is analysed in the same way. This time, the denominator in (4.2.56) is larger than unity for all $\tilde{t} > 0$, but it still tends to unity as $\tilde{t} \to \infty$. This means that the solution again approaches the neutral oscillations as $\tilde{t} \to \infty$, now 'from below'.

Subcritical case. For a subcritically unstable flow, the Landau parameter l_r is negative, and the neutral Reynolds number decreases with the amplitude of perturbations as shown in Figure 4.5(b). We shall first consider two points, 4 and 5, that lie above and below the neutral solution, and assume that they represent the initial state of the perturbations. Both at point 4 and at point 5, the Reynolds number deviation parameter r is negative, which allows us to use the solution of the Landau equation in the form given by (4.2.56). At point 4, the ratio $A_*/|A_0|$ is confined to the interval $(0, 1)$, which makes the denominator in (4.2.56) initially positive. But, as the time $\tilde{t}$ increases, the exponential function $e^{-2r\kappa_r\tilde{t}}$ becomes progressively larger, and the denominator turns zero at

$$\tilde{t}_c = \frac{1}{2r\kappa_r} \ln\left(1 - \frac{A_*^2}{|A_0|^2}\right). \tag{4.2.57}$$

As a result, the amplitude of the perturbations $|A|$ displays an unbounded growth as $\tilde{t} \to \tilde{t}_c - 0$.

However, if we use point 5 as a starting point (see Figure 4.5b), then the denominator in (4.2.56) will remain positive for all $\tilde{t} > 0$. Moreover, it grows exponentially, so that $|A| \to 0$ as $\tilde{t} \to \infty$.

Finally, we shall consider point 6. At this point, the Reynolds number deviation parameter r is positive, and since for subcritically unstable flows $l_r < 0$, the equilibrium amplitude given in (4.2.55) does not exist. This precludes us from using (4.2.56) as the solution of the Landau equation. Instead, we have to return to (4.2.53) and (4.2.54). Applying equation (4.2.54) to point 6, we see that at this point $Cl_r \in (0, 1)$. Therefore, the denominator in (4.2.53) is initially negative, but $e^{-2r\kappa_r\tilde{t}}$ decays with time, and there exists a finite value of $\tilde{t}$ at which $Cl_r - e^{-2r\kappa_r\tilde{t}}$ becomes zero, and the amplitude $|A|$ of the perturbations displays an unbounded growth.[7]

To summarize, we see that the neutral solution (4.2.55) of the Landau equation is stable for supercritical flows, where any deviation from this solution will bring the perturbations back to (4.2.55). For subcritical flows however, the neutral solution is unstable. In this case, a deviation from (4.2.55) will lead to either a finite time singularity or to a complete disappearance of the perturbations in the limit $\tilde{t} \to \infty$. Interestingly enough, in subcritical flows, perturbations can grow at values of the Reynolds number

---

[7]Of course, one should keep in mind that these predictions are restricted by the assumptions of weakly nonlinear theory. In reality, when the amplitude of the perturbations becomes sufficiently large, additional physical effects will come into play and bound the growth of the perturbations (see Section 4.3).

smaller than that predicted by linear stability theory. For such growth to be observed, the initial amplitude of the perturbations should exceed a certain threshold.

Our final comment concerns Landau's description of laminar-turbulent transition (see Section 4.1). Remember that in this scenario, the boundary layer is considered to be initially laminar, but loses its stability when the Reynolds number reaches the first critical value Re_{*1}, allowing a Tollmien–Schlichting wave to form. This wave grows in the boundary layer until its amplitude reaches the value given by (4.2.55). As a result, a new flow state is achieved, where the boundary layer is superimposed with the Tollmien–Schlichting wave. This new flow state becomes unstable at a higher Reynolds number, Re_{*2}, and the process repeats itself. Clearly, this scenario requires not only the first, but also subsequent flow states to develop instability in a supercritical manner. No such flows are known to exist.

Exercises 13

1. In Problem 2, Exercises 2, the linear stability properties of the Eckhaus equation

$$\frac{1}{R}\left(\frac{\partial^2 u}{\partial y^2}+\frac{\partial^2 u}{\partial x^2}\right)-\frac{\partial^4 u}{\partial x^4}-\frac{\partial u}{\partial t}=\frac{\partial u}{\partial y}\frac{\partial^2 u}{\partial x^2} \tag{4.2.58}$$

were studied. Here, your task is to apply weakly nonlinear theory to (4.2.58).

Equation (4.2.58) should be solved in the infinite strip $x \in (-\infty, \infty)$, $y \in (0, 1)$ subject to the boundary conditions

$$u\Big|_{y=0}=0, \qquad u\Big|_{y=1}=1. \tag{4.2.59}$$

You may perform your analysis in the following steps:

(a) Consider the first root $(n = 1)$ of the dispersion equation (1.3.29). For a wave number α_0 from the interval $(0, 1)$, choose the neutral value R_0 of the parameter R by setting $\sigma = 0$ in (1.3.29), which yields

$$R_0 = \frac{\pi^2 + \alpha_0^2}{\alpha_0^2 - \alpha_0^4}. \tag{4.2.60}$$

To study the behaviour of the perturbations in the vicinity of the neutral point (α_0, R_0), represent the solution to (4.2.58), (4.2.59) in the form

$$u = y + \varepsilon u_1(\tilde{t}, x, y) + \varepsilon^2 u_2(\tilde{t}, x, y) + \varepsilon^3 u_3(\tilde{t}, x, y) + \cdots,$$

where

$$\tilde{t} = \epsilon^2 t, \qquad R = R_0 + \varepsilon^2 r.$$

Show that the functions u_1, u_2, and u_3 satisfy the equations

$$\frac{\partial^2 u_i}{\partial y^2}-(R_0-1)\frac{\partial^2 u_i}{\partial x^2}-R_0\frac{\partial^4 u_i}{\partial x^4}=Q_i, \qquad i=1,\,2,\,3, \tag{4.2.61}$$

with

$$Q_1 = 0,$$

$$Q_2 = R_0 \frac{\partial u_1}{\partial y} \frac{\partial^2 u_1}{\partial x^2}, \tag{4.2.62}$$

$$Q_3 = R_0 \frac{\partial u_1}{\partial \tilde{t}} + r\left(\frac{\partial^4 u_1}{\partial x^4} + \frac{\partial^2 u_1}{\partial x^2}\right)$$

$$+ R_0\left(\frac{\partial u_1}{\partial y}\frac{\partial^2 u_2}{\partial x^2} + \frac{\partial u_2}{\partial y}\frac{\partial^2 u_1}{\partial x^2}\right). \tag{4.2.63}$$

What are the boundary conditions for u_1, u_2, and u_3?

(b) Seek the solution of the linear problem in the form

$$u_1 = A(\tilde{t})e^{i\alpha_0 x}\phi_1(y) + \overline{A}(\tilde{t})e^{-i\alpha_0 x}\overline{\phi}_1(y), \tag{4.2.64}$$

and show that, for the first root (4.2.60) of the dispersion equation, a suitably normalized function $\phi_1(y)$ may be written as

$$\phi_1 = \sin(\pi y).$$

(c) Next, consider the quadratic perturbations. Using (4.2.64) in (4.2.62) show that

$$Q_2 = A\overline{A}\mathcal{H}(y) + A^2 e^{i2\alpha_0 x}\mathcal{F}(y) + \overline{A}^2 e^{-i2\alpha_0 x}\overline{\mathcal{F}}(y),$$

where

$$\mathcal{H}(y) = -R_0\alpha_0^2\left(\phi_1 \frac{d\overline{\phi}_1}{dy} + \overline{\phi}_1 \frac{d\phi_1}{dy}\right), \qquad \mathcal{F}(y) = -R_0\alpha_0^2\phi_1 \frac{d\phi_1}{dy}.$$

Correspondingly, seek the solution for u_2 in the form

$$u_2 = A\overline{A}\,h(y) + A^2 e^{i2\alpha_0 x}\phi_2(y) + \overline{A}^2 e^{-i2\alpha_0 x}\overline{\phi}_2(y). \tag{4.2.65}$$

and deduce that

$$h = \frac{R_0\alpha_0^2}{4\pi}\sin(2\pi y), \qquad \phi_2 = \frac{\pi}{24\alpha_0^2}\sin(2\pi y).$$

(d) Finally, consider the cubic approximation. For this purpose, return to (4.2.61) and this time set $i = 3$. Using (4.2.64) and (4.2.65) in (4.2.63), deduce that

$$Q_3 = \left[R_0\phi_1 \frac{dA}{d\tilde{t}} + r(\alpha_0^4 - \alpha_0^2)\phi_1 A\right.$$

$$- R_0\alpha_0^2\left(4\phi_2 \frac{d\overline{\phi}_1}{dy} + \phi_1 \frac{dh}{dy} + \overline{\phi}_1 \frac{d\phi_2}{dy}\right)A^2\overline{A}\right]e^{i\alpha_0 x}$$

$$- R_0\alpha_0^2\left[\left(4\phi_2 \frac{d\phi_1}{dy} + \phi_1 \frac{d\phi_2}{dy}\right)A^3\right]e^{i3\alpha_0 x} + (c.c.).$$

Hence, seek u_3 in the form

$$u_3 = A^3 e^{i3\alpha_0 x}\phi_3(y) + e^{i\alpha_0 x}\Phi(\tilde{t}, y) + (c.c.).$$

Show that the function $\Phi(\tilde{t}, y)$ satisfies the equation

$$\frac{\partial^2 \Phi}{\partial y^2} + \pi^2 \Phi = \left[R_0 \frac{dA}{d\tilde{t}} + r(\alpha_0^4 - \alpha_0^2)A + \left(\frac{R_0^2 \alpha_0^4}{4} - \frac{\pi^2 R_0}{24}\right)A^2\overline{A}\right]\sin(\pi y)$$

$$- \left[\left(\frac{R_0^2 \alpha_0^4}{4} + \frac{3\pi^2 R_0}{24}\right)A^2\overline{A}\right]\sin(3\pi y).$$

Argue that the solution to this equation, satisfying the boundary conditions

$$\Phi = 0 \quad \text{at} \quad y = 0 \text{ and } y = 1,$$

exists only if

$$\frac{dA}{d\tilde{t}} + r\frac{\alpha_0^4 - \alpha_0^2}{R_0}A + \left(\frac{R_0 \alpha_0^4}{4} - \frac{\pi^2}{24}\right)A^2\overline{A} = 0. \qquad (4.2.66)$$

(e) Can you obtain the coefficient of A in (4.2.66) directly from the linear dispersion equation (1.3.29)?

2. Consider a fourth order ordinary differential equation

$$A(y)\frac{d^4 \Phi}{dy^4} + B(y)\frac{d^3 \Phi}{dy^3} + C(y)\frac{d^2 \Phi}{dy^2} + D(y)\frac{d\Phi}{dy} + E(y)\Phi = F(y) \qquad (4.2.67)$$

that should be solved on the interval $y \in [0, 1]$ subject to the boundary conditions

$$\Phi = \frac{d\Phi}{dy} = 0 \quad \text{at} \quad y = 0 \text{ and } y = 1. \qquad (4.2.68)$$

Your task is to formulate the boundary-value problem that is adjoint to (4.2.67), (4.2.68). You may perform this task in the following steps:

(a) Multiply all the terms in equation (4.2.67) by the function $\Phi^*(y)$ that represents a solution of the adjoint problem, and start with the integral

$$\int_0^1 A\Phi^* \frac{d^4 \Phi}{dy^4}\, dy. \qquad (4.2.69)$$

Perform integration by parts in (4.2.69) to show that

$$\int_0^1 A\Phi^* \frac{d^4 \Phi}{dy^4}\, dy = A\Phi^* \frac{d^3 \Phi}{dy^3}\bigg|_0^1 - \int_0^1 \frac{d^3 \Phi}{dy^3}\frac{d}{dy}\left(A\Phi^*\right) dy.$$

Continue to integrate by parts until all the derivatives are 'transferred' from Φ to Φ^*. You will see that

$$\int_0^1 A\Phi^* \frac{d^4 \Phi}{dy^4}\, dy = \left[A\Phi^* \frac{d^3 \Phi}{dy^3} - \frac{d}{dy}\left(A\Phi^*\right)\frac{d^2 \Phi}{dy^2}\right]_0^1 + \int_0^1 \Phi \frac{d^4}{dy^4}\left(A\Phi^*\right) dy.$$

(b) Similarly, show that

$$\int\limits_0^1 B\Phi^* \frac{d^3\Phi}{dy^3}\, dy = B\Phi^* \frac{d^2\Phi}{dy^2}\Big|_0^1 - \int\limits_0^1 \Phi \frac{d^3}{dy^3}(B\Phi^*)\, dy,$$

$$\int\limits_0^1 C\Phi^* \frac{d^2\Phi}{dy^2}\, dy = \int\limits_0^1 \Phi \frac{d^2}{dy^2}(C\Phi^*)\, dy,$$

$$\int\limits_0^1 D\Phi^* \frac{d\Phi}{dy}\, dy = -\int\limits_0^1 \Phi \frac{d}{dy}(D\Phi^*)\, dy.$$

(c) Combine the above integral relations to show that

$$\underbrace{\left[A\Phi^* \frac{d^3\Phi}{dy^3} - \frac{d}{dy}(A\Phi^*)\frac{d^2\Phi}{dy^2} + B\Phi^* \frac{d^2\Phi}{dy^2}\right]_0^1}_{\text{term 1}}$$

$$+ \underbrace{\int\limits_0^1 \Phi\mathcal{L}(\Phi^*)\, dy}_{\text{term 2}} = \int\limits_0^1 F\Phi^*\, dy, \qquad (4.2.70)$$

where

$$\mathcal{L}(\Phi^*) = \frac{d^4}{dy^4}(A\Phi^*) - \frac{d^3}{dy^3}(B\Phi^*) + \frac{d^2}{dy^2}(C\Phi^*) - \frac{d}{dy}(D\Phi^*) + E\Phi^*.$$

Argue that the second term on the left-hand side of equation (4.2.70) vanishes for an arbitrary function $\Phi(y)$ if

$$\mathcal{L}(\Phi^*) = 0. \qquad (4.2.71)$$

Argue further that for the first term to vanish, one has to pose the following boundary conditions for (4.2.71):

$$\Phi^* = \frac{d\Phi^*}{dy} = 0 \quad \text{at} \quad y = 0 \text{ and } y = 1. \qquad (4.2.72)$$

(d) Consider the particular case when the left-hand side of (4.2.67) is represented by the Orr–Sommerfeld equation as in (4.2.45). Show that in this case the adjoint equation is written as

$$\frac{d^4\Phi^*}{dy^4} - \left(2\alpha_0^2 + i\alpha_0 Re_0 U - i\omega_0 Re_0\right)\frac{d^2\Phi^*}{dy^2}$$

$$- 2i\alpha_0 Re_0 \frac{dU}{dy}\frac{d\Phi^*}{dy} + \left(\alpha_0^4 + i\alpha_0^3 Re_0 U - i\omega_0\alpha_0^2 Re_0\right)\Phi^* = 0. \qquad (4.2.73)$$

(e) How should the boundary conditions (4.2.72) be modified for a boundary-layer flow?

3. Show that the general solution of the Landau equation

$$\frac{d|A|^2}{d\tilde{t}} = 2r\kappa_r|A|^2 - l_r|A|^4 \tag{4.2.74}$$

may be written as

$$|A|^2 = \frac{2r\kappa_r C}{Cl_r - e^{-2r\kappa_r\tilde{t}}},$$

where C is an arbitrary constant.

Suggestion: Denote $|A|^2$ in (4.2.74) by Z, and perform separation of variables. You should find that

$$\frac{dZ}{(2r\kappa_r - l_rZ)Z} = d\tilde{t}. \tag{4.2.75}$$

Integrate equation (4.2.75) using a partial fraction decomposition on the left-hand side.

4. Use the Landau–Stuart equation

$$\frac{dA}{d\tilde{t}} = r\kappa A - \frac{l}{2}A^2\overline{A} \tag{4.2.76}$$

to show that the amplitude A_* of neutral oscillations is given by

$$A_* = \sqrt{\frac{2r\kappa_r}{l_r}}.$$

Suggestion: Seek the solution of (4.2.76) in the form

$$A = A_*e^{i\Omega\tilde{t}}, \tag{4.2.77}$$

where Ω is a real constant. Substitute (4.2.77) into (4.2.76) and consider the real part of the resulting equation.

4.3   Finite-Amplitude Nonlinear Travelling Wave Solutions

We have seen in Section 4.2 that weakly nonlinear stability analysis in the vicinity of a point on the linear neutral curve leads to the Landau-Stuart equation which possesses finite-amplitude equilibrium solutions in the form of travelling waves.[8] The analysis also allows us to predict the nature of the bifurcation (subcritical or supercritical) from linear to nonlinear travelling wave solutions. In this section we explore beyond the weakly nonlinear regime and compute fully nonlinear travelling wave solutions of the Navier–Stokes equations. Again for simplicity we shall concentrate on plane Poiseuille flow, but the methods involved are easily applied to other parallel flows. The first solutions of this type were computed by Herbert (1977) and over the years have been gradually refined and calculated at ever larger Reynolds number Re as computing

---

[8]These are obtained by substituting (4.2.77) into (4.2.33) and (4.2.35).

resources have become more powerful and numerical algorithms more sophisticated and efficient.

As demonstrated in Section 4.2.4, for nonlinear calculations we can either fix the streamwise pressure gradient or the fluid flux across the channel: in the calculations presented here we choose the former of these two possibilities. Correspondingly, we seek the pressure $p(t, x, y)$ in the form

$$p(t, x, y) = -\frac{2}{Re}x + \tilde{p}(\xi, y). \qquad (4.3.1)$$

The first term on the right-hand side of (4.3.1) is given by the laminar solution (1.2.11), while the second term $\tilde{p}(\xi, y)$ represents the perturbations. We seek solutions to $\tilde{p}(\xi, y)$ that depend not on x and t independently but on the phase variable

$$\xi = x - ct, \qquad (4.3.2)$$

where the phase speed c is taken to be purely real and to be determined as part of the solution. Corresponding to (4.3.1), we seek the velocity components (u, v) in the form

$$u(t, x, y) = U(y) + \tilde{u}(\xi, y), \qquad v(t, x, y) = \tilde{v}(\xi, y), \qquad (4.3.3)$$

where $U(y) = 1 - y^2$ is the velocity profile in the basic laminar flow (1.2.11).

Substitution of (4.3.1)–(4.3.3) into the Navier–Stokes equations (4.2.1) yields the following nonlinear governing equations

$$\left.\begin{aligned}
(U + \tilde{u} - c)\frac{\partial \tilde{u}}{\partial \xi} + \tilde{v}\left(\frac{dU}{dy} + \frac{\partial \tilde{u}}{\partial y}\right) &= -\frac{\partial \tilde{p}}{\partial \xi} + \frac{1}{Re}\left(\frac{\partial^2 \tilde{u}}{\partial \xi^2} + \frac{\partial^2 \tilde{u}}{\partial y^2}\right), \\
(U + \tilde{u} - c)\frac{\partial \tilde{v}}{\partial \xi} + \tilde{v}\frac{\partial \tilde{v}}{\partial y} &= -\frac{\partial \tilde{p}}{\partial y} + \frac{1}{Re}\left(\frac{\partial^2 \tilde{v}}{\partial \xi^2} + \frac{\partial^2 \tilde{v}}{\partial y^2}\right), \\
\frac{\partial \tilde{u}}{\partial \xi} + \frac{\partial \tilde{v}}{\partial y} &= 0,
\end{aligned}\right\} \qquad (4.3.4)$$

which are subject to the usual no-slip conditions on the channel walls, i.e.

$$\tilde{u} = \tilde{v} = 0 \quad \text{on} \quad y = \pm 1. \qquad (4.3.5)$$

A trivial solution $\tilde{u} - \tilde{v} = \tilde{p} = 0$ to (4.3.4), (4.3.5) is the basic plane Poiseuille flow state. Here we seek alternative nontrivial solutions assuming $\tilde{u}$, $\tilde{v}$, and $\tilde{p}$ to be periodic in ξ with period $2\pi/\alpha$ where α is a prescribed real wave number. This allows us to represent the solution in Fourier series form as

$$\tilde{u} = \sum_{n=-\infty}^{\infty} u_n(y)e^{i\alpha n\xi}, \qquad \tilde{v} = \sum_{n=-\infty}^{\infty} v_n(y)e^{i\alpha n\xi}, \qquad \tilde{p} = \sum_{n=-\infty}^{\infty} p_n(y)e^{i\alpha n\xi}. \qquad (4.3.6)$$

Here, due to nonlinearity, all harmonics should be retained whereas in the weakly nonlinear theory of Section 4.2 it proves necessary to only retain the first three harmonics to determine the equilibrium amplitudes. Of course, in weakly nonlinear theory we only perturb the linear neutral values of wave number α and phase speed c by a small amount so that the resulting amplitude is also small. Now we are considering an $O(1)$ perturbation to α and c and therefore anticipate the corresponding equilibrium amplitude to also be $O(1)$. This means that formally we require all the harmonics in (4.3.6). Of course, when performing the calculations, we can take advantage of the fact that the functions $\tilde{u}$, $\tilde{v}$, and $\tilde{p}$ are real, and therefore, the Fourier coefficients with negative n may be obtained by simply taking the complex conjugates of the coefficients with positive n, i.e.

$$u_{-n}(y) = \bar{u}_n(y), \quad v_{-n}(y) = \bar{v}_n(y), \quad p_{-n}(y) = \bar{p}_n(y) \quad \text{for all} \quad n \geq 1.$$

The terms $u_0(y)$, $p_0(y)$ representing distortion of the basic flow state are real, and $v_0(y) \equiv 0$; see Problem 1 in Exercises 14.

Substitution of (4.3.6) into (4.3.4) leads to the formulation of an infinite set of ordinary differential equations for the Fourier coefficients $u_n(y)$, $v_n(y)$, and $p_n(y)$; see Problem 1 in Exercises 14. This system of equations is then truncated at some suitably large $n = N$, and solved iteratively using the weakly nonlinear solution established in Section 4.2 as an initial guess. Once a converged nonlinear solution is obtained at a given amplitude it can be used as an initial seed for a Newton iteration at a larger disturbance level. The results of the calculations are summarized in Figure 4.6. The boundary of the shaded region in the (Re, α)-plane represents the linear neutral curve as seen in Figure 1.8(a) in Chapter 1. On Figure 4.6, a third axis is introduced that measures the amplitude $\mathcal{A}$ of the disturbance. Here it is defined in terms of the normal velocity component as

$$\mathcal{A} = \sqrt{\sum_{n=1}^{N} |v_n(y_0)|^2}, \tag{4.3.7}$$

with y_0 being a representative point in the channel. In this case we took $y_0 = 0$.

As was mentioned earlier, when performing the calculations, various strategies may be adopted. In Section 4.2 we fixed the wave number and moved away from the linear neutral curve by perturbing the Reynolds number (see Figure 4.2). The amplitude A_* of the neutral perturbations was then found from equation (4.2.51). Here, the Reynolds number is kept fixed and the amplitude $\mathcal{A}$ is increased in small steps, while the wave number α is found for each value of $\mathcal{A}$ as part of the solution of the problem. This produces a path in the $(Re, \alpha, \mathcal{A})$-space that lies in a plane which is drawn perpendicular to the Re-axis and crosses this axis at a chosen value of the Reynolds number. If Re is large enough, then the nonlinear solution path links the lower and upper branches of the linear neutral curve. In Figure 4.6, three such paths are shown, for $Re = 9000$, 7500 and 6000. We can see that at Reynolds numbers $Re = 9000$ and $Re = 7500$ the bifurcation from the lower branch neutral point is in the direction of increasing wave number α, and hence is supercritical, while on the upper branch the bifurcation is subcritical. However as the Reynolds number is decreased, the bifurcation from the lower branch neutral point changes to subcritical as can be seen from the solution curve for $Re = 6000$. These observations are consistent with the conclusions of weakly nonlinear stability theory and in particular the results presented in Figure 4.3.

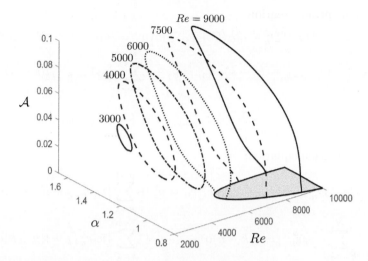

Fig. 4.6: Nonlinear neutral curves for plane Poiseuille flow at various values of the Reynolds number. Figure courtesy of Dr. R. Kumar.

A consequence of the subcritical bifurcation near the nose of the linear neutral curve is that nonlinear solutions exist below the critical Reynolds number $Re_c \simeq$ 5772.2 predicted by linear theory. A nonlinear solution path at a subcritical Reynolds number forms a closed loop in the $(\alpha, \mathcal{A})$-plane as can be seen on Figure 4.6 for Reynolds numbers $Re = 5000, 4000$, and 3000. Of course, solutions of this type may be constructed for other values of the Reynolds number. Considered together they form what may be termed a *nonlinear neutral surface* in the $(Re, \alpha, \mathcal{A})$-space. The minimum Reynolds number on this surface is approximately 2900, roughly half the linear critical Re.

In conclusion we remind the reader that in the above analysis the perturbations in (4.3.1), (4.3.3) are assumed two-dimensional. If this restriction is lifted and three-dimensionality is considered, many more closed loop solutions, totally unconnected to the linear neutral state, and with even lower critical Reynolds numbers, can be shown to exist (e.g. see Gibson and Brand, 2014). It transpires that the combined effects of nonlinearity and three-dimensionality lead to the formation of so-called *coherent structures* within the flow, with these providing the focus for Chapter 5.

Exercises 14

1. *The computation of neutral surfaces for plane Poiseuille flow*

   (a) Show that if we substitute (4.3.6) truncated at N terms into (4.3.4) then we obtain the following set of ordinary differential equations for $n = 1, \ldots, N$.

   Continuity equation:

   $$in\alpha u_n + v_n' = 0, \tag{4.3.8}$$

x-momentum equation:

$$\left[Re^{-1}D_n^2 - in\alpha(U + u_0 - c)\right]u_n = in\alpha p_n + v_n(U + u_0)'$$

$$+ \sum_{k=1}^{n-1}\left[v_k u'_{n-k} + i(n-k)\alpha u_{n-k}u_k\right] + \sum_{k=n+1}^{N}\left[v_k \overline{u}'_{k-n} - i(k-n)\alpha \overline{u}_{k-n}u_k\right]$$

$$+ \sum_{k=1}^{N-n}\left[\overline{v}_k u'_{n+k} + i(n+k)\alpha u_{n+k}\overline{u}_k\right],$$

y-momentum equation:

$$\left[Re^{-1}D_n^2 - in\alpha(U + u_0 - c)\right]v_n = p'_n + \sum_{k=1}^{n-1}\left[v_k v'_{n-k} + i(n-k)\alpha v_{n-k}u_k\right]$$

$$+ \sum_{k=n+1}^{N}\left[v_k \overline{v}'_{k-n} - i(k-n)\alpha \overline{v}_{k-n}u_k\right] + \sum_{k=1}^{N-n}\left[\overline{v}_k v'_{n+k} + i(n+k)\alpha v_{n+k}\overline{u}_k\right],$$

and the mean flow distortion equation:

$$\mathrm{Re}^{-1}D_0^2 u_0 = \sum_{k=1}^{N}\left(\overline{v}_k u'_k + c.c.\right). \tag{4.3.9}$$

Here $\prime$ denotes d/dy and $D_n^2 = d^2/dy^2 - n^2\alpha^2$.

In addition, write down an equation involving $p_0(y)$ and show that it uncouples from the other equations. Use the continuity equation (4.3.8) to explain why there is no term $v_0(y)$ in the expansion (4.3.6).

(b) Show that if (u_n, v_n, p_n) is a solution of the system of equations derived in part (a), then $(u_n, v_n, p_n)\exp(in\Theta)$ is also a solution for any real value of Θ.

(c) Show that the system written down in (a) together with the amplitude measure (4.3.7) constitutes $6N + 2$ real equations for $6N + 3$ real unknowns (for a prescribed value of Re). Explain how the property discussed in (b) allows us to specify an extra condition, thus ensuring that the number of equations matches the number of unknowns.

2. *Properties of the mean flow distortion*

Given a travelling wave expression of the form (4.3.6), the mean flow distortion u_0 satisfies

$$Re^{-1}u_0'' = \sum_{k=1}^{\infty}\overline{v}_k u'_k + c.c., \tag{4.3.10}$$

which is simply a rewrite of equation (4.3.9) but with all the harmonics retained, rather than a numerical truncation. Suppose that the travelling perturbation is linear and inviscid in nature, i.e. it satisfies the equations

$$ik\alpha u_k + v'_k = 0, \tag{4.3.11a}$$

$$ik\alpha(U + u_0 - c)u_k + v_k(U + u_0)' = -ik\alpha p_k, \tag{4.3.11b}$$

$$ik\alpha(U + u_0 - c)v_k = -p'_k, \tag{4.3.11c}$$

for each value of k. Show that, provided $U + u_0 \neq c$, the right-hand side of (4.3.10) is identically zero.

You can accomplish this task in the following steps:

(a) Eliminate the pressure between (4.3.11b) and (4.3.11c), and deduce that v_k satisfies

$$(U + u_0 - c)(v_k'' - k^2\alpha^2 v_k) = (U + u_0)'' v_k. \qquad (4.3.12)$$

(b) Use (4.3.12) and the continuity equation (4.3.11a) to deduce that

$$\bar{v}_k u_k' = i\left[\frac{(U + u_0)''}{k\alpha(U + u_0 - c)} + k\alpha\right]\bar{v}_k v_k.$$

(c) Hence deduce that $\bar{v}_k u_k'$ is purely imaginary, and that the right-hand side of (4.3.10) is therefore zero. What form does the mean flow distortion take in this case?

5

Coherent Structures and Self-Sustaining Processes in Shear Flows

We have seen earlier in this volume that one well-established route to transition to turbulence is via a sequence of steps, beginning with the generation of weak disturbances in the flow by means of the receptivity to external perturbations (Chapter 3), the linear growth of disturbances described by the Orr–Sommerfeld equation (Chapter 1), followed by weakly nonlinear growth, and finally the entering of a nonlinear stage governed by the full Navier–Stokes equations (Chapter 4). At each step of the process the flow becomes gradually more complicated, resulting ultimately in the apparently random motions characteristic of a turbulent flow. In reality the route to transition depends on the disturbance environment and the specific flow involved, and for some flows this classical description is not appropriate due to three main difficulties. The first is that the Reynolds number at which transition typically occurs in experiments and direct numerical simulations is observed to often be much less than the linear critical Reynolds number that is obtained from calculations of the Orr–Sommerfeld equation. The second unavoidable fact is that, as we have seen in Chapter 1, certain canonical flows, such as plane Couette flow and pipe Poiseuille flow, are linearly stable at all Reynolds numbers and in these cases there is therefore no linear instability to provide the second step of the ladder on this stairway to transition. Evidently, any theory which claims to explain the route to transition in such flows must be inherently nonlinear and somehow incorporate a way of reaching the 'higher rungs' of the ladder. Such a process, which involves removing the linear stage of transition, is often known as a *bypass mechanism* and is particularly appropriate when the disturbance level is relatively high. The final observation which calls into question the appropriateness of the classical linear route to transition, and one that we focus upon in this chapter, is the experimental revelation that within a turbulent flow it is often possible to discern the existence of *coherent structures* despite the apparent randomness of the overall motion. For example, streaky structures possessing spanwise variation have been seen in many experimental visualizations of shear flows such as that presented in Figure 5.1(a), which is for flow within a boundary layer. In addition these streaky structures often appear to be accompanied by a *longitudinal vortex structure*

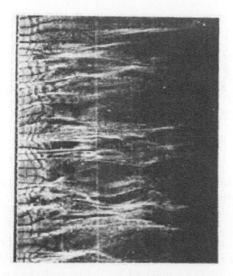

(a) Streaky structure in the boundary layer (image from Kline *et al.*, 1967); the flow is from left to right.

(b) Cross-flow visualization showing the existence of longitudinal vortices or streamwise rolls in a boundary-layer flow (from Lee and Lee, 2001).

Fig. 5.1: Experimental visualization of turbulent boundary layers.

(Figure 5.1b) in the cross-stream plane.[1] Motivated by these experimental observations, it has proved possible to identify the individual building blocks of a turbulent flow computationally: this was achieved in plane Couette flow by beginning with a fully turbulent flow simulation and then gradually reducing both the dimensions of the computational domain and the Reynolds number to eliminate much of the randomness associated with the location of the turbulent structures without destroying the structures themselves. It is then possible to observe an interaction between three distinct structures in the flow: a flow in the cross-stream plane which is independent of downstream distance, known as a *roll flow*, a streamwise component possessing some cross-stream variation (a *streak*), and finally *three-dimensional travelling waves* propagating in the streamwise direction, but with a complicated cross-stream structure. Numerical results from such a turbulent simulation are shown in Figure 5.2 at a particular instant in time: these cross-stream visualizations clearly show the streak and roll parts of the interaction.

In a flow such as plane Couette flow, if only two of these three components were present they would both inevitably decay in time. There clearly must therefore be some mechanism that operates only if all three components are present and which mutually sustains them against viscous decay. In this chapter we will investigate the nature of that feedback mechanism.

---

[1]In Section 1.7 we saw that in the presence of surface curvature such a structure could occur in the form of Taylor–Görtler vortices. It is noteworthy here that no such curvature is required for this structure to be observed.

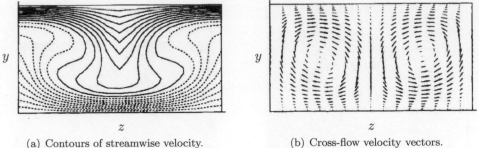

(a) Contours of streamwise velocity.   (b) Cross-flow velocity vectors.

Fig. 5.2: An instantaneous numerical visualization of turbulent plane Couette flow in which only the part of the flow which is independent of the downstream coordinate is plotted (from Hamilton *et al.*, 1995).

The theory was developed independently for asymptotically large Reynolds number (where exact sets of interaction equations can be formulated) and for finite Reynolds number where a number of heuristic approximations need to be made and the theory essentially forms the foundation for a successful numerical approach to generating three-dimensional solutions of the governing Navier–Stokes equations. The high-Reynolds-number theory is often referred to as *vortex–wave interaction* (VWI) and was formulated by Hall and Smith (1988, 1991) and co-workers in a series of papers in the late 1980s and early 1990s, while the finite-Reynolds-number approach or theory of *self-sustaining processes* (SSP) was the result of work by Waleffe and co-workers in the mid- to late 1990s (e.g. see Hamilton *et al.*, 1995; Waleffe, 1997). Although these studies were carried out independently, both were motivated by the failings of linear theory described above and experiments demonstrating the existence of coherent structures in turbulent shear flows such as those shown in Figure 5.1. In addition both were heavily-influenced by the mean-flow-wave interaction theory proposed by Benney (1984).

Our aim in this chapter is to introduce the reader to the ideas behind the SSP and VWI theories, to demonstrate that they involve the interaction of precisely the same structures, and to show that the fundamental ideas underpinning the instability mechanism are identical. It is only the detailed mathematical nature of the interaction which varies, according to whether the Reynolds number is taken to be finite or asymptotically large, and also whether the wave involved in the process is influenced predominantly by viscous or inviscid effects. For simplicity we will concentrate upon flow in a channel, although many of the ideas expounded here have also been successfully applied to pipe and boundary-layer flows.

We will tackle the finite-Reynolds-number SSP first and follow this with a formulation of the VWI interaction equations for the two cases of an inviscid and viscous travelling wave. Before we do this it is instructive to set out the fundamental processes underlying the interaction, which are common to both approaches.

5.1 The Fundamental Building Blocks of a Self-Sustaining Process

Throughout this chapter we will assume the fluid to be incompressible. The three-dimensional unsteady Navier–Stokes equations will be used in their dimensionless form:

$$\frac{\partial u}{\partial x} + \frac{\partial v}{\partial y} + \frac{\partial w}{\partial z} = 0, \tag{5.1.1a}$$

$$\frac{\partial u}{\partial t} + u\frac{\partial u}{\partial x} + v\frac{\partial u}{\partial y} + w\frac{\partial u}{\partial z} = -\frac{\partial p}{\partial x} + Re^{-1}\left(\frac{\partial^2 u}{\partial x^2} + \frac{\partial^2 u}{\partial y^2} + \frac{\partial^2 u}{\partial z^2}\right), \tag{5.1.1b}$$

$$\frac{\partial v}{\partial t} + u\frac{\partial v}{\partial x} + v\frac{\partial v}{\partial y} + w\frac{\partial v}{\partial z} = -\frac{\partial p}{\partial y} + Re^{-1}\left(\frac{\partial^2 v}{\partial x^2} + \frac{\partial^2 v}{\partial y^2} + \frac{\partial^2 v}{\partial z^2}\right), \tag{5.1.1c}$$

$$\frac{\partial w}{\partial t} + u\frac{\partial w}{\partial x} + v\frac{\partial w}{\partial y} + w\frac{\partial w}{\partial z} = -\frac{\partial p}{\partial z} + Re^{-1}\left(\frac{\partial^2 w}{\partial x^2} + \frac{\partial^2 w}{\partial y^2} + \frac{\partial^2 w}{\partial z^2}\right), \tag{5.1.1d}$$

where the Cartesian coordinates (x, y, z) represent the streamwise, wall-normal, and spanwise directions respectively with corresponding velocity components (u, v, w) and associated pressure field p. The precise form that the Reynolds number Re takes depends on the flow geometry and will be defined specifically for each problem we consider below.

As we already mentioned, there are three essential components that together comprise a self-sustained interaction. The first of these is the so-called *roll flow*. This is a flow in the cross-stream (y, z)-plane which has dependence solely on the cross-stream coordinates. Mathematically, a roll flow can be represented as

$$u = 0, \quad v = v_R(y, z, t), \quad w = w_R(y, z, t), \tag{5.1.2}$$

with the dependence on t allowing for the rolls to be unsteady. We will now derive equations governing the behaviour of such an unsteady roll flow under the action of viscosity. It is convenient to introduce a scaled time

$$\tau = t/Re, \tag{5.1.3}$$

and to write the roll flow in the form

$$v_R = Re^{-1}V_R(y, z, \tau), \quad w_R = Re^{-1}W_R(y, z, \tau), \tag{5.1.4}$$

together with an associated roll pressure

$$p = p_R = Re^{-2}P_R(y, z, \tau). \tag{5.1.5}$$

If we then substitute these expressions into the non-dimensional incompressible Navier–Stokes equations (5.1.1) we see that the nonlinear governing equations for the unsteady roll flow are

$$\frac{\partial V_R}{\partial y} + \frac{\partial W_R}{\partial z} = 0, \tag{5.1.6a}$$

$$\frac{\partial V_R}{\partial \tau} + V_R\frac{\partial V_R}{\partial y} + W_R\frac{\partial V_R}{\partial z} = -\frac{\partial P_R}{\partial y} + \frac{\partial^2 V_R}{\partial y^2} + \frac{\partial^2 V_R}{\partial z^2}, \tag{5.1.6b}$$

$$\frac{\partial W_R}{\partial \tau} + V_R\frac{\partial W_R}{\partial y} + W_R\frac{\partial W_R}{\partial z} = -\frac{\partial P_R}{\partial z} + \frac{\partial^2 W_R}{\partial y^2} + \frac{\partial^2 W_R}{\partial z^2}. \tag{5.1.6c}$$

Typically, these would be solved subject to the usual no-slip boundary conditions on the boundary S of the flow domain, i.e.

$$V_R = W_R = 0 \quad \text{on} \quad S. \tag{5.1.7}$$

There are two important observations to be made here: one is that the roll flow satisfies (5.1.6) exactly, in other words no approximations have been made in arriving at the governing roll equations; secondly, since equations (5.1.6) are independent of Re, the only dependence on Reynolds number is purely an algebraic one arising from the scalings given in (5.1.3), (5.1.4). This means that if a solution to (5.1.6) exists, it does so for all Reynolds numbers. However, it is readily established using energy stability theory (see Problem 3 in Exercises 15) that no steady solutions of (5.1.6), (5.1.7) exist and thus a roll flow cannot be maintained against viscous decay without the existence of a forcing mechanism arising from the presence of some other structure in the flow field.

Leaving aside for the moment the issue of how the roll field is maintained, the next building block in our self-sustained process is the *streak*. This is a flow in the streamwise direction possessing some cross-stream structure, i.e.

$$u = U_S(y, z, \tau).$$

Substitution of this expression into the longitudinal momentum equation (5.1.1b) alongside the roll form (5.1.4), shows that the streak satisfies the linear equation

$$\frac{\partial U_S}{\partial \tau} + V_R(y, z, \tau)\frac{\partial U_S}{\partial y} + W_R(y, z, \tau)\frac{\partial U_S}{\partial z} = K + \frac{\partial^2 U_S}{\partial y^2} + \frac{\partial^2 U_S}{\partial z^2} \tag{5.1.8}$$

exactly. Here K is a constant corresponding to the uniform pressure gradient that drives the underlying basic flow velocity which itself depends purely on the normal coordinate y. For example $K = 0$ for plane Couette flow where the flow is driven by the movement of the walls, and is non-zero for plane Poiseuille flow, where the walls are at rest.[2] It is already possible to see from (5.1.8) that the roll flow will automatically generate a streaky flow with complicated cross-stream dependence.

These simple arguments based on the Navier–Stokes equations had of course been appreciated for many years: what was missing from the jigsaw was the final piece that explained how the roll flow could be sustained and therefore continue to drive the streaky flow. The answer to the puzzle came in the form of *travelling waves*, which form the third and final essential component of the self-sustaining process. The waves are generated as a classical linear instability of the spanwise-modulated streaky flow, but then self-interact in a nonlinear fashion (along the lines anticipated by Benney, 1984, in his mean-field theory) to reinforce the roll flow and to keep the interaction alive.

We now look at the mathematical details of the theory for the finite-Reynolds-number SSP and for the large-Reynolds-number VWI. In the latter case we will consider interactions involving both inviscid waves, where it acts as a bypass mechanism,

---

[2]A detailed discussion of Couette and Poiseuille flows may be found in Section 2.1 in Part 1 of this book series.

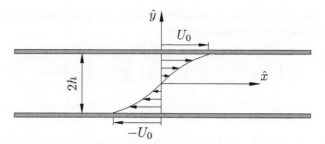

Fig. 5.3: The geometrical configuration for plane Couette flow.

and for viscous waves where the nonlinear interaction develops from a classical linear instability of the basic flow.

5.2   The Self-Sustaining Process (SSP) at Finite Reynolds Number

We will consider each part of the interaction in turn, before fitting the pieces together. For definiteness here we consider the classical plane Couette flow (see Figure 5.3) in which walls separated by a distance $2h$ move in the streamwise direction with equal and opposite constant velocities $\pm U_0$. We shall use h to make the coordinates dimensionless, and U_0 to scale the velocity components. Then the Reynolds number for this flow can then be defined as

$$Re = U_0 h/\nu, \tag{5.2.1}$$

and within the non-dimensional geometry, the walls lie at the scaled locations $y = \pm 1$.

5.2.1   The roll flow

In order to gain an insight into the nature of the roll solution in an analytical way, we suppose that the roll amplitude is small. This allows us to linearize equations (5.1.6) and determine solutions using elementary mathematical methods. Specifically, we write

$$V_R = \Delta \widetilde{V}_R, \qquad W_R = \Delta \widetilde{W}_R, \qquad P_R = \Delta \widetilde{P}_R, \tag{5.2.2}$$

with Δ representing the small roll amplitude. Substitution into the roll equations (5.1.6), followed by the discarding of terms of $O(\Delta^2)$, leads us to the linear equations

$$0 = \frac{\partial \widetilde{V}_R}{\partial y} + \frac{\partial \widetilde{W}_R}{\partial z}, \tag{5.2.3a}$$

$$\frac{\partial \widetilde{V}_R}{\partial \tau} = -\frac{\partial \widetilde{P}_R}{\partial y} + \frac{\partial^2 \widetilde{V}_R}{\partial y^2} + \frac{\partial^2 \widetilde{V}_R}{\partial z^2}, \tag{5.2.3b}$$

$$\frac{\partial \widetilde{W}_R}{\partial \tau} = -\frac{\partial \widetilde{P}_R}{\partial z} + \frac{\partial^2 \widetilde{W}_R}{\partial y^2} + \frac{\partial^2 \widetilde{W}_R}{\partial z^2}. \tag{5.2.3c}$$

The no-slip condition of viscous flow then implies that the appropriate boundary conditions are

$$\widetilde{V}_R = \widetilde{W}_R = 0 \quad \text{on} \quad y = \pm 1, \tag{5.2.4}$$

for all values of spanwise coordinate z and scaled time τ.

Eliminating the pressure by cross-differentiating (5.2.3b) and (5.2.3c) we find that the normal component of the roll satisfies

$$\left(\frac{\partial^2}{\partial y^2} + \frac{\partial^2}{\partial z^2}\right)^2 \widetilde{V}_R = \frac{\partial}{\partial \tau}\left(\frac{\partial^2}{\partial y^2} + \frac{\partial^2}{\partial z^2}\right)\widetilde{V}_R, \qquad (5.2.5)$$

where use has been made of the continuity equation, (5.2.3a), to eliminate the spanwise component $\widetilde{W}_R$.

We can proceed analytically by separating the variables, and seeking solutions in the normal-mode form

$$\widetilde{V}_R = \breve{v}_R(y)\, e^{i\beta z - \Lambda^2 \tau}. \qquad (5.2.6)$$

It follows that $\breve{v}_R$ satisfies the fourth order ordinary differential equation

$$\frac{d^4\breve{v}_R}{dy^4} - 2\beta^2\frac{d^2\breve{v}_R}{dy^2} + \beta^4\breve{v}_R + \Lambda^2\left(\frac{d^2\breve{v}_R}{dy^2} - \beta^2\breve{v}_R\right) = 0, \qquad (5.2.7)$$

and the no-slip boundary conditions

$$\breve{v}_R = \frac{d\breve{v}_R}{dy} \quad \text{on} \quad y = \pm 1, \qquad (5.2.8)$$

which follow from (5.2.4) and the continuity equation (5.2.3a).

Equation (5.2.7), considered together with the boundary conditions (5.2.8), constitute an eigen-value problem which determines the possible values of the roll growth/decay rate Λ^2. Defining

$$q^2 = \Lambda^2 - \beta^2, \qquad (5.2.9)$$

we can write the four independent solutions of (5.2.7) as

$$\cosh \beta y, \qquad \sinh \beta y, \qquad \cos qy, \qquad \sin qy.$$

Applying the boundary conditions (5.2.8) to a linear combination of these solutions, it becomes evident that the resulting solution is either odd or even about the channel centreline $y = 0$. For the even solution we have

$$\breve{v}_R^{(E)} = A \cosh \beta y + C \cos qy,$$

with

$$\begin{pmatrix} \cosh \beta & \cos q \\ \beta \sinh \beta & -q \sin q \end{pmatrix} \begin{pmatrix} A \\ C \end{pmatrix} = \begin{pmatrix} 0 \\ 0 \end{pmatrix} \qquad (5.2.10)$$

in view of (5.2.8). For non-trivial solutions the determinant of the above 2×2 matrix must vanish, leading to the eigenrelation

$$q \tan q + \beta \tanh \beta = 0. \qquad (5.2.11)$$

The quantity q could be complex-valued. However, by splitting (5.2.11) into two equations for the real and imaginary parts of q, it can be established that for given real values of the spanwise wave number β, all possible solutions for q are real. This implies, in view of (5.2.9), that the quantity Λ^2 is real and positive, and hence the even roll solutions are all damped. A list of the smallest three values of Λ^2 for $\beta = 5/3$ is given in Table 5.1. (The reason for choosing this particular value of spanwise wave number will be elaborated on presently). After solving (5.2.10) for the eigen-vector $\begin{pmatrix} A \\ C \end{pmatrix}$ we can write the even roll solution as

$$\widetilde{V}_R^{(E)} = A_0 \left(\frac{\cos qy}{\cos q} - \frac{\cosh \beta y}{\cosh \beta} \right) \cos(\beta z) \, e^{-\Lambda^2 \tau}, \tag{5.2.12a}$$

$$\widetilde{W}_R^{(E)} = A_0 \left(\frac{q \sin qy}{\beta \cos q} + \frac{\sinh \beta y}{\cosh \beta} \right) \sin(\beta z) \, e^{-\Lambda^2 \tau}, \tag{5.2.12b}$$

with the form for the spanwise component following from the continuity equation (5.2.3a). The constant of proportionality A_0, corresponding to the roll amplitude, is inevitably left undetermined due to the linearity of the original problem (5.2.3).

We now turn to the odd solution of (5.2.7) which takes the form

$$\breve{v}_R^{(O)} = B \sinh \beta y + D \sin qy.$$

This time, application of the boundary conditions (5.2.8) leads to the eigenrelation

$$q \tanh \beta - \beta \tan q = 0. \tag{5.2.13}$$

Again, for real β, the corresponding solutions for q are real and the rolls once more decay in time. A list of the corresponding three smallest decay rates for Λ^2 for the odd roll solution is given in Table 5.1, again for $\beta = 5/3$. It can be seen that all the modes are relatively heavily-damped. The odd roll solution with arbitrary amplitude B_0 can be expressed as

$$\widetilde{V}_R^{(O)} = B_0 \left(\frac{\sinh \beta y}{\sinh \beta} - \frac{\sin qy}{\sin q} \right) \cos(\beta z) \, e^{-\Lambda^2 \tau}, \tag{5.2.14a}$$

$$\widetilde{W}_R^{(O)} = B_0 \left(\frac{q \cos qy}{\beta \sin q} - \frac{\cosh \beta y}{\sinh \beta} \right) \sin(\beta z) \, e^{-\Lambda^2 \tau}. \tag{5.2.14b}$$

The streamlines in the (y, z)-plane corresponding to the least-decaying even and odd roll solutions are pictured instantaneously in Figure 5.4 for $\beta = 5/3$. Of course, these modes are damped and so the fluid motion will diminish over time. It is evident from these figures why this type of streamline pattern is often referred to as a 'roll flow'.

Table 5.1: The smallest three decay rates Λ^2 for the even and odd roll solutions with spanwise wave number $\beta = 5/3$, from the numerical solution of the eigenrelations (5.2.11), (5.2.13) with Λ^2 given in (5.2.9).

Λ^2 (Even solution)	Λ^2 (Odd solution)
9.5596	21.4362
39.1550	60.8934
88.5015	120.1057

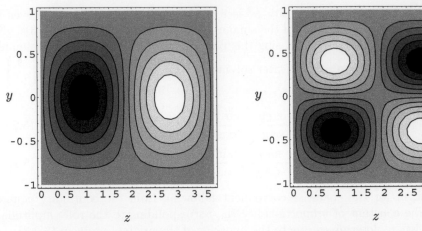

(a) The least-damped even roll mode.   (b) The least-damped odd roll mode.

Fig. 5.4: Streamlines for spanwise wave number $\beta = 5/3$.

As mentioned earlier, it can be proven (see Problem 3 in Exercises 15) that all non-trivial solutions of the *nonlinear* system (5.1.6), (5.1.7) decay in time: in other words there are no steady roll solutions of (5.1.6) satisfying no-slip boundary conditions. The implications of this result are profound: it means that if a roll flow is to be sustained against viscous decay, some other forcing needs to be introduced into the roll equations, necessarily complicating the flow field. We will see later that in our self-sustaining process the forcing will take the form of naturally-occurring three-dimensional travelling waves. Before we consider this aspect of the process however, we turn our attention to the streamwise streak that is generated by the two-dimensional roll flow.

5.2.2 The streamwise streak

As mentioned already, the second component of the SSP is a flow $u = U_S(y, z, \tau)$ in the streamwise direction possessing some cross-stream structure, and satisfying equation (5.1.8). If we assume a linearized roll field of the form (5.2.2), then (5.1.8) becomes

$$\frac{\partial U_S}{\partial \tau} + \Delta \tilde{V}_R(y, z, \tau) \frac{\partial U_S}{\partial y} + \Delta \widetilde{W}_R(y, z, \tau) \frac{\partial U_S}{\partial z} = \frac{\partial^2 U_S}{\partial y^2} + \frac{\partial^2 U_S}{\partial z^2}, \qquad (5.2.15)$$

for the present case of plane Couette flow where the applied streamwise pressure gradient $K = 0$. No approximations have been made at arriving at (5.2.15) and we see that the form of the streak is independent of Reynolds number and is driven explicitly by the rolls. The streak flow incorporates the basic parallel flow that exists for this specific geometrical configuration and boundary conditions. The appropriate steady, parallel solution to the Navier–Stokes equations (5.1.1) is simply[3]

$$u = y, \qquad (5.2.16)$$

[3]See Section 2.1.1 in Part 1 of this book series.

which, of course, corresponds to an exact solution $U_S = y$ of the streak equation (5.2.15) with no roll forcing, i.e. $\tilde{V}_R = \widetilde{W}_R = 0$. The reasons for studying our self-sustaining process in the context of plane Couette flow are two-fold. Firstly, the simple nature of the basic flow (5.2.16) leads to algebraic simplifications that facilitate our discussion; secondly, and more importantly, it was remarked upon in Section 1.2.2 of this volume that the profile (5.2.16) is linearly stable for all values of Re. In contrast, experiments and numerical simulations of the full Navier–Stokes equations (5.1.1) indicate that alternative nonlinear three-dimensional solutions to the plane Couette flow problem exist in the form of travelling waves, some with zero phase speed. Our aim here is to show how such solutions can be constructed using a roll/streak/wave interaction. Since we are concentrating on the plane Couette flow problem, the appropriate boundary conditions on (5.2.15) are

$$U_S = \pm 1 \quad \text{on} \quad y = \pm 1.$$

In the present situation, with the roll flow described by decaying modes of the form (5.2.12) or (5.2.14), solutions of (5.2.15) will inevitably revert, as $\tau \to \infty$, to the simple solution $U_S = y$.

Up to this point, all our analysis has been mathematically rigorous. In order to make further analytical progress it is now necessary to make some heuristic approximations that appear to hold good from a practical point of view when considering transitional flows. The model we obtain after making these assumptions will allow us to demonstrate that a numerical approach based upon a roll/streak/wave interaction could plausibly generate alternative solutions to the plane Couette flow problem. Furthermore the arguments can be easily extended to apply to other shear flows. Later, in sections 5.3, 5.4, we will be able to place the theory on a rigorous mathematical foundation, but only in the limit of large Reynolds number where, among other things, the roll/streak combination becomes a parallel flow at leading order in view of the Reynolds number dependence in the roll scaling (5.1.4).

We suppose now that by some means to be discussed presently, the roll flow has been stabilized so that a steady roll solution exists with the cross-stream dependence derived above. For definiteness we will choose the least-damped even mode which corresponds to (5.2.12). We therefore 'freeze' the roll flow at some arbitrary time ($\tau = 0$ say) so that the roll field is described by

$$\tilde{V}_R(y, z) = A_0 \left(\frac{\cos q_0 y}{\cos q_0} - \frac{\cosh \beta y}{\cosh \beta} \right) \cos \beta z, \tag{5.2.17a}$$

$$\widetilde{W}_R(y, z) = A_0 \left(\frac{q_0 \sin q_0 y}{\beta \cos q_0} + \frac{\sinh \beta y}{\cosh \beta} \right) \sin \beta z, \tag{5.2.17b}$$

with q_0 denoting the smallest root of (5.2.11) for a given spanwise wave number β. We now seek a steady solution for the streak flow satisfying (5.2.15) with the roll field given by (5.2.17). We can exploit the assumed smallness of the roll amplitude parameter Δ by seeking a solution in the form

$$U_S(y, z) = y + \Delta \breve{u}_{S1}(y) \cos \beta z + \Delta^2 \left[\breve{u}_{S0}(y) + \breve{u}_{S2}(y) \cos 2\beta z \right] + \cdots. \tag{5.2.18}$$

Here we anticipate the crucial occurrence of a spanwise-independent perturbation to the basic Couette flow (which we refer to as a *mean flow distortion*) in addition to harmonics in the spanwise coordinate. From substitution of (5.2.18) together with (5.2.17) into (5.2.15) we find that $\breve{u}_{S1}$ satisfies the equation

$$\frac{d^2\breve{u}_{S1}}{dy^2} - \beta^2\breve{u}_{S1} = A_0\left(\frac{\cos q_0 y}{\cos q_0} - \frac{\cosh \beta y}{\cosh \beta}\right), \tag{5.2.19a}$$

which should be solved subject to the boundary conditions

$$\breve{u}_{S1} = 0 \quad \text{on} \quad y = \pm 1. \tag{5.2.19b}$$

The solution of (5.2.19) is written as

$$\breve{u}_{S1} = A_0\left[\left(\frac{1}{q_0^2+\beta^2} + \frac{\tanh\beta}{2\beta}\right)\frac{\cosh\beta y}{\cosh\beta} - \frac{1}{q_0^2+\beta^2}\frac{\cos q_0 y}{\cos q_0} - \frac{y\sinh\beta y}{2\beta\cosh\beta}\right], \tag{5.2.20}$$

where A_0 is chosen such that $\breve{u}_{S1} = -1$ on $y = 0$.

A similar procedure establishes that the mean flow distortion and the second harmonic satisfy the forced equations

$$\frac{d^2\breve{u}_{S0}}{dy^2} = \frac{1}{2}\left(\breve{v}_R\frac{d\breve{u}_{S1}}{dy} - \beta\breve{w}_R\breve{u}_{S1}\right), \tag{5.2.21}$$

$$\frac{d^2\breve{u}_{S2}}{dy^2} - 4\beta^2\breve{u}_{S2} = \frac{1}{2}\left(\breve{v}_R\frac{d\breve{u}_{S1}}{dy} + \beta\breve{w}_R\breve{u}_{S1}\right), \tag{5.2.22}$$

where $\breve{w}_R$ is the coefficient of $\sin\beta z$ in (5.2.17b). The solutions of (5.2.21), (5.2.22) satisfying the no-slip conditions

$$\breve{u}_{S0}(\pm 1) = \breve{u}_{S2}(\pm 1) = 0,$$

can be determined analytically and they are plotted in Figure 5.5 alongside the corresponding form for $\breve{u}_{S1}$ calculated from (5.2.20).

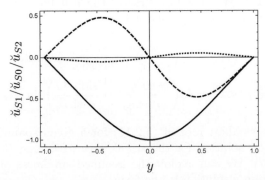

Fig. 5.5: The streak flow components for $\beta = 5/3$. First harmonic $\breve{u}_{S1}$ (solid line), mean flow distortion $\breve{u}_{S0}$ (dashed line), second harmonic $\breve{u}_{S2}$ (dotted line).

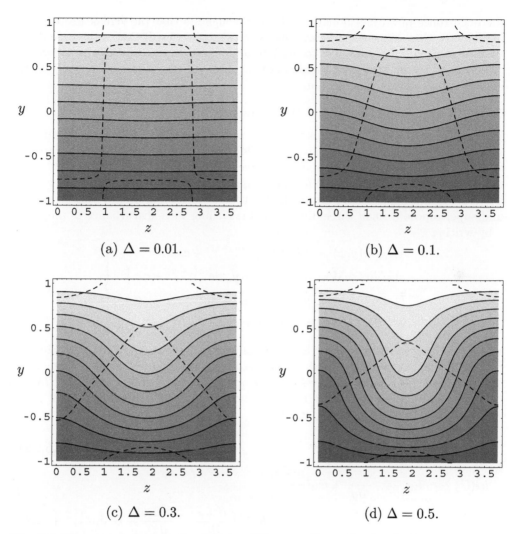

(a) $\Delta = 0.01$.   (b) $\Delta = 0.1$.

(c) $\Delta = 0.3$.   (d) $\Delta = 0.5$.

Fig. 5.6: The streak field at $\beta = 5/3$ for different roll amplitudes Δ. The dashed lines denote the paths along which the streak profile is inflexional.

It may be observed that the second harmonic term is significantly smaller numerically than the other terms: we will exploit this fact later. Combining these terms together with the basic Couette flow in (5.2.18) we can obtain an expression for an approximation of the streak response to the imposed roll field. This is shown in Figure 5.6 as a contour plot for various values of the roll amplitude Δ, again for the case $\beta = 5/3$. The streak profile looks remarkably similar to the structure found in the turbulent flow simulation referred to earlier (Figure 5.2). We can see that as Δ is increased the streak field becomes more disturbed and develops numerous points of inflexion. These are represented on the contour plots in Figure 5.6 by the dashed curves. Although not a sufficient condition for instability (see Theorem 1.2 on page 35), the occurrence of an inflexional streak profile suggests that the streak itself may be unstable to three-dimensional travelling wave disturbances. If this is the case it may then be possible for these waves to self-interact in such a nonlinear way as to force the roll flow, re-energizing it and sustaining it against viscous decay. We investigate this idea more fully in the next section.

5.2.3   The three-dimensional travelling wave

In Section 1.2 we saw how viscous travelling wave disturbances to a basic parallel flow satisfy the Orr–Sommerfeld equation (1.2.24) in which the growth rate of the wave depends on its streamwise and spanwise wave numbers (α, β) and the Reynolds number Re. We now wish to perform a similar analysis on the streak profile $U_S(y, z)$ given by (5.2.18) and depicted in Figure 5.6. We therefore take our steady roll/streak flow and superimpose a small three-dimensional travelling wave disturbance of amplitude ε by writing

$$
\left.
\begin{aligned}
u &= U_S(y, z) + \varepsilon\big[\tilde{u}_W(y, z)e^{i\alpha(x-ct)} + c.c.\big], \\
v &= Re^{-1}\Delta\widetilde{V}_R(y, z) + \varepsilon\big[\tilde{v}_W(y, z)e^{i\alpha(x-ct)} + c.c.\big], \\
w &= Re^{-1}\Delta\widetilde{W}_R(y, z) + \varepsilon\big[\tilde{w}_W(y, z)e^{i\alpha(x-ct)} + c.c.\big], \\
p &= Re^{-2}\Delta\widetilde{P}_R(y, z) + \varepsilon\big[\tilde{p}_W(y, z)e^{i\alpha(x-ct)} + c.c.\big],
\end{aligned}
\right\}
\tag{5.2.23}
$$

where *c.c.* denotes (as before) the complex conjugate of the term in front of it. Since the roll amplitude Δ is assumed small, the flow is dominated by the streak at leading order. The equations governing the travelling wave are obtained by substituting (5.2.23) into the Navier–Stokes equations (5.1.1) and linearizing, under the assumption that $\varepsilon \ll 1$. At $O(\varepsilon)$ we obtain the following equations:

$$
i\alpha\tilde{u}_W + \frac{\partial\tilde{v}_W}{\partial y} + \frac{\partial\tilde{w}_W}{\partial z} = 0,
\tag{5.2.24a}
$$

$$
i\alpha(U_S - c)\tilde{u}_W + \tilde{v}_W\frac{\partial U_S}{\partial y} + \tilde{w}_W\frac{\partial U_S}{\partial z}
$$

$$
= -i\alpha\tilde{p}_W + Re^{-1}\left(\frac{\partial^2\tilde{u}_W}{\partial y^2} + \frac{\partial^2\tilde{u}_W}{\partial z^2} - \alpha^2\tilde{u}_W\right),
\tag{5.2.24b}
$$

$$
i\alpha(U_S - c)\tilde{v}_W = -\frac{\partial\tilde{p}_W}{\partial y} + Re^{-1}\left(\frac{\partial^2\tilde{v}_W}{\partial y^2} + \frac{\partial^2\tilde{v}_W}{\partial z^2} - \alpha^2\tilde{v}_W\right),
\tag{5.2.24c}
$$

$$
i\alpha(U_S - c)\tilde{w}_W = -\frac{\partial\tilde{p}_W}{\partial z} + Re^{-1}\left(\frac{\partial^2\tilde{w}_W}{\partial y^2} + \frac{\partial^2\tilde{w}_W}{\partial z^2} - \alpha^2\tilde{w}_W\right).
\tag{5.2.24d}
$$

The usual no-slip condition is applied to the wave, namely

$$
\tilde{u}_W = \tilde{v}_W = \tilde{w}_W = 0 \quad \text{on} \quad y = \pm 1.
\tag{5.2.25}
$$

If the basic streak flow U_S were just to depend on the normal coordinate y, equations (5.2.24) would reduce to the Orr–Sommerfeld ordinary differential equation (1.2.24) describing the linear instability of a parallel flow to a three-dimensional disturbance. In the present more complicated situation where the streak profile is also dependent on the spanwise coordinate z, the wave components must be expressed in terms of Fourier series expansions in z, and the problem is a challenging numerical one best solved by spectral methods. We can however gain significant insight into the nature of the solution of (5.2.24) if we recall that the second harmonic term in the streak profile (5.2.18) is numerically small (Figure 5.5). If we neglect this term and the corresponding higher harmonics in the expansion of the wave, we can seek a solution of (5.2.24) in which

$$U_S(y,z) = y + \Delta \ddot{u}_{S1}(y)\cos\beta z + \Delta^2 \ddot{u}_{S0}(y), \qquad (5.2.26)$$

and the wave has the form

$$\tilde{u}_W = \ddot{u}_{W1}(y)\sin(\beta z), \quad \tilde{v}_W = \ddot{v}_{W1}(y)\sin(\beta z), \quad \tilde{p}_W = \ddot{p}_{W1}(y)\sin(\beta z), \qquad (5.2.27\text{a})$$
$$\tilde{w}_W = \ddot{w}_{W0}(y) + \ddot{w}_{W1}(y)\cos(\beta z), \qquad (5.2.27\text{b})$$

so that the overall solution (5.2.23) possesses the shift-reflect symmetry

$$(u,v,w,p)\Big|_{(x,y,z)} = (u,v,-w,p)\Big|_{(x+\pi/\alpha,y,-z)}. \qquad (5.2.28)$$

Substituting the ansatz consisting of (5.2.26) and (5.2.27) into (5.2.24) and neglecting higher harmonics, we obtain the eigen-value problem

$$i\alpha \ddot{u}_{W1} + \frac{d\ddot{v}_{W1}}{dy} - \beta \ddot{w}_{W1} = 0, \qquad (5.2.29\text{a})$$

$$i\alpha(y + \Delta^2 \ddot{u}_{S0} - c)\ddot{u}_{W1} + \ddot{v}_{W1} + \Delta^2 \ddot{v}_{W1}\frac{d\ddot{u}_{S0}}{dy}$$
$$-\beta\Delta \ddot{u}_{S1}\ddot{w}_{W0} = -i\alpha \ddot{p}_{W1} + Re^{-1}\left(\frac{d^2 \ddot{u}_{W1}}{dy^2} - \gamma^2 \ddot{u}_{W1}\right), \qquad (5.2.29\text{b})$$

$$i\alpha(y + \Delta^2 \ddot{u}_{S0} - c)\ddot{v}_{W1} = -\frac{d\ddot{p}_{W1}}{dy} + Re^{-1}\left(\frac{d^2 \ddot{v}_{W1}}{dy^2} - \gamma^2 \ddot{v}_{W1}\right), \qquad (5.2.29\text{c})$$

$$i\alpha(y + \Delta^2 \ddot{u}_{S0} - c)\ddot{w}_{W0} + \frac{1}{2}i\alpha\Delta \ddot{u}_{S1}\ddot{w}_{W1}$$
$$= Re^{-1}\left(\frac{d^2 \ddot{w}_{W0}}{dy^2} - \alpha^2 \ddot{w}_{W0}\right), \qquad (5.2.29\text{d})$$

$$i\alpha(y + \Delta^2 \ddot{u}_{S0} - c)\ddot{w}_{W1} + i\alpha\Delta \ddot{u}_{S1}\ddot{w}_{W0}$$
$$= -\beta \ddot{p}_{W1} + Re^{-1}\left(\frac{d^2 \ddot{w}_{W1}}{dy^2} - \gamma^2 \ddot{w}_{W1}\right), \qquad (5.2.29\text{e})$$

with

$$\ddot{u}_{W1} = \ddot{v}_{W1} = \ddot{w}_{W0} = \ddot{w}_{W1} = 0 \quad \text{on} \quad y = \pm 1,$$

and the streak perturbations $\ddot{u}_{S0}$, $\ddot{u}_{S1}$ given in (5.2.21), (5.2.20). In (5.2.29) we have defined $\gamma^2 \equiv \beta^2 + \alpha^2$. The key terms here are the effect of the *mean flow distortion* $\ddot{u}_{S0}$ and the term proportional to $\ddot{w}_{W0}$ in (5.2.29b). Without this latter term, the last two

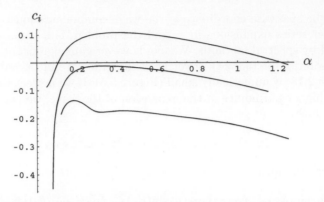

Fig. 5.7: Imaginary part of the phase speed versus streamwise wave number α for roll amplitudes (bottom to top) $\Delta = 0.3,\ 0.4,\ 0.5$ and $(\beta, Re) = (5/3, 400)$ from the solution of the eigen-value problem (5.2.29).

equations would uncouple from the first three. For given values of Δ, the solution to (5.2.29) and the resulting complex eigen-values for $c(\alpha)$ can be found by an elementary numerical technique such as a Runge–Kutta method. The disturbances are proportional to $e^{\alpha c_i t}$, where c_i is the imaginary part of c, and in Figure 5.7 we plot the mode with the largest value of c_i versus α for various values of the roll amplitude Δ, and with $(\beta, Re) = (5/3, 400)$. The motivation behind this choice of parameters was that computations of turbulent flows with these parameter values, in domains of decreasing dimension, provided clear evidence for the existence of a self-sustaining mechanism of the type investigated here (see, for example Hamilton *et al.*, 1995). The instability wave observed in our calculations is in fact stationary as the real part of c is zero for this particular mode for all values of the streamwise wave number α. As Δ is increased, we observe that the disturbance becomes unstable over a range of values of α with the disturbance passing through neutral (i.e. $c_i = 0$) for a second time at some specific value α_s say, such that the flow is stable for $\alpha > \alpha_s$. The wave contribution to the flow field can be quantified by considering the *inertial forcing* $\mathbf{F} = (F_y, F_z)$ where

$$F_y = i\alpha \tilde{u}_W^* \tilde{v}_W + \tilde{v}_W^* \frac{\partial \tilde{v}_W}{\partial y} + \tilde{w}_W^* \frac{\partial \tilde{v}_W}{\partial z} + c.c., \qquad (5.2.30a)$$

$$F_z = i\alpha \tilde{u}_W^* \tilde{w}_W + \tilde{v}_W^* \frac{\partial \tilde{w}_W}{\partial y} + \tilde{w}_W^* \frac{\partial \tilde{w}_W}{\partial z} + c.c.. \qquad (5.2.30b)$$

Here, and throughout this chapter, we use $*$ to represent complex conjugate. In Figure 5.8 we show contours of $|\mathbf{F}|$ for the case where $\alpha = \alpha_s$. The light colours indicate the higher values of this quantity and indicate that the effect of the wave is at its strongest at various locations along a dashed curve. This curve is the contour along which the streak profile (5.2.26) is equal to the phase speed of the neutral wave disturbance (zero in this case). Such a curve is known as a *critical curve* and it will be shown in the next subsection that the flow dynamics in the vicinity of this curve play a crucial role in the *roll/streak/wave interaction*.

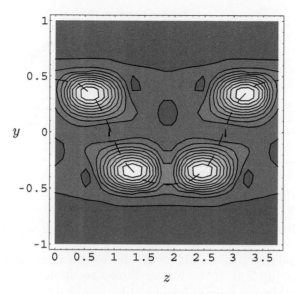

Fig. 5.8: Contour plot of the quantity $|\mathbf{F}|$ for the neutral wave with parameters $\Delta = 0.5$, $(\beta, Re) = (5/3, 400)$ and $\alpha = \alpha_s \simeq 1.21$, from the solution of the eigen-value problem (5.2.29).

We have therefore established that once the streak profile has sufficient spanwise variation, a three-dimensional neutral wave with a complicated cross-stream structure can be supported within the flow. We now need to demonstrate that this neutral disturbance can interact in a nonlinear way to reinforce and re-energize the roll flow that drove the streak originally.

5.2.4 The nonlinear feedback on the rolls

If we once again substitute the roll-streak-wave expansion (5.2.23) into the Navier–Stokes equations (5.1.1), but this time consider the components in the normal and spanwise directions which are independent of the streamwise coordinate x, we can derive an equation for the roll field that is nonlinearly-forced by the inertial response of the wave. The continuity equation

$$\frac{\partial \widetilde{V}_R}{\partial y} + \frac{\partial \widetilde{W}_R}{\partial z} = 0, \tag{5.2.31}$$

remains unchanged, but the normal and spanwise momentum balances are now

$$\varepsilon^2 F_y = \frac{\Delta}{Re^2} \left\{ -\frac{\partial \widetilde{P}_R}{\partial y} + \frac{\partial^2 \widetilde{V}_R}{\partial y^2} + \frac{\partial^2 \widetilde{V}_R}{\partial z^2} \right\}, \tag{5.2.32a}$$

$$\varepsilon^2 F_z = \frac{\Delta}{Re^2} \left\{ -\frac{\partial \widetilde{P}_R}{\partial z} + \frac{\partial^2 \widetilde{W}_R}{\partial y^2} + \frac{\partial^2 \widetilde{W}_R}{\partial z^2} \right\}, \tag{5.2.32b}$$

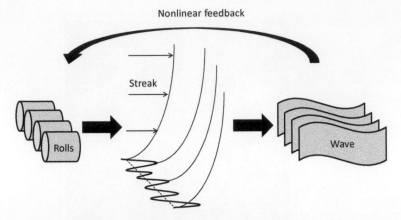

Fig. 5.9: A sketch of the self-sustaining process involving roll/streak/wave interaction.

where the left-hand sides (defined in (5.2.30) above) arise from the nonlinear convective acceleration term $(\mathbf{u} \cdot \nabla)\mathbf{u}$ in the Navier–Stokes equations. The roll equations will therefore be forced by the wave at leading order provided

$$\varepsilon = Re^{-1}\Delta^{1/2}\varepsilon_0, \tag{5.2.33}$$

with ε_0 of $O(1)$, indicating that only a tiny three-dimensional travelling wave disturbance is necessary to generate this strong feedback, especially in view of the relative smallness of the roll amplitude and the fact that in practice transitional Reynolds numbers often tend to be moderately large numerically. A schematic diagram of the SSP process is given in Figure 5.9.

In order to demonstrate the suitability of the three-pronged *roll/streak/wave interaction* for generating nonlinear solutions in plane Couette flow, we will show that the roll field generated by the wave interaction is similar to the original roll field (5.2.17) that was used to generate a streak and hence support the neutral wave in the first instance. To facilitate the solution we perform some manipulations on (5.2.32) which involve eliminating the roll pressure $\widetilde{P}_R$ by cross-differentiation and use of the continuity balances (5.2.31) and (5.2.24a) to eliminate the streamwise wave component $\widetilde{u}_W$ and spanwise roll component $\widetilde{W}_R$. This leaves the normal roll component governed by the equation

$$\left(\frac{\partial^2}{\partial y^2} + \frac{\partial^2}{\partial z^2}\right)^2 \widetilde{V}_R = \varepsilon_0^2 \mathcal{F}(y, z), \tag{5.2.34}$$

with

$$\mathcal{F}(y, z) = \frac{\partial^3}{\partial y \partial z^2}\left(\tilde{v}_W^* \tilde{v}_W - \tilde{w}_W^* \tilde{w}_W + c.c.\right)$$
$$+ \left(\frac{\partial^2}{\partial z^2} - \frac{\partial^2}{\partial y^2}\right)\frac{\partial}{\partial z}\left(\tilde{v}_W^* \tilde{w}_W + c.c.\right). \tag{5.2.35}$$

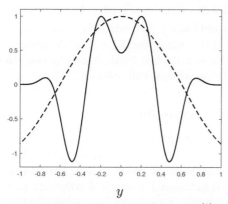 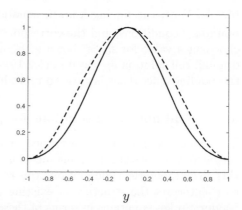

(a) Solid curve: forcing function $\overline{\mathcal{F}}^{(1)}(y)$; dashed curve: the original forcing given by $-\Lambda_0^2(\breve{v}_R'' - \beta^2\breve{v}_R)$.

(b) Solid curve: the regenerated roll solution $\breve{v}_R^{(1)}$; dashed curve: the original roll solution $\breve{v}_R$.

Fig. 5.10: Roll regeneration for spanwise wave number $\beta = 5/3$, Reynolds number $Re = 400$, axial wave number $\alpha = \alpha_s$ and roll amplitude $\Delta = 0.5$.

This is a forced, steady version of (5.2.5). Of particular interest here is the $\cos\beta z$ component of $\mathcal{F}(y, z)$ as this reinforces the original roll flow. If we denote this component by $\overline{\mathcal{F}}^{(1)}(y)$ and the corresponding $\cos\beta z$ component of the normal roll velocity as $\breve{v}_R^{(1)}(y)$, then by definition:

$$\overline{\mathcal{F}}^{(1)} = \frac{1}{\pi}\int_{-\pi}^{\pi}\mathcal{F}(y, z)\cos\beta z\, dz, \qquad \breve{v}_R^{(1)} = \frac{1}{\pi}\int_{-\pi}^{\pi}\widetilde{V}_R(y, z)\cos\beta z\, dz,$$

with these two components related via the expression

$$\left(\frac{\partial^2}{\partial y^2} - \beta^2\right)^2 \breve{v}_R^{(1)} = \varepsilon_0^2\overline{\mathcal{F}}^{(1)}. \tag{5.2.36}$$

Strictly speaking, all harmonics in z need to be retained when calculating the forcing function, but we will use our simplified low-dimensional approximation for the wave to check whether the solution to (5.2.36) for $\breve{v}_R^{(1)}$ is close to the original roll form $\breve{v}_R$ in (5.2.19). Using expansions (5.2.27) for the wave, and after some algebra, the forcing function reduces to

$$\overline{\mathcal{F}}^{(1)}(y) = -\beta^3\breve{v}_{W1}^*\breve{w}_{W0} - \beta\left(\breve{v}_{W1}''\breve{w}_{W0}^* + \breve{v}_{W1}^*\breve{w}_{W0}'' + 2\breve{v}_{W1}^{*\prime}\breve{w}_{W0}'\right)$$
$$+ 2\beta^2\left(\breve{w}_{W0}'\breve{w}_{W1}^* + \breve{w}_{W0}^*\breve{w}_{W1}'\right) + c.c.. \tag{5.2.37}$$

where 'prime' denotes differentiation with respect to y.

After solving (5.2.29) for the wave components, the forcing can be determined from (5.2.37) and, suitably normalized, is shown in Figure 5.10(a) for the neutral case of $(\beta, Re) = (5/3, 400)$ and $\alpha = \alpha_s$ for roll amplitude $\Delta = 0.5$. We also plot as a dashed curve the original effective time-dependent forcing given by the right-hand side of

(5.2.5). With the forcing known, equation (5.2.36) can be solved subject to no-slip boundary conditions, and the corresponding solution for $\breve{v}_R^{(1)}$, suitably normalized by choosing a value for ε_0, is shown in Figure 5.10(b), again for $\Delta = 0.5$, alongside the original roll solution $\breve{v}_R$ given in (5.2.19) (shown as a dashed line). It can be seen that the nonlinear feedback is close to reproducing the original roll solution.

5.2.5 Full numerical solutions for plane Couette flow

The calculations we have performed here act as a form of 'proof of concept' for the roll/streak/wave self-sustaining mechanism that we have outlined. Having established this firm theoretical foundation, the theory can then be used in two ways: firstly to re-interpret the structure of existing three-dimensional numerical solutions of the Navier–Stokes equations in terms of these three basic ingredients, and secondly to use the SSP as the basis of a numerical method to discover new alternative Navier–Stokes solutions. This has been achieved for plane Couette flow and for other important shear flows such as plane Poiseuille flow in a channel and Hagen–Poiseuille flow in a circular pipe. In the plane Couette flow case, many different solutions of both steady and travelling wave form, possessing widely-varying symmetry properties, have been found using this technique over a large range of wave number space, with solutions found down to Reynolds numbers as low as $Re = 150$. We finish this section by presenting the results of some of these calculations for plane Couette flow. The solutions are typically obtained by adopting a spectral approach involving Chebyshev polynomials in the y-direction and a Fourier series expansion in z, with continuation from existing states facilitated by using a multi-dimensional Newton iteration.

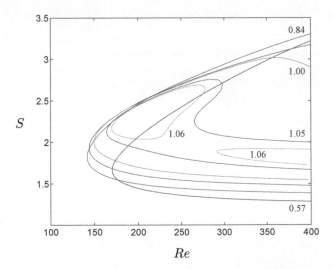

Fig. 5.11: Bifurcation diagram showing numerically-generated alternative steady solutions to plane Couette flow for spanwise wave number $\beta = 5/3$. Here S is the spanwise-averaged wall shear due to the streak. The values of axial wave number α are shown next to the curves. Figure courtesy of Waleffe (2002).

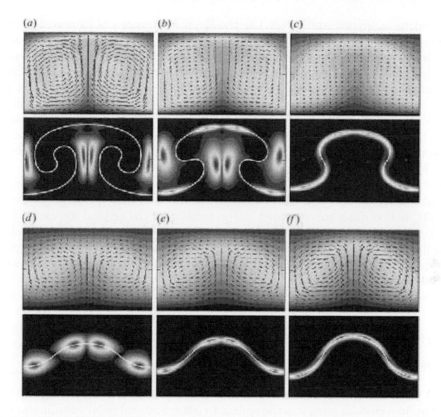

Fig. 5.12: Three-dimensional steady solutions of plane Couette flow at spanwise wave number $\beta = 2$ and Reynolds number $Re = 16000$. In each panel the upper plot shows contours of the streak, with arrows depicting the roll flow superimposed, while the lower plot depicts the wave forcing $|\mathbf{F}|$, defined in (5.2.30), with the contour of zero streak velocity denoted by a white curve. (a,b,c): upper branch solutions with streamwise wave numbers $\alpha = 0.4, 0.8, 1.2$; (d,e,f): lower branch solutions with $\alpha = 0.05, 0.8, 1.2$. From Deguchi and Hall (2014).

In Figure 5.11 we show a bifurcation diagram taken from Waleffe (2002) that illustrates the existence of some of these solutions for the value of $\beta = 5/3$ that we have concentrated on in our discussions. The diagram clearly shows the existence of three-dimensional steady solutions for various values of streamwise wave number α for Reynolds numbers up to 400. It can be seen that this particular class of solutions ceases to exist at $Re = 400$ when $\alpha > \alpha_c$ with $\alpha_c \simeq 1.1$. This wave number cut-off is very close to the value α_s predicted by our simple low-dimensional solution of the wave eigen-value problem (5.2.29) (see Figure 5.7).

Solutions have now been shown to exist for a wide range of values of β and at much higher Reynolds number. In Figure 5.12 we show the detailed flow fields for some plane Couette flow solutions for $(\beta, Re) = (2, 16000)$ computed by Deguchi and Hall (2014). For this spanwise wave number it proves possible to compute two branches of solutions (referred to as 'upper' and 'lower') as the Reynolds number is increased, apparently without bound. The solutions possess the same characteristic shapes of roll and streak fields seen earlier, although the spanwise variations of the streak field are much more pronounced for the upper branch solutions. It has recently been discovered that these two branches of solutions are both associated with turbulent flows but in slightly different ways. The lower branch solutions have been shown by Itano and Toh (2001), and Skufca *et al.* (2006) to act as so-called *edge states* which separate disturbances which become turbulent from those that return to the laminar state. The upper branch solutions appear to be associated with attractors for turbulent flows (Gibson *et al.*, 2008; Kreilos and Eckhardt, 2012).

Returning again to Figure 5.12, it is interesting to observe at this relatively large Reynolds number how the wave activity, measured by the quantity $|\mathbf{F}|$ defined in (5.2.30), is becoming concentrated in the vicinity of the critical location where the streak velocity is equal to the phase speed of the wave (denoted by a white curve in these figures). This is suggestive of the emergence of a distinct, simplified structure as $Re \to \infty$ and forms part of the motivation for the high-Reynolds-number version of the SSP presented in the next section.

Exercises 15

1. *The roll/streak equations in cylindrical geometry*

   When dealing with fluid flows in cylindrical geometry, such as Hagen–Poiseuille flow in a circular pipe, it is convenient to use cylindrical polar coordinates (x, r, ϕ). Here x is measured along the pipe axis, r is the radial coordinate, and ϕ is the azimuthal angle; see Figure 5.13. The corresponding velocity components are denoted as (u, v, w).

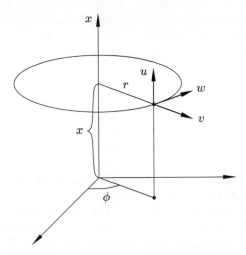

Fig. 5.13: Cylindrical polar coordinates.

The non-dimensional Navier–Stokes equations are written in these variables as[4]

$$\frac{\partial v}{\partial t} + v\frac{\partial v}{\partial r} + \frac{w}{r}\frac{\partial v}{\partial \phi} + u\frac{\partial v}{\partial x} - \frac{w^2}{r} = -\frac{\partial p}{\partial r}$$
$$+ Re^{-1}\left(\frac{\partial^2 v}{\partial x^2} + \frac{1}{r^2}\frac{\partial^2 v}{\partial \phi^2} - \frac{2}{r^2}\frac{\partial w}{\partial \phi} + \frac{\partial^2 v}{\partial r^2} + \frac{1}{r}\frac{\partial v}{\partial r} - \frac{v}{r^2}\right),$$

$$\frac{\partial w}{\partial t} + v\frac{\partial w}{\partial r} + \frac{w}{r}\frac{\partial w}{\partial \phi} + u\frac{\partial w}{\partial x} + \frac{vw}{r} = -\frac{1}{r}\frac{\partial p}{\partial \phi}$$
$$+ Re^{-1}\left(\frac{\partial^2 w}{\partial x^2} + \frac{1}{r^2}\frac{\partial^2 w}{\partial \phi^2} + \frac{2}{r^2}\frac{\partial v}{\partial \phi} + \frac{\partial^2 w}{\partial r^2} + \frac{1}{r}\frac{\partial w}{\partial r} - \frac{w}{r^2}\right),$$

$$\frac{\partial u}{\partial t} + v\frac{\partial u}{\partial r} + \frac{w}{r}\frac{\partial u}{\partial \phi} + u\frac{\partial u}{\partial x} = -\frac{\partial p}{\partial x}$$
$$+ Re^{-1}\left(\frac{\partial^2 u}{\partial x^2} + \frac{1}{r^2}\frac{\partial^2 u}{\partial \phi^2} + \frac{\partial^2 u}{\partial r^2} + \frac{1}{r}\frac{\partial u}{\partial r}\right),$$

$$\frac{\partial v}{\partial r} + \frac{1}{r}\frac{\partial w}{\partial \phi} + \frac{\partial u}{\partial x} + \frac{v}{r} = 0.$$

Starting from the above equations, use similar arguments to those presented in Section 5.2.1 to show that the unsteady roll satisfies

$$\frac{\partial V_R}{\partial r} + \frac{V_R}{r} + \frac{1}{r}\frac{\partial W_R}{\partial \phi} = 0, \tag{5.2.38}$$

$$\frac{\partial V_R}{\partial \tau} + V_R\frac{\partial V_R}{\partial r} + \frac{W_R}{r}\frac{\partial V_R}{\partial \phi} - \frac{W_R^2}{r}$$
$$= -\frac{\partial P_R}{\partial r} + \frac{\partial^2 V_R}{\partial r^2} + \frac{1}{r}\frac{\partial V_R}{\partial r} + \frac{1}{r^2}\frac{\partial^2 V_R}{\partial \phi^2} - \frac{V_R}{r^2} - \frac{2}{r^2}\frac{\partial W_R}{\partial \phi}, \tag{5.2.39}$$

$$\frac{\partial W_R}{\partial \tau} + V_R\frac{\partial W_R}{\partial r} + \frac{W_R}{r}\frac{\partial W_R}{\partial \phi} + \frac{V_R W_R}{r}$$
$$= -\frac{1}{r}\frac{\partial P_R}{\partial \phi} + \frac{\partial^2 W_R}{\partial r^2} + \frac{1}{r}\frac{\partial W_R}{\partial r} + \frac{1}{r^2}\frac{\partial^2 W_R}{\partial \phi^2} - \frac{W_R}{r^2} + \frac{2}{r^2}\frac{\partial V_R}{\partial \phi}, \tag{5.2.40}$$

while the streak equation is given by

$$\frac{\partial U_S}{\partial \tau} + V_R\frac{\partial U_S}{\partial r} + \frac{W_R}{r}\frac{\partial U_S}{\partial \phi} = K + \frac{\partial^2 U_S}{\partial r^2} + \frac{1}{r}\frac{\partial U_S}{\partial r} + \frac{1}{r^2}\frac{\partial^2 U_S}{\partial \phi^2}. \tag{5.2.41}$$

Suppose that the associated Reynolds number is based on the pipe radius and the centreline velocity. Show that in the absence of a roll flow the steady streak velocity is simply the Hagen–Poiseuille profile

$$U_S = 1 - r^2.$$

[4]See equations (1.8.45) in Part 1 of this book series.

2. *The form of the rolls in pipe flow*

Using equations (5.2.38)–(5.2.41), show that for pipe flow the linearized roll solutions can be expressed in the form

$$V_R = N\left[\frac{J_N(\mu r)}{r} - J_N(\mu)r^{N-1}\right]\cos(N\phi)\,e^{-\mu^2\tau}, \qquad (5.2.42)$$

$$W_R = \left[Nr^{N-1}J_N(\mu) - \mu J_N'(\mu r)\right]\sin(N\phi)\,e^{-\mu^2\tau}, \qquad (5.2.43)$$

where the possible decay rates μ are roots of the equation

$$\mu J_N'(\mu) = N J_N(\mu), \qquad (5.2.44)$$

with J_N a Bessel function of the first kind, and 'prime' denoting differentiation.
You can accomplish this task in the following steps:

(a) Based on the continuity equation (5.2.38) introduce the stream function ψ such that

$$V_R = \frac{1}{r}\frac{\partial\psi}{\partial\phi}, \qquad W_R = -\frac{\partial\psi}{\partial r}. \qquad (5.2.45)$$

Assuming the rolls to be small in amplitude, neglect the nonlinear terms in the momentum equations (5.2.39), (5.2.40). Then eliminate the pressure P_R by cross-differentiating the linearized momentum equations. Express the resulting equation in terms of ψ by using (5.2.45) and show that ψ satisfies

$$\nabla^2\left(\nabla^2\psi - \frac{\partial\psi}{\partial\tau}\right) = 0,$$

where

$$\nabla^2 = \frac{\partial^2}{\partial r^2} + \frac{1}{r}\frac{\partial}{\partial r} + \frac{1}{r^2}\frac{\partial^2}{\partial\phi^2}.$$

(b) Seek a normal-mode solution for ψ in the form

$$\psi = \widetilde{\psi}(r)e^{iN\phi}e^{-\mu^2\tau},$$

where N is an integer, and deduce that

$$\left(\frac{d^2}{dr^2} + \frac{1}{r}\frac{d}{dr} - \frac{N^2}{r^2}\right)F = 0, \qquad (5.2.46)$$

where

$$F(r) = \frac{d^2\widetilde{\psi}}{dr^2} + \frac{1}{r}\frac{d\widetilde{\psi}}{dr} + \mu^2\widetilde{\psi} - \frac{N^2}{r^2}\widetilde{\psi}. \qquad (5.2.47)$$

Show that the solution of (5.2.46) which is regular at $r = 0$ is

$$F(r) = Cr^N, \qquad (5.2.48)$$

where C is an arbitrary constant.

(c) Show that on the pipe wall ($r = 1$) the no-slip boundary conditions dictate that

$$\widetilde{\psi} = \frac{d\widetilde{\psi}}{dr} = 0. \tag{5.2.49}$$

Given the solution (5.2.48) for F, show that the solution of (5.2.47) for $\widetilde{\psi}$ that is regular at the axis of the pipe and satisfies the first condition in (5.2.49) is

$$\widetilde{\psi} = D\big[J_N(\mu r) - r^N J_N(\mu)\big]. \tag{5.2.50}$$

Explain why we can take the constant $D = 1$ without loss of generality.
   Hint: Remember that the Bessel equation of order N is written as

$$x^2 \frac{d^2 y}{dx^2} + x \frac{dy}{dx} + (x^2 - N^2)y = 0.$$

You may use without proof the fact that the two complementary solutions to this equation are the Bessel function of the first kind, $J_N(x)$, and the Bessel function of the second kind, $Y_N(x)$. The former is regular at $x = 0$, while the latter develops a singularity: $Y_N(x) = O(x^{-N})$.
(d) Show that the decay rate μ must satisfy the relation given in (5.2.44).
(e) Use (5.2.45) and your expression for the stream function (5.2.50) to deduce the components of the linearized roll and show that they agree with those given in (5.2.42), (5.2.43). These expressions are the cylindrical equivalents of the planar versions given in (5.2.12), (5.2.14).

3. *Energy analysis of the viscous decay of the roll equations*

   Consider the unsteady roll flow

$$u = 0, \qquad v = V(y, z, t), \qquad w = W(y, z, t).$$

The governing equations are

$$\frac{\partial V}{\partial y} + \frac{\partial W}{\partial z} = 0, \tag{5.2.51a}$$

$$\frac{\partial V}{\partial t} + V\frac{\partial V}{\partial y} + W\frac{\partial V}{\partial z} = -\frac{\partial P}{\partial y} + \frac{\partial^2 V}{\partial y^2} + \frac{\partial^2 V}{\partial z^2}, \tag{5.2.51b}$$

$$\frac{\partial W}{\partial t} + V\frac{\partial W}{\partial y} + W\frac{\partial W}{\partial z} = -\frac{\partial P}{\partial z} + \frac{\partial^2 W}{\partial y^2} + \frac{\partial^2 W}{\partial z^2}. \tag{5.2.51c}$$

These are to be solved subject to no-slip on the boundaries at $y = \pm 1$:

$$V = W = 0 \quad \text{on} \quad y = \pm 1, \tag{5.2.52}$$

and the condition of periodicity of $2\pi/\beta$ in z.

Multiply (5.2.51b) by V, (5.2.51c) by W and add the resulting equations together. Show that after integrating over the domain $-1 \leq y \leq 1$, $0 \leq z \leq 2\pi/\beta$, the left-hand side of your expression may be written in the form

$$\frac{d}{dt}\left[\int\limits_{0}^{2\pi/\beta} \int\limits_{-1}^{1} \frac{1}{2}(V^2 + W^2)\, dy\, dz \right].$$

By using integration by parts, the boundary conditions (5.2.52), and the spanwise periodicity, establish that the corresponding right-hand side can be expressed as

$$-\int\limits_{0}^{2\pi/\beta} \int\limits_{-1}^{1} \left[\left(\frac{\partial V}{\partial y}\right)^2 + \left(\frac{\partial W}{\partial y}\right)^2 + \left(\frac{\partial V}{\partial z}\right)^2 + \left(\frac{\partial W}{\partial z}\right)^2 \right] dy\, dz.$$

Hence deduce that no steady non-trivial spanwise-periodic solutions to (5.2.51), (5.2.52) exist.

5.3   Self-Sustaining Processes at High Reynolds Number: Vortex-Inviscid Wave Interaction

Although the finite-Reynolds-number description presented above provides a successful starting point for the computation of nonlinear solutions in plane Couette flow, and indeed other linearly stable flows such as pipe Poiseuille flow, the description of the interaction is not entirely satisfactory from a strict mathematical viewpoint. The obvious criticism is that because the roll field is assumed small in (5.2.2), its effect on the streak can inevitably only be a small perturbation to the undisturbed Couette flow, as is evident explicitly from expression (5.2.26). However, in order to generate a neutral wave it is obviously necessary, since plane Couette flow is linearly stable for all Re, that the streak flow represents a finite perturbation to the uniform shear flow. This can be seen in Figure 5.7 where a neutral wave is only found to exist provided Δ is in excess of 0.4 for the parameters chosen. On the other hand, if Δ is assumed $O(1)$ from the outset, the roll field will be comparable in size with the streak and the wave instability problem can no longer be formulated in terms of a small perturbation to a parallel, spanwise-modulated flow as in (5.2.23). The consequence of all of this is that if Δ is $O(1)$ there is no simplification at all to the governing Navier–Stokes equations (5.1.1). Thus, the amplitude of the rolls possesses something of a schizophrenic nature in the present finite-Reynolds-number theory, required to be asymptotically small in some places in the analysis and finite in others. The theory can however be put on a firm mathematical foundation if we assume that the Reynolds number is asymptotically large, i.e.

$$Re \gg 1,$$

for then the assumption that Δ is small will no longer be necessary as we shall see below. For definiteness we will once again consider the plane Couette flow problem (see Figure 5.3) for which we can define a Reynolds number

$$Re = U_0 h/\nu,$$

as in (5.2.1), where the walls move at velocities $\pm U_0$ in the x-direction and are separated by a distance $2h$. Virtually all the details we present here remain unchanged for channel flows subject to pressure gradients and flows in more complicated geometries such as pipe flows. The theory can also be extended fairly easily to non-parallel flows such as boundary-layer flow.

5.3.1 The core flow

We begin by posing the roll/streak/wave expansion first considered in (5.2.23), which we repeat here for convenience, but with the scaled roll amplitude Δ set equal to unity:

$$u = U_S(y, z) + \varepsilon\left[\tilde{u}_W(y, z)e^{i\alpha(x-ct)} + c.c.\right], \tag{5.3.1a}$$

$$v = Re^{-1}V_R(y, z) + \varepsilon\left[\tilde{v}_W(y, z)e^{i\alpha(x-ct)} + c.c.\right], \tag{5.3.1b}$$

$$w = Re^{-1}W_R(y, z) + \varepsilon\left[\tilde{w}_W(y, z)e^{i\alpha(x-ct)} + c.c.\right], \tag{5.3.1c}$$

$$p = Re^{-2}P_R(y, z) + \varepsilon\left[\tilde{p}_W(y, z)e^{i\alpha(x-ct)} + c.c.\right]. \tag{5.3.1d}$$

An important distinction from the earlier analysis is that since $Re \gg 1$, the flow is parallel to the x-direction at leading order without requiring the assumption of a small scaled roll amplitude: hence the reason that Δ can be set to unity without loss of generality. In the finite-Reynolds-number SSP, the wave amplitude ε was related to Δ in order to provide a strong wave feedback upon the rolls; see equation (5.2.33). In the present high-Reynolds-number approach, the philosophy is the same, namely that nonlinear wave interaction forces the roll flow, but now our analysis will identify ε in terms of the Reynolds number. This nonlinear forcing will be confined to a small region of the channel but will nevertheless exert a strong global effect on the roll flow in a manner that will become evident as we proceed with the analysis.

Substituting (5.3.1) into the Navier–Stokes equations (5.1.1) we obtain the nonlinear equations for the roll

$$\frac{\partial V_R}{\partial y} + \frac{\partial W_R}{\partial z} = 0, \tag{5.3.2a}$$

$$V_R\frac{\partial V_R}{\partial y} + W_R\frac{\partial V_R}{\partial z} = -\frac{\partial P_R}{\partial y} + \frac{\partial^2 V_R}{\partial y^2} + \frac{\partial^2 V_R}{\partial z^2}, \tag{5.3.2b}$$

$$V_R\frac{\partial W_R}{\partial y} + W_R\frac{\partial W_R}{\partial z} = -\frac{\partial P_R}{\partial z} + \frac{\partial^2 W_R}{\partial y^2} + \frac{\partial^2 W_R}{\partial z^2}, \tag{5.3.2c}$$

together with the streak equation

$$V_R(y, z)\frac{\partial U_S}{\partial y} + W_R(y, z)\frac{\partial U_S}{\partial z} = \frac{\partial^2 U_S}{\partial y^2} + \frac{\partial^2 U_S}{\partial z^2}. \tag{5.3.3}$$

Equations (5.3.2), (5.3.3) are the steady versions of equations (5.1.6), (5.2.15) with $\Delta = 1$, and are derived under the assumption that

$$\varepsilon \ll Re^{-1}, \tag{5.3.4}$$

which ensures that the roll equations are not forced directly by the wave at leading order. We will show later that this assumption on the size of ε is consistent with the asymptotic structure that emerges. In vortex-wave interaction terminology the roll/streak combination is referred to as a vortex flow (mainly because this interaction was first considered in curved channels where the Görtler vortex instability mechanism considered in Section 1.7.2 operates; see Hall and Smith, 1988). Note also that these equations are exact as the Reynolds number scaling cancels throughout. We have remarked already that these equations, when subject to no-slip boundary conditions, possess no solution other than the basic plane Couette flow $U_S = y$, $V_R = W_R = 0$. Some form of forcing is therefore required in order to generate alternative solutions: this will be provided by the wave in a novel manner, which is a much more subtle effect than the global forcing proposed in the finite-Reynolds-number version we have examined; see for example the forcing in equations (5.2.32).

The key difference between the finite-Reynolds-number and asymptotic descriptions arises when we consider the equation governing the evolution of the wave. Since we are now assuming that the Reynolds number is asymptotically large, across the majority of the pipe the dynamics are inviscid to leading order. Substitution of the expansions (5.3.1) into the Navier–Stokes equations (5.1.1) leads to the inviscid version of (5.2.24), namely:

$$i\alpha \tilde{u}_W + \frac{\partial \tilde{v}_W}{\partial y} + \frac{\partial \tilde{w}_W}{\partial z} = 0, \tag{5.3.5a}$$

$$i\alpha \big[U_S(y,z) - c \big] \tilde{u}_W + \tilde{v}_W \frac{\partial U_S}{\partial y} + \tilde{w}_W \frac{\partial U_S}{\partial z} = -i\alpha \tilde{p}_W, \tag{5.3.5b}$$

$$i\alpha \big[U_S(y,z) - c \big] \tilde{v}_W = -\frac{\partial \tilde{p}_W}{\partial y}, \tag{5.3.5c}$$

$$i\alpha \big[U_S(y,z) - c \big] \tilde{w}_W = -\frac{\partial \tilde{p}_W}{\partial z}. \tag{5.3.5d}$$

It is possible to eliminate the wave velocity components from these equations, leaving the wave pressure driven by the streak profile, and governed by the Rayleigh equation (see Problem 1 in Exercises 16):

$$\big[U_S(y,z) - c \big] \left(\frac{\partial^2 \tilde{p}_W}{\partial y^2} + \frac{\partial^2 \tilde{p}_W}{\partial z^2} - \alpha^2 \tilde{p}_W \right) = 2 \frac{\partial U_S}{\partial y} \frac{\partial \tilde{p}_W}{\partial y} + 2 \frac{\partial U_S}{\partial z} \frac{\partial \tilde{p}_W}{\partial z}. \tag{5.3.6}$$

This equation is to be solved subject to the usual inviscid impermeability condition on the boundaries:

$$\tilde{v}_W = 0 \quad \text{on} \quad y = \pm 1.$$

Using (5.3.5c), this condition may be expressed in terms of the pressure

$$\frac{\partial \tilde{p}_W}{\partial y} = 0 \quad \text{on} \quad y = \pm 1. \tag{5.3.7}$$

If U_S were independent of the spanwise coordinate z then equation (5.3.6) would reduce to the familiar one-dimensional Rayleigh equation (1.4.1) now written in terms of the pressure $\tilde{p}_W$. The nature of the wave involved here has led to this self-sustaining process often being referred to as *vortex-Rayleigh wave interaction*.

The basis of the interaction is that the streak profile $U_S(y, z)$ can support a neutral wave disturbance, i.e. one for which the phase speed c is real. A necessary consequence of this is that equation (5.3.6) possesses a singularity along the critical location $y = f(z)$ where[5]

$$U_S(f(z), z) = c. \tag{5.3.8}$$

We will now obtain asymptotic forms for the wave velocity components as this critical curve is approached.

5.3.2 Asymptotic behaviour near the critical curve

Close to the critical location we have that

$$U_S(y, z) - c = \lambda(z)\big[y - f(z)\big] + \cdots \quad \text{as} \quad y \to f(z), \tag{5.3.9}$$

where

$$\lambda(z) = \left.\frac{\partial U_S}{\partial y}\right|_{y=f(z)} \tag{5.3.10}$$

is the streak shear stress evaluated along the critical curve. It therefore follows from the normal and spanwise momentum balances (5.3.5c), (5.3.5d) that the corresponding wave velocity components adopt the singular forms

$$\left.\begin{aligned}
\tilde{v}_W &= -\big[i\alpha\lambda(z)\big]^{-1}\left.\frac{\partial \tilde{p}_W}{\partial y}\right|_{y=f(z)} \big[y - f(z)\big]^{-1} + \cdots, \\
\tilde{w}_W &= -\big[i\alpha\lambda(z)\big]^{-1}\left.\frac{\partial \tilde{p}_W}{\partial z}\right|_{y=f(z)} \big[y - f(z)\big]^{-1} + \cdots.
\end{aligned}\right\} \tag{5.3.11}$$

Next, assuming that the pressure at the critical level is finite, we define

$$\tilde{p}_{WC}(z) = \tilde{p}_W(f(z), z). \tag{5.3.12}$$

It follows from differentiation of (5.3.12), using the chain rule, that

$$\frac{d\tilde{p}_{WC}}{dz} = \left.\frac{\partial \tilde{p}_W}{\partial y}\right|_{y=f(z)} f_z + \left.\frac{\partial \tilde{p}_W}{\partial z}\right|_{y=f(z)}, \tag{5.3.13}$$

where the notation f_z is used to represent df/dz. Since the streak velocity U_S remains constant along the critical curve, a further application of the chain rule yields

$$\left.\frac{\partial U_S}{\partial y}\right|_{y=f(z)} f_z + \left.\frac{\partial U_S}{\partial z}\right|_{y=f(z)} = 0,$$

and hence

$$\left.\frac{\partial U_S}{\partial z}\right|_{y=f(z)} = -\lambda(z)f_z. \tag{5.3.14}$$

---

[5]For a detailed discussion of the critical layer theory the reader is referred to Chapter 2.

Evaluating the pressure equation (5.3.6) on the critical curve, it is evident that

$$\frac{\partial U_S}{\partial y}\frac{\partial \tilde{p}_W}{\partial y} + \frac{\partial U_S}{\partial z}\frac{\partial \tilde{p}_W}{\partial z} = 0 \quad \text{on} \quad y = f(z),$$

and upon substituting for $\partial U_S/\partial z$ from (5.3.14) and $\partial U_S/\partial y$ from (5.3.10), this expression simplifies to

$$\frac{\partial \tilde{p}_W}{\partial y} - f_z\frac{\partial \tilde{p}_W}{\partial z} = 0 \quad \text{on} \quad y = f(z).$$

Finally, using this result in (5.3.13), we can deduce from the asymptotic forms (5.3.11) that

$$\left.\begin{array}{l} \tilde{v}_W = -\left[i\alpha\lambda(z)\right]^{-1}\dfrac{f_z}{1+f_z^2}\dfrac{d\tilde{p}_{WC}}{dz}\left[y - f(z)\right]^{-1} + \cdots, \\[3mm] \tilde{w}_W = -\left[i\alpha\lambda(z)\right]^{-1}\dfrac{1}{1+f_z^2}\dfrac{d\tilde{p}_{WC}}{dz}\left[y - f(z)\right]^{-1} + \cdots \end{array}\right\} \quad \text{as} \quad y \to f(z). \quad (5.3.15)$$

The form for $\tilde{u}_W$ then immediately follows from the continuity equation (5.3.5a) as

$$\tilde{u}_W = -\alpha^{-2}\frac{d}{dz}\left[\frac{1}{\lambda(1+f_z^2)}\frac{d\tilde{p}_{WC}}{dz}\right]\left[y - f(z)\right]^{-1} + \cdots. \quad (5.3.16)$$

Thus all three components of wave velocity have an algebraic singularity along the critical curve. This singular behaviour needs to be smoothed out by processes occurring within a thin layer centred at the location $y = f(z)$. This is referred to as the *critical layer* and its consideration has arisen earlier in this volume, particularly when considering the high-Reynolds-number linear stability of boundary-layer and channel flows (see Chapter 2). In those applications the critical layer had no spanwise dependence and was therefore flat, rather than the present complicated unknown shape, determined as part of the solution. The essential principle remains the same however, namely that by reintroducing viscosity within a thin region it is possible to regularize the behaviour evident in (5.3.15), (5.3.16). In the critical layers of Section 2.4–2.6, particular emphasis was placed on the phase shift induced across the layer: while such a phenomenon still exists, it is very much a higher order effect within the highly-curved critical layer under consideration here.

The asymptotic forms (5.3.15) are quite revealing as they indicate that although all three components of wave velocity are singular at the critical level, the combination $\tilde{v}_W - f_z\tilde{w}_W$ remains finite, which means that the projection of the velocity vector on the direction normal to the critical curve is bounded, with the singularity being confined to the tangential component. This result suggests that the details of the critical layer may be uncovered more easily if we work in a coordinate system in which we consider velocity components locally normal and tangential to the critical curve. A suitable choice is the body-fitted coordinates,[6] with the critical curve playing the role of the body contour; see Figure 5.14.

---

[6]A detailed description of the body-fitted coordinates is given in Section 1.8 in Part 1 of this book series.

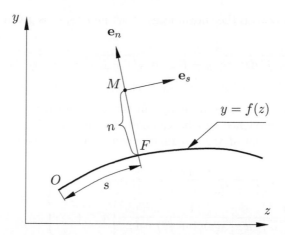

Fig. 5.14: The body-fitted coordinate system used to describe the flow within the critical layer.

Remember that in these coordinates the position of the observation point M is given by three coordinates (x, s, n). The first of these is measured, as before, in the streamwise direction, i.e. perpendicular to the sketch in Figure 5.14. The second coordinate is the arc length s measured between an arbitrarily chosen initial point O to the foot F of the perpendicular dropped from the point M to the critical curve. The third coordinate n is the distance between points M and F. The unit vector triad $(\mathbf{i}, \mathbf{e}_s, \mathbf{e}_n)$ is composed of the vector $\mathbf{i}$ directed along the x-axis, vector $\mathbf{e}_s$ tangential to the critical curve at point M, and the normal vector $\mathbf{e}_n$. Using these, the velocity vector for the roll motion is represented as

$$Re^{-1}\left[V_R^{(s)}(s, n)\,\mathbf{e_s} + V_R^{(n)}(s, n)\,\mathbf{e_n}\right],$$

with the components $V_R^{(s)}$, $V_R^{(n)}$ related to their counterparts in the Cartesian system by

$$V_R^{(s)} = \frac{V_R f_z + W_R}{\sqrt{1 + f_z^2}}, \qquad V_R^{(n)} = \frac{V_R - W_R f_z}{\sqrt{1 + f_z^2}}. \tag{5.3.17}$$

It should be noted that the streak velocity and roll pressure remain invariant with respect to the coordinate transformation.

A similar decomposition of the velocity vector for the wave motion is written as

$$\digamma\left[\tilde{u}_W(s, n)\,\mathbf{i} + \tilde{v}_W^{(s)}(s, n)\,\mathbf{e_s} + \tilde{v}_W^{(n)}(s, n)\,\mathbf{e_n}\right]e^{i\alpha(x - ct)} + \text{c.c.}.$$

Again, the tangential and normal components $(\tilde{v}_W^{(s)}, \tilde{v}_W^{(n)})$ are linked to their Cartesian counterparts $(\tilde{v}_W, \tilde{v}_W)$ via the relations

$$\tilde{v}_W^{(s)} = \frac{\tilde{v}_W f_z + \tilde{w}_W}{\sqrt{1 + f_z^2}}, \qquad \tilde{v}_W^{(n)} = \frac{\tilde{v}_W - \tilde{w}_W f_z}{\sqrt{1 + f_z^2}}. \tag{5.3.18}$$

The relationship between the coordinates (s, n) and (y, z) is given by

$$s = \int_0^z \sqrt{1 + f_z^2}\, dz; \qquad y - f(z) = \sqrt{1 + f_z^2}\, n + \cdots \quad \text{as} \quad n \to 0. \qquad (5.3.19)$$

In terms of this new coordinate system the corresponding wave behaviour as the critical curve is approached can be calculated from (5.3.15), (5.3.16), (5.3.18), and (5.3.19). For the tangential and streamwise components of the wave velocity we have

$$\left.\begin{aligned}
\tilde{v}_W^{(s)} &= -\left[\frac{1}{i\alpha\lambda(z)(1 + f_z^2)}\frac{d\tilde{p}_{WC}}{dz}\right]n^{-1} + \cdots, \\[2mm]
\tilde{u}_W &= -\frac{\alpha^{-2}}{\sqrt{1 + f_z^2}}\frac{d}{dz}\left[\frac{1}{\lambda(z)(1 + f_z^2)}\frac{d\tilde{p}_{WC}}{dz}\right]n^{-1} + \cdots
\end{aligned}\right\} \quad \text{as} \quad n \to 0. \qquad (5.3.20)$$

We will now investigate the dynamics of the critical layer using the (x, s, n) coordinate system with the aim of smoothing out this singular behaviour.

5.3.3   Inside the critical layer

A detailed discussion of the procedure that allowed us to express the Navier–Stokes equations in the body-fitted coordinates may be found in Section 1.8 in Part 1 of this book series. For ease of use the equations are repeated here:

$$\frac{1}{H_1}\frac{\partial v^{(s)}}{\partial s} + \frac{\partial v^{(n)}}{\partial n} + \frac{\kappa v^{(n)}}{H_1} + \frac{\partial u}{\partial x} = 0, \qquad (5.3.21a)$$

$$\frac{\partial v^{(s)}}{\partial t} + \frac{v^{(s)}}{H_1}\frac{\partial v^{(s)}}{\partial s} + v^{(n)}\frac{\partial v^{(s)}}{\partial n} + u\frac{\partial v^{(s)}}{\partial x} + \frac{\kappa}{H_1}v^{(n)}v^{(s)} = $$
$$-\frac{1}{H_1}\frac{\partial p}{\partial s} + Re^{-1}\left[\frac{1}{H_1}\frac{\partial}{\partial s}\left(\frac{1}{H_1}\frac{\partial v^{(s)}}{\partial s}\right) + \frac{\partial^2 v^{(s)}}{\partial n^2} + \frac{\partial^2 v^{(s)}}{\partial x^2}\right.$$
$$\left. + \kappa\frac{\partial}{\partial n}\left(\frac{v^{(s)}}{H_1}\right) + \frac{1}{H_1}\frac{\partial}{\partial s}\left(\frac{\kappa v^{(n)}}{H_1}\right) + \frac{\kappa}{H_1^2}\frac{\partial v^{(n)}}{\partial s}\right], \qquad (5.3.21b)$$

$$\frac{\partial v^{(n)}}{\partial t} + \frac{v^{(s)}}{H_1}\frac{\partial v^{(n)}}{\partial s} + v^{(n)}\frac{\partial v^{(n)}}{\partial n} + u\frac{\partial v^{(n)}}{\partial x} - \frac{\kappa}{H_1}\left(v^{(s)}\right)^2 = $$
$$-\frac{\partial p}{\partial n} + Re^{-1}\left[\frac{1}{H_1}\frac{\partial}{\partial s}\left(\frac{1}{H_1}\frac{\partial v^{(n)}}{\partial s}\right) + \frac{\partial^2 v^{(n)}}{\partial n^2} + \frac{\partial^2 v^{(n)}}{\partial x^2}\right.$$
$$\left. + \kappa\frac{\partial}{\partial n}\left(\frac{v^{(n)}}{H_1}\right) - \frac{1}{H_1}\frac{\partial}{\partial s}\left(\frac{\kappa v^{(s)}}{H_1}\right) - \frac{\kappa}{H_1^2}\frac{\partial v^{(s)}}{\partial s}\right], \qquad (5.3.21c)$$

$$\frac{\partial u}{\partial t} + \frac{v^{(s)}}{H_1}\frac{\partial u}{\partial s} + v^{(n)}\frac{\partial u}{\partial n} + u\frac{\partial u}{\partial x} = $$
$$-\frac{\partial p}{\partial x} + Re^{-1}\left[\frac{1}{H_1}\frac{\partial}{\partial s}\left(\frac{1}{H_1}\frac{\partial u}{\partial s}\right) + \frac{\partial^2 u}{\partial n^2} + \frac{\partial^2 u}{\partial x^2} + \frac{\kappa}{H_1}\frac{\partial u}{\partial n}\right]. \qquad (5.3.21d)$$

It should be noted that the notation differs slightly here from that in Part 1 where the coordinate z was used, rather than x. The quantity H_1 is the Lamé coefficient. It is related to the local curvature $\kappa(s)$ of the critical curve by

$$H_1 = 1 + \kappa(s)n. \tag{5.3.22}$$

We define the curvature as

$$\kappa(s) = -\frac{f_{zz}}{(1 + f_z^2)^{3/2}}, \tag{5.3.23}$$

such that κ is positive for a convex critical curve and negative for concave critical curve. Here, and in what follows, we denote the first and second derivatives of the critical curve shape function $f(z)$ by f_z and f_{zz}, respectively.

Since the critical layer is thin we introduce a scaled normal coordinate N by

$$n = \delta N, \tag{5.3.24}$$

where the parameter $\delta \ll 1$ and will be identified subsequently in terms of the Reynolds number.

We now need to write down the asymptotic expansions for the velocity components in the critical layer. These take the form

$$u = c + \delta \mu(s) N + \cdots$$
$$+ (\varepsilon/\delta)\left[\overline{u}_W(s, N)\, e^{i\alpha(x-ct)} + c.c.\right] + \cdots, \tag{5.3.25a}$$

$$v^{(s)} = Re^{-1}\left[\overline{V}_{R0}^{(s)}(s, N) + \delta \overline{V}_{R1}^{(s)}(s, N) + \cdots\right]$$
$$+ (\varepsilon/\delta)\left[\overline{v}_W^{(s)}(s, N)\, e^{i\alpha(x-ct)} + c.c.\right] + \cdots, \tag{5.3.25b}$$

$$v^{(n)} = Re^{-1}\left[\overline{V}_{R0}^{(n)}(s, N) + \delta \overline{V}_{R1}^{(n)}(s, N) + \cdots\right]$$
$$+ \varepsilon\left[\overline{v}_W^{(n)}(s, N)\, e^{i\alpha(x-ct)} + c.c.\right] + \cdots. \tag{5.3.25c}$$

The first two terms in the streamwise velocity u are the Taylor expansion of the streak velocity $U_S(s, n)$ about the critical curve where, from comparison of (5.3.25a) with (5.3.9) and use of (5.3.19) we have

$$\mu(s) = \sqrt{1 + f_z^2}\, \lambda(z). \tag{5.3.26}$$

The scalings for the wave reflect the fact that the streamwise and tangential components will be larger here than in the rest of the flow field by a factor δ^{-1} due to the singular behaviour evident in (5.3.20). Note that, since the wave velocity normal to the critical curve remains finite, it retains the same $O(\varepsilon)$ size in both the core region and the critical layer and is formally smaller than the other components. This difference in size between the wave velocity components leads to a considerable simplification in the governing equations as we shall see presently. The corresponding expansion for the pressure is

$$p = Re^{-2}\left[\overline{P}_{R0}(s, N) + \delta \overline{P}_{R1}(s, N) + \cdots\right]$$
$$+ \varepsilon\left[\overline{p}_W(s, N)\, e^{i\alpha(x-ct)} + c.c.\right] + \cdots. \tag{5.3.27}$$

Wave motion

The velocity and pressure expansions (5.3.25), (5.3.27) within the critical layer are substituted into the Navier–Stokes equations (5.3.21) together with the normal coordinate scaling (5.3.24). A consequence of this scaling is that the Lamé coefficient (5.3.22) is $1 + O(\delta)$, and can be taken to be unity to the order of magnitude we need in our analysis. To leading order, the equation (5.3.21b) governing the tangential wave component can be seen to be

$$\varepsilon i \alpha \mu N \overline{v}_W^{(s)} = -\varepsilon \frac{\partial \overline{p}_W}{\partial s} + Re^{-1}(\varepsilon/\delta^3) \frac{\partial^2 \overline{v}_W^{(s)}}{\partial N^2}. \tag{5.3.28}$$

Clearly, the viscous forces restore their action in the critical layer provided

$$\delta = Re^{-1/3}, \tag{5.3.29}$$

which fixes the thickness of the critical layer in terms of the Reynolds number.

Similarly, the other equations in (5.3.21) assume the form

$$i\alpha \overline{u}_W + \frac{\partial \overline{v}_W^{(n)}}{\partial N} + \frac{\partial \overline{v}_W^{(s)}}{\partial s} = 0, \tag{5.3.30a}$$

$$\frac{\partial \overline{p}_W}{\partial N} = 0, \tag{5.3.30b}$$

$$i\alpha \mu N \overline{u}_W + \mu \overline{v}_W^{(n)} + \frac{d\mu}{ds} N \overline{v}_W^{(s)} = -i\alpha \overline{p}_W + \frac{\partial^2 \overline{u}_W}{\partial N^2}. \tag{5.3.30c}$$

The boundary conditions for equations (5.3.28) and (5.3.30) are formulated by matching with the solution (5.3.20) in the core of the flow field. We have

$$\left. \begin{array}{r} \overline{u}_W \to 0, \\ \overline{v}_W^{(s)} \to 0 \end{array} \right\} \quad \text{as} \quad N \to \pm\infty. \tag{5.3.31}$$

Equation (5.3.30b) signifies that the wave pressure is independent of the normal coordinate N measured across the critical layer, and so $\overline{p}_W = \overline{p}_W(s)$. From matching of (5.3.27) with the core flow pressure expansion (5.3.1d) we see that the wave pressure in the critical layer is simply the core flow wave pressure evaluated along the critical curve, i.e.

$$\overline{p}_W(s) = \overline{p}_W(s, 0) = \widetilde{p}_{WC}(z), \tag{5.3.32}$$

with $\widetilde{p}_{WC}(z)$ introduced in (5.3.12). It is worth noting that this quantity is known already in principle, from the solution of the Rayleigh equation (5.3.6) in the core, and it is this pressure that drives the wave motion.

Although all the velocity components of the wave can be calculated it will prove sufficient for our purposes to concentrate on the tangential component. With the scaling (5.3.29) in place, it can be shown by the method of Fourier transforms that the solution of (5.3.28) which satisfies the conditions for $\overline{v}_W^{(s)}$ in (5.3.31) is given by[7]

---

[7] See Problem 3 in Exercises 16.

$$\overline{v}_W^{(s)}(s, N) = -[\alpha\mu(s)]^{-2/3}\frac{d\overline{p}_W}{ds}\int_0^\infty \exp\left\{-t^3/3 - i[\alpha\mu(s)]^{1/3}tN\right\}dt. \qquad (5.3.33)$$

It transpires that the most important property of this solution is that

$$\int_{-\infty}^\infty \left|\overline{v}_W^{(s)}\right|^2 dN = \frac{1}{2}n_0[\alpha\mu(s)]^{-5/3}\left|\frac{d\overline{p}_W}{ds}\right|^2, \qquad (5.3.34)$$

with

$$n_0 = 2\pi(2/3)^{2/3}\Gamma(1/3). \qquad (5.3.35)$$

Roll motion

Having solved for the wave in the critical layer, we now turn to the corresponding solution for the roll flow. From substitution of (5.3.25b), (5.3.25c) into the continuity equation (5.3.21a), we find, that at leading order

$$\frac{\partial\left(\overline{V}_{R0}^{(n)}\right)}{\partial N} = 0,$$

indicating that

$$\overline{V}_{R0}^{(n)} = \overline{V}_{R0}^{(n)}(s, 0),$$

i.e. the normal component of the roll velocity is continuous across the critical layer. At $O(Re^{-2}\delta^{-2})$ in the s-momentum equation (5.3.21b) we find

$$\frac{\partial^2\left(\overline{V}_{R0}^{(s)}\right)}{\partial N^2} = 0,$$

which implies, upon matching the critical layer expansion (5.3.25b) to the core flow expansions (5.3.1b), (5.3.1c), projected on the s-axis, that the tangential component is given by

$$\overline{V}_{R0}^{(s)} = \overline{V}_{R0}^{(s)}(s, 0).$$

It therefore emerges that the entire leading-order roll flow contribution within the critical layer is simply the core roll field evaluated on the critical curve.

At next order however, wave forcing comes into play, and from the normal momentum equation (5.3.21c) at order $O(Re^{-2}\delta^{-1})$ we obtain the following balance between the roll pressure gradient and the curvature-induced centrifugal terms arising from nonlinear interactions of the wave:

$$-\kappa(\varepsilon/\delta)^2\left(\overline{v}_W^{(s)}\overline{v}_W^{(s)*} + c.c.\right) = -Re^{-2}\delta^{-1}\frac{\partial\overline{P}_{R0}}{\partial N}. \qquad (5.3.36)$$

The terms are in balance provided $(\varepsilon/\delta)^2 \sim Re^{-2}\delta^{-1}$, with $\delta = Re^{-1/3}$ from (5.3.29). This uniquely identifies the wave amplitude parameter ε as

$$\varepsilon = Re^{-7/6}. \qquad (5.3.37)$$

In passing we note that this value of ε is consistent with the inequality (5.3.4) assumed earlier.

Integrating (5.3.36) across the critical layer, we see that there is a jump in the roll pressure, with

$$\left[\overline{P}_{R0}\right]_{-}^{+} \equiv \lim_{N\to\infty} \overline{P}_{R0} - \lim_{N\to-\infty} \overline{P}_{R0} = 2\kappa(s) \int_{-\infty}^{+\infty} \left|\overline{v}_W^{(s)}\right|^2 dN. \tag{5.3.38}$$

The integral on the right-hand side of (5.3.38) is given by (5.3.34). Hence, using the principle of matching of asymptotic expansions, we can see that in the core flow, the pressure jump across the critical curve is

$$\left[P_R\right]_{-}^{+} = \lim_{n\to0+} P_R(s,n) - \lim_{n\to0-} P_R(s,n) = n_0\kappa(s)\left[\alpha\mu(s)\right]^{-5/3}\left|\frac{d\overline{p}_W}{ds}\right|^2. \tag{5.3.39}$$

Next, if we examine the $O(Re^{-2}\delta^{-1})$ contribution from the s-momentum equation (5.3.21b), we find

$$\overline{v}_W^{(s)*}\frac{\partial \overline{v}_W^{(s)}}{\partial s} + \overline{v}_W^{(n)*}\frac{\partial \overline{v}_W^{(s)}}{\partial N} + i\alpha \overline{u}_W^* \overline{v}_W^{(s)} + c.c. = \frac{\partial^2 \overline{V}_{R1}^{(s)}}{\partial N^2}. \tag{5.3.40}$$

This equation is reminiscent of the finite-Reynolds-number SSP equations (5.2.30), (5.2.32) which describe the nonlinear wave forcing of the roll flow, the difference here being that explicit forcing is confined to the $O(Re^{-1/3})$ thick critical layer. Integrating (5.3.40) across the critical layer as we did with the normal momentum component, we obtain the result that the normal derivative of the tangential roll component undergoes the jump

$$\left[\frac{\partial \overline{V}_{R1}^{(s)}}{\partial N}\right]_{-}^{+} \equiv \lim_{N\to\infty} \frac{\partial \overline{V}_{R1}^{(s)}}{\partial N} - \lim_{N\to-\infty} \frac{\partial \overline{V}_{R1}^{(s)}}{\partial N}$$

$$= \int_{-\infty}^{+\infty} \left(\overline{v}_W^{(s)*}\frac{\partial \overline{v}_W^{(s)}}{\partial s} + \overline{v}_W^{(n)*}\frac{\partial \overline{v}_W^{(s)}}{\partial N} + i\alpha \overline{u}_W^* \overline{v}_W^{(s)} + c.c.\right) dN. \tag{5.3.41}$$

We apply integration by parts to the second term in the integral on the right-hand side of (5.3.41). This gives

$$\int_{-\infty}^{\infty} \overline{v}_W^{(n)*}\frac{\partial \overline{v}_W^{(s)}}{\partial N} dN = \left[\overline{v}_W^{(n)*}\overline{v}_W^{(s)}\right]_{-\infty}^{\infty} - \int_{-\infty}^{\infty} \overline{v}_W^{(s)}\frac{\partial \overline{v}_W^{(n)*}}{\partial N} dN. \tag{5.3.42}$$

Here, the integrated part vanishes in view of the second condition in (5.3.31), and to evaluate the integral on the right-hand side of (5.3.42) we use the continuity equation (5.3.30a). We find

$$\int_{-\infty}^{\infty} \overline{v}_W^{(n)*}\frac{\partial \overline{v}_W^{(s)}}{\partial N} dN = \int_{-\infty}^{\infty} \left(\overline{v}_W^{(s)}\frac{\partial \overline{v}_W^{(s)*}}{\partial s} - i\alpha \overline{u}_W^* \overline{v}_W^{(s)}\right) dN.$$

Using this result, (5.3.41) simplifies to

$$
\left[\frac{\partial \overline{V}_{R1}^{(s)}}{\partial N}\right]_{-}^{+} = 2\int_{-\infty}^{+\infty} \left(\overline{v}_W^{(s)*}\frac{\partial \overline{v}_W^{(s)}}{\partial s} + c.c.\right)dN = 2\frac{d}{ds}\int_{-\infty}^{\infty} \overline{v}_W^{(s)}\overline{v}_W^{(s)*}dN. \qquad (5.3.43)
$$

The formalism of matching of asymptotic expansions allows us to write (5.3.43) in terms of the variable of the core flow region. Using again (5.3.34) for the integral on the right-hand side of (5.3.43) we have

$$
\left[\frac{\partial V_R^{(s)}}{\partial n}\right]_{-}^{+} = \lim_{n\to0+}\frac{\partial V_R^{(s)}}{\partial n} - \lim_{n\to0-}\frac{\partial V_R^{(s)}}{\partial n} = n_0\alpha^{-5/3}\frac{d}{ds}\left[\mu(s)^{-5/3}\left|\frac{d\overline{p}_W}{ds}\right|^2\right]. \qquad (5.3.44)
$$

Additionally, we know that the normal component of the roll velocity vector in the core region is non-singular. Consequently,

$$
\left[\frac{\partial V_R^{(n)}}{\partial n}\right]_{-}^{+} = 0. \qquad (5.3.45)
$$

5.3.4   The full nonlinear interaction

Our formulation of the roll/streak/wave interaction at high Reynolds number is complete. It consists of solving the nonlinear roll equations (5.3.2) subject to no-slip conditions at the walls and the jump conditions (5.3.39), (5.3.44), (5.3.45) at the (unknown) location $y = f(z)$ where $U_S = c$. The value of c is found from the solution of the Rayleigh eigen-value problem (5.3.6), (5.3.7), while the streak flow is found from solving the equation (5.3.3). The system is fully interactive in the sense that the Rayleigh equation is forced by the streak shear, while the streak is driven by the roll in the usual way, and the roll itself is forced by the wave pressure via the jump conditions.

Although the formulation in the critical layer in terms of tangential and normal components facilitated the derivation of the jump conditions, to solve the full problem for a flow such as plane Couette flow, it is convenient to transform the jump conditions into their more complicated Cartesian form. Firstly, in view of the zero jump (5.3.45) and the form for $V_R^{(n)}$ given in (5.3.17), we must have

$$
\left[\frac{\partial}{\partial n}(V_R - W_R f_z)\right]_{-}^{+} = 0. \qquad (5.3.46)
$$

Suppose G represents any of the quantities that experience a jump in their normal derivative across the critical layer, while maintaining a continuous tangential derivative. It can be shown that[8]

$$
\left[\frac{\partial G}{\partial n}\right]_{-}^{+} = (1 + f_z^2)^{1/2}\left[\frac{\partial G}{\partial y}\right]_{-}^{+}. \qquad (5.3.47)
$$

[8]See Problem 4 in Exercises 16.

Therefore (5.3.46) implies

$$\left[\frac{\partial V_R}{\partial y}\right]_-^+ = f_z \left[\frac{\partial W_R}{\partial y}\right]_-^+. \tag{5.3.48}$$

In addition, it follows from (5.3.19) that along the critical curve:

$$\frac{d}{ds} = (1 + f_z^2)^{-1/2}\frac{d}{dz}. \tag{5.3.49}$$

Finally, we see that

$$\left[\frac{\partial V_R^{(s)}}{\partial n}\right]_-^+ = \left[\frac{\partial}{\partial n}\left(\frac{V_R f_z + W_R}{(1 + f_z^2)^{1/2}}\right)\right]_-^+$$

$$= \left[\frac{\partial}{\partial y}(V_R f_z + W_R)\right]_-^+ = (1 + f_z^2)\left[\frac{\partial W_R}{\partial y}\right]_-^+, \tag{5.3.50}$$

where we have made use of (5.3.17), (5.3.47), and (5.3.48). Now we perform the replacements (5.3.50), (5.3.49), (5.3.26) in the jump conditions (5.3.39), (5.3.44). Using expression (5.3.23) for the curvature and recalling the equality of wave pressures in (5.3.32), we obtain

$$\left[P_R\right]_-^+ = \mathcal{J}_p, \qquad \left[\frac{\partial W_R}{\partial y}\right]_-^+ = \frac{1}{f_z}\left[\frac{\partial V_R}{\partial y}\right]_-^+ = \mathcal{J}_w,$$

with the jumps given by

$$\mathcal{J}_p = -n_0 f_{zz}[\alpha\lambda(z)]^{-5/3}(1 + f_z^2)^{-10/3}\left|\frac{d\widetilde{p}_{WC}}{dz}\right|^2, \tag{5.3.51a}$$

$$\mathcal{J}_w = n_0\alpha^{-5/3}(1 + f_z^2)^{-3/2}\frac{d}{dz}\left\{(1 + f_z^2)^{-11/6}[\lambda(z)]^{-5/3}\left|\frac{d\widetilde{p}_{WC}}{dz}\right|^2\right\}. \tag{5.3.51b}$$

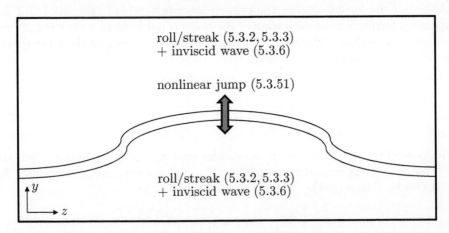

Fig. 5.15: A sketch of the vortex-Rayleigh wave interaction for plane Couette flow. The curved strip represents the $O(Re^{-1/3})$ thick critical layer centred at $y = f(z)$.

We shall use these as the conditions to be applied at the critical curve $y = f(z)$. A sketch of the interaction is given in Figure 5.15.

In Figure 5.16 we plot the results of numerical solutions of the interaction equations (5.3.2) subject to no-slip conditions at the walls and the jump conditions (5.3.51), together with the solution of the streak equation (5.3.3) and the Rayleigh equation (5.3.6) subject to the impermeability condition (5.3.7). The spanwise wave number was chosen to be $\beta = 2$ in order to compare with the Navier–Stokes solution of Deguchi and Hall (2014) presented in Figure 5.12. In Figure 5.16(a) we show a solution from the upper branch in which the critical layer is highly distorted: this should be compared with the Navier–Stokes solution in Figure 5.12(b) which is computed at a lower Reynolds number and slightly different streamwise wave number. One can see that the flow fields are very similar. In Figure 5.16(b,c) we show lower branch solutions at a Reynolds number of 16000 and two different streamwise wave numbers $\alpha = 0.8, 1.2$. These parameter values are identical to those used in the Navier–Stokes computations shown in Figures 5.12(e,f) and it can be seen that the flow fields are almost indistinguishable. It is therefore clear that both branches of self-sustaining solutions for plane Couette flow are indeed described accurately by the vortex-Rayleigh interaction equations at large Reynolds number.

A similarly successful agreement has been established for self-sustaining solutions in other shear flows such as pipe flow. Often the agreement with a Navier–Stokes generated flow field proves to be excellent down to even moderate Reynolds numbers, making the vortex-Rayleigh approach to computing self-sustaining solutions an at-

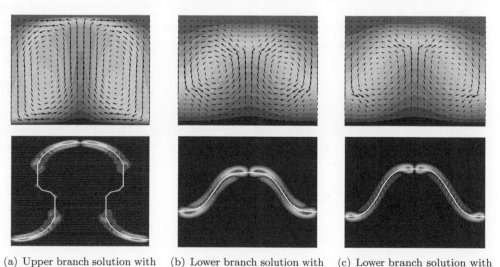

(a) Upper branch solution with $(\alpha, Re) = (0.7, 50000)$.

(b) Lower branch solution with $(\alpha, Re) = (0.8, 16000)$.

(c) Lower branch solution with $(\alpha, Re) = (1.2, 16000)$.

Fig. 5.16: Three-dimensional vortex-Rayleigh solutions for plane Couette flow with $\beta = 2$. In each panel the upper plot shows contours of the streak, with arrows depicting the superimposed roll flow, while the lower plot shows the wave forcing $|\mathbf{F}|$ with $\mathbf{F}$ given in (5.2.30); the critical layer is denoted by the white curve. Figure courtesy of Dr. J. Maestri.

tractive one in view of its less demanding computational requirements in comparison with a full Navier–Stokes simulation. In addition it produces a better insight into the physical nature of the flows.

Exercises 16

1. *Derivation of the streak-driven Rayleigh equation*

   Starting from the inviscid wave disturbance equations given in (5.3.5), namely

   $$i\alpha \tilde{u}_W + \frac{\partial \tilde{v}_W}{\partial y} + \frac{\partial \tilde{w}_W}{\partial z} = 0, \tag{5.3.52a}$$

   $$i\alpha \big[U_S(y,z) - c\big]\tilde{u}_W + \tilde{v}_W \frac{\partial U_S}{\partial y} + \tilde{w}_W \frac{\partial U_S}{\partial z} = -i\alpha \tilde{p}_W, \tag{5.3.52b}$$

   $$i\alpha \big[U_S(y,z) - c\big]\tilde{v}_W = -\frac{\partial \tilde{p}_W}{\partial y}, \tag{5.3.52c}$$

   $$i\alpha \big[U_S(y,z) - c\big]\tilde{w}_W = -\frac{\partial \tilde{p}_W}{\partial z}, \tag{5.3.52d}$$

   show that the wave pressure $\tilde{p}_W$ satisfies the Rayleigh equation

   $$\big[U_S(y,z) - c\big]\left(\frac{\partial^2 \tilde{p}_W}{\partial y^2} + \frac{\partial^2 \tilde{p}_W}{\partial z^2} - \alpha^2 \tilde{p}_W\right) = 2\frac{\partial U_S}{\partial y}\frac{\partial \tilde{p}_W}{\partial y} + 2\frac{\partial U_S}{\partial z}\frac{\partial \tilde{p}_W}{\partial z}, \tag{5.3.53}$$

   as quoted in (5.3.6).

   You may accomplish this task in the following steps:

   (a) Multiply equation (5.3.52b) by $i\alpha(U_S - c)$ and then use (5.3.52c), (5.3.52d) to obtain

   $$-\alpha^2 (U_S - c)^2 \tilde{u}_W - \frac{\partial U_S}{\partial y}\frac{\partial \tilde{p}_W}{\partial y} - \frac{\partial U_S}{\partial z}\frac{\partial \tilde{p}_W}{\partial z} = \alpha^2 (U_S - c)\tilde{p}_W. \tag{5.3.54}$$

   (b) Divide (5.3.52c) by $(U_S - c)$ and then differentiate with respect to y. Divide (5.3.52d) by $(U_S - c)$ and differentiate with respect to z. Add the resulting expressions together and multiply through by $(U_S - c)^2$ to obtain

   $$i\alpha(U_S - c)^2 \left(\frac{\partial \tilde{v}_W}{\partial y} + \frac{\partial \tilde{w}_W}{\partial z}\right)$$

   $$= -(U_S - c)\left(\frac{\partial^2 \tilde{p}_W}{\partial y^2} + \frac{\partial^2 \tilde{p}_W}{\partial z^2}\right) + \frac{\partial U_S}{\partial y}\frac{\partial \tilde{p}_W}{\partial y} + \frac{\partial U_S}{\partial z}\frac{\partial \tilde{p}_W}{\partial z}.$$

   (c) Use the continuity equation (5.3.52a) on the left-hand side of the above equation to rewrite it in the form

   $$\alpha^2 (U_S - c)^2 \tilde{u}_W = -(U_S - c)\left(\frac{\partial^2 \tilde{p}_W}{\partial y^2} + \frac{\partial^2 \tilde{p}_W}{\partial z^2}\right)$$

   $$+ \frac{\partial U_S}{\partial y}\frac{\partial \tilde{p}_W}{\partial y} + \frac{\partial U_S}{\partial z}\frac{\partial \tilde{p}_W}{\partial z}. \tag{5.3.55}$$

(d) Finally, add equations (5.3.54) and (5.3.55) together to eliminate $\tilde{u}_W$. You should be left with the Rayleigh equation (5.3.53).

2. *Streak-driven flow through a pipe*

   Consider flow through a circular pipe driven by a streak velocity $U_S(r, \phi)$ with (r, ϕ) being polar coordinates. Use similar manipulations to those indicated above to show that the corresponding Rayleigh equation takes the form

$$(U_S - c)\left(\frac{\partial^2 \tilde{p}_W}{\partial r^2} + \frac{1}{r} \frac{\partial \tilde{p}_W}{\partial r} + \frac{1}{r^2} \frac{\partial^2 \tilde{p}_W}{\partial \phi^2} - \alpha^2 \tilde{p}_W \right) = 2\frac{\partial U_S}{\partial r}\frac{\partial \tilde{p}_W}{\partial r} + \frac{2}{r^2}\frac{\partial U_S}{\partial \phi}\frac{\partial \tilde{p}_W}{\partial \phi}.$$

3. *Vortex-Rayleigh wave interaction critical layer dynamics*

   The function $\mathcal{B}(y)$ satisfies the equation

$$\frac{d^2 \mathcal{B}}{dy^2} - iay\mathcal{B} = 1, \qquad (5.3.56)$$

   and the boundary conditions

$$\mathcal{B} \to 0 \quad \text{as} \quad y \to \pm\infty,$$

   where a is a real positive constant.

   You have to perform the following two tasks:

   (a) Define the Fourier transform of $\mathcal{B}$ by

$$\breve{\mathcal{B}}(\omega) = \int\limits_{-\infty}^{\infty} \mathcal{B}(y)e^{-i\omega y}\, dy, \qquad (5.3.57)$$

   and show that

$$\breve{\mathcal{B}}(\omega) = \begin{cases} 0 & \text{if} \quad \omega > 0, \\ -2\pi a^{-1}e^{\omega^3/3a} & \text{if} \quad \omega < 0. \end{cases}$$

   Hence deduce that the solution for $\mathcal{B}$ may be written in the form

$$\mathcal{B}(y) = -a^{-2/3} \int\limits_{0}^{\infty} e^{-t^3/3 - ia^{1/3}ty}\, dt,$$

   thus confirming the validity of the expression (5.3.33).

   Suggestions: When applying Fourier transforms to (5.3.56) note that the derivative of (5.3.57) with respect to ω is given by

$$\frac{d\breve{\mathcal{B}}}{d\omega} = -\int\limits_{-\infty}^{\infty} iy\mathcal{B}(y)e^{-i\omega y}\, dy.$$

   You may also use without proof the fact that the Fourier transform of unity is $2\pi\delta(\omega)$, where $\delta(\omega)$ is the Dirac delta function.

(b) Show further that

$$\int_{-\infty}^{\infty} \left| \mathcal{B}(y) \right|^2 dy = \left(\frac{2}{3} \right)^{2/3} \frac{\pi}{a^{5/3}} \Gamma \left(\frac{1}{3} \right),$$

which is the result quoted in (5.3.34), (5.3.35) with the gamma function defined by

$$\Gamma(x) = \int_0^{\infty} t^{x-1} e^{-t} dt.$$

Suggestion: You might find it helpful to use the 'sifting property' of the delta function, namely that

$$\int_{-\infty}^{\infty} f(t)\delta(t-s)\, dt = f(s)$$

for any continuous function f.

4. *Conversion of the jump formulae*

Suppose $G(y, z)$ is a function which experiences a jump in its normal derivative across the curve $y = f(z)$, while its tangential derivative remains continuous. Show that if $[\ \ ]_-^+$ denotes such a jump, then

$$\left[\frac{\partial G}{\partial n} \right]_-^+ = \sqrt{1 + f_z^2} \left[\frac{\partial G}{\partial y} \right]_-^+, \qquad (5.3.58)$$

where $f_z = df/dz$. You can achieve this result in the following steps:

(a) Show that if $\mathbf{j}, \mathbf{k}$ are unit vectors in the y, z directions respectively, then unit vectors $\mathbf{e}_s, \mathbf{e}_n$ in the tangential and normal directions to the curve $y = f(z)$ can be written as

$$\mathbf{e}_s = (f_z \mathbf{j} + \mathbf{k})/\sqrt{1 + f_z^2}, \qquad \mathbf{e}_n = (\mathbf{j} - f_z \mathbf{k})/\sqrt{1 + f_z^2}.$$

(b) Use the formulae

$$\frac{\partial G}{\partial s} = \mathbf{e}_s \cdot \nabla G, \qquad \frac{\partial G}{\partial n} = \mathbf{e}_n \cdot \nabla G,$$

where the gradient vector

$$\nabla G = \frac{\partial G}{\partial y} \mathbf{j} + \frac{\partial G}{\partial z} \mathbf{k},$$

to establish that

$$\frac{\partial G}{\partial n} = \frac{1}{\sqrt{1 + f_z^2}} \frac{\partial G}{\partial y} - \frac{f_z}{\sqrt{1 + f_z^2}} \frac{\partial G}{\partial z},$$

$$\frac{\partial G}{\partial s} = \frac{f_z}{\sqrt{1 + f_z^2}} \frac{\partial G}{\partial y} + \frac{1}{\sqrt{1 + f_z^2}} \frac{\partial G}{\partial z}.$$

(c) Eliminate $\partial G/\partial z$ between the last two equations to show that

$$\frac{\partial G}{\partial n} + f_z \frac{\partial G}{\partial s} = \sqrt{1 + f_z^2}\frac{\partial G}{\partial y}.$$

(d) Using the properties of G given at the beginning of the question, deduce the jump condition (5.3.58).

5. *The jump across the critical layer*

Consider expression (5.3.51b) for the jump $\mathcal{J}_w$ in the shear of the spanwise roll component across the critical layer. Show that by defining the quantities

$$\Delta(z) = 1 + f_z^2, \qquad a(z) = \alpha\lambda/\Delta,$$

the jump can be rewritten in the form

$$\mathcal{J}_w = \frac{n_0}{\Delta^5 a^{5/3}}\left\{\left[-\frac{7\Delta'(z)}{2\Delta} - \frac{5a'(z)}{3a}\right]\left|\frac{d\widetilde{p}_{WC}}{dz}\right|^2 + \frac{d}{dz}\left(\left|\frac{d\widetilde{p}_{WC}}{dz}\right|^2\right)\right\},$$

as given in Brown *et al.* (1993) and Hall and Sherwin (2010).

Hint: In order to derive this expression you might find it useful to observe that

$$a^{-5/3}\Delta^{-5} = (\alpha\lambda)^{-5/3}(1 + f_z^2)^{-10/3}, \qquad \frac{\lambda'(z)}{\lambda} = \frac{a'(z)}{a} + \frac{\Delta'(z)}{\Delta}.$$

5.4 Self-Sustaining Processes at High Reynolds Number: Vortex-Viscous Wave Interaction

The waves in the vortex-Rayleigh VWI described in Section 5.3 are governed by inviscid dynamics and in particular by the Rayleigh equation (5.3.6). In this section we investigate the mathematics and physics behind a similar self-sustaining mechanism, but involving waves governed by viscous effects at leading order. A classical example of this type of instability wave are the Tollmien–Schlichting (TS) waves observed in boundary layers (see Section 1.3). Experimentally, roll/streak formation in boundary layers has been achieved from a strong TS input by Klebanoff *et al.* (1962), for example. In such a situation the mechanism involved latches on to the linear instability of the basic unperturbed motion. In view of this, the mechanism to be described is not applicable to plane Couette flow, where the basic profile is known to be linearly stable at all Reynolds numbers. Instead we will illustrate the ideas using plane Poiseuille flow (see Section 1.2.1). Similar arguments can be put forward to describe the analogous interactions in boundary layers: for this reason, this type of self-sustaining process is often referred to as a *vortex-TS wave interaction*.

Returning to our canonical problem, we will again suppose that the channel is of width $2h$, but now the channel walls are stationary and the flow is driven by a constant streamwise pressure gradient. To study the flow we shall use dimensionless variables introduced through the scalings (1.2.10). This leads once again to the equations (5.1.1). Throughout this section we will focus our attention on the high-Reynolds-number limit

$$Re \gg 1.$$

5.4.1   The unforced roll/streak flow

As before we start by considering the roll flow but now we allow for a longer spanwise scaling by writing

$$z = \beta^{-1}\bar{z}, \tag{5.4.1}$$

where we shall assume that

$$O(Re^{-1}) \ll \beta \ll 1. \tag{5.4.2}$$

The significance of the lower bound on β will become clear later in this section. Our ultimate aim is to show that a strong roll/streak/wave interaction occurs when β is a particular inverse power of the Reynolds number to be determined subsequently.

Leaving aside the wave contribution for now, we pose a roll/streak expansion of the form

$$\left.\begin{aligned} u = U_S(y,\bar{z}), \quad v = Re^{-1}V_R(y,\bar{z}), \quad w = Re^{-1}\beta^{-1}\overline{W}_R(y,\bar{z}), \\ p = -\frac{2}{Re}x + Re^{-2}\beta^{-2}\overline{P}_R(y,\bar{z}), \end{aligned}\right\} \tag{5.4.3}$$

which is the same as the roll/streak part of the vortex-Rayleigh wave expansion (5.3.1) except for the extra factors involving the small parameter β. These are used in (5.4.3) to ensure a continuity balance and a pressure gradient-inertial-viscous balance in the spanwise momentum equation (5.1.1d). The first term in the pressure expansion represents the constant streamwise pressure gradient. If we substitute (5.4.3) into the governing Navier–Stokes equations (5.1.1) we obtain, with no approximation:

$$\frac{\partial V_R}{\partial y} + \frac{\partial \overline{W}_R}{\partial \bar{z}} = 0,$$

$$V_R\frac{\partial V_R}{\partial y} + \overline{W}_R\frac{\partial V_R}{\partial \bar{z}} = -\beta^{-2}\frac{\partial \overline{P}_R}{\partial y} + \frac{\partial^2 V_R}{\partial y^2} + \beta^2\frac{\partial^2 V_R}{\partial \bar{z}^2},$$

$$V_R\frac{\partial \overline{W}_R}{\partial y} + \overline{W}_R\frac{\partial \overline{W}_R}{\partial \bar{z}} = -\frac{\partial \overline{P}_R}{\partial \bar{z}} + \frac{\partial^2 \overline{W}_R}{\partial y^2} + \beta^2\frac{\partial^2 \overline{W}_R}{\partial \bar{z}^2},$$

for the roll, and

$$V_R\frac{\partial U_S}{\partial y} + \overline{W}_R\frac{\partial U_S}{\partial \bar{z}} = 2 + \frac{\partial^2 U_S}{\partial y^2} + \beta^2\frac{\partial^2 U_S}{\partial \bar{z}^2}$$

for the streak. Thus, under the assumptions on β given in (5.4.2), the roll/streak equations at leading order are:

$$\frac{\partial V_R}{\partial y} + \frac{\partial \overline{W}_R}{\partial \bar{z}} = 0, \tag{5.4.4a}$$

$$\frac{\partial \overline{P}_R}{\partial y} = 0, \tag{5.4.4b}$$

$$V_R\frac{\partial \overline{W}_R}{\partial y} + \overline{W}_R\frac{\partial \overline{W}_R}{\partial \bar{z}} = -\frac{\partial \overline{P}_R}{\partial \bar{z}} + \frac{\partial^2 \overline{W}_R}{\partial y^2}, \tag{5.4.4c}$$

$$V_R\frac{\partial U_S}{\partial y} + \overline{W}_R\frac{\partial U_S}{\partial \bar{z}} = 2 + \frac{\partial^2 U_S}{\partial y^2}. \tag{5.4.4d}$$

We notice, in the now familiar way, how the nonlinear roll field drives the linear streak equation. These equations are typically subject to the no-slip boundary conditions

$$U_S = V_R = \overline{W}_R = 0 \quad \text{on} \quad y = \pm 1. \tag{5.4.5}$$

Again, energy theory arguments (similar to those in Problem 3, Exercises 15) can be used to prove that the unforced roll/streak equations (5.4.4) subject to no-slip conditions (5.4.5) admit only the uni-directional solution

$$U_S = 1 - y^2, \quad V_R = \overline{W}_R = \overline{P}_R = 0, \tag{5.4.6}$$

which is, of course, the famous parabolic profile of laminar Poiseuille flow (1.2.11).

Our aim in this section is to show how the introduction of a small travelling wave disturbance governed predominantly by viscous dynamics (rather than the inviscid wave of Section 5.3) can provide a strong feedback on the roll/streak flow and generate alternative solutions to (5.4.6). To demonstrate this, we first need to investigate the regions of the flow field near the channel walls where, given that the Reynolds number is large, viscous effects are at their most significant.

5.4.2   The viscous wall layers

As mentioned above, in order to generate a non-zero roll flow, we need to incorporate some wave forcing into the problem. Again we will seek travelling waves that are proportional to

$$E \equiv e^{i\alpha(x - ct)}, \tag{5.4.7}$$

but now, in order to ensure that the waves are three-dimensional at leading order near the wall, we will suppose that the streamwise wavelength of the wave is comparable with the long spanwise length scale. Hence, we choose

$$\alpha = \beta \bar{\alpha}, \tag{5.4.8}$$

where $\bar{\alpha}$ is assumed to be an order one quantity. In addition, we expect the flow in the wall layers to be unsteady at leading order, such that the first two terms in (5.1.1b) remain in balance with one another:

$$\frac{\partial u}{\partial t} \sim u \frac{\partial u}{\partial x}. \tag{5.4.9}$$

Let us consider, to be definite, the lower wall layer adjacent to $y = -1$. In this layer

$$y = -1 + \delta Y_-, \tag{5.4.10}$$

where $\delta \ll 1$ is the characteristic thickness of the layer, and Y_- is an order one variable measured across the lower wall layer. We assume that the streak flow behaves regularly as the wall is approached, i.e.

$$U_S = \lambda_-(\bar{z})(y + 1) + \cdots \quad \text{as} \quad y \to -1+, \tag{5.4.11}$$

where

$$\lambda_-(\bar{z}) = \frac{\partial U_S}{\partial y}\Big|_{y=-1} \tag{5.4.12}$$

is the streak shear on the lower wall. It follows from (5.4.10) and (5.4.11) that in the wall layer $u \sim \delta$ which allows us to write (5.4.9) as

$$\frac{\partial u}{\partial t} \sim \delta \frac{\partial u}{\partial x}. \tag{5.4.13}$$

Calculating the derivatives in (5.4.13) with the help of (5.4.7), we see that the required balance holds provided that $c \sim \delta$. Hence, we write

$$c = \delta \bar{c}. \tag{5.4.14}$$

Similar to the streak (5.4.11) we suppose that the roll flow

$$V_R \propto (y+1), \quad \overline{W}_R \propto (y+1) \quad \text{as} \quad y \to -1+, \tag{5.4.15}$$

so that the no-slip conditions (5.4.5) are satisfied.

Based on the above arguments we express the fluid-dynamic functions in the lower wall layer in the form

$$u = \delta\lambda_-(\bar{z})Y_- + \cdots + \varepsilon\big[u_W^{(-)}(Y_-,\bar{z})E + c.c.\big] + \cdots, \tag{5.4.16a}$$

$$v = Re^{-1}\delta V_R^{(-)}(Y_-,\bar{z}) + \cdots + \varepsilon\delta\beta\big[v_W^{(-)}(Y_-,\bar{z})E + c.c.\big] + \cdots, \tag{5.4.16b}$$

$$w = Re^{-1}\beta^{-1}\delta W_R^{(-)}(Y_-,\bar{z}) + \cdots + \varepsilon\big[w_W^{(-)}(Y_-,\bar{z})E + c.c.\big] + \cdots, \tag{5.4.16c}$$

$$p = -\frac{2}{Re}x + Re^{-2}\beta^{-2}P_R^{(-)}(Y_-,\bar{z}) + \cdots + \varepsilon\delta\big[p_W^{(-)}(Y_-,\bar{z})E + c.c.\big] + \cdots. \tag{5.4.16d}$$

Here

$$E = e^{i\beta\bar{\alpha}(x-\delta\bar{c}t)}, \tag{5.4.17}$$

while ε is the unknown wave amplitude which will be determined later in the analysis in order to ensure a strong feedback on the rolls. The magnitude of the roll flow has been reduced by a factor δ in comparison to the core flow (5.4.3), in view of the assumptions given in (5.4.15).

We start with the continuity equation (5.1.1a). Substituting (5.4.16a)–(5.4.16c) into (5.1.1a) we obtain at $O(\varepsilon\beta)$:

$$i\bar{\alpha}u_W^{(-)} + \frac{\partial v_W^{(-)}}{\partial Y_-} + \frac{\partial w_W^{(-)}}{\partial \bar{z}} = 0,$$

indicating that the wave is fully three-dimensional in this layer.

Next we consider the streamwise momentum equation (5.1.1b). Substituting (5.4.16a)–(5.4.16d) into (5.1.1b), we find

$$\delta\beta\varepsilon\bigg[i\bar{\alpha}(\lambda_-Y_- - \bar{c})u_W^{(-)} + \lambda_-v_W^{(-)} + \frac{d\lambda_-}{d\bar{z}}Y_-w_W^{(-)} + i\bar{\alpha}p_W^{(-)}\bigg]$$

$$= Re^{-1}\delta^{-2}\varepsilon\frac{\partial^2 u_W^{(-)}}{\partial Y_-^2} + Re^{-1}\varepsilon\beta^2\bigg(\frac{\partial^2 u_W^{(-)}}{\partial \bar{z}^2} - \bar{\alpha}^2 u_W^{(-)}\bigg).$$

Since $\beta \ll 1$, the first term on the right-hand side dominates over the final two terms. For there to be a balance between this leading-order viscous term and the effects of inertia, we then require

$$\delta\beta\varepsilon \sim Re^{-1}\delta^{-2}\varepsilon,$$

which means that the thickness of the wall layer should be chosen to be

$$\delta = (\beta Re)^{-1/3}, \tag{5.4.18}$$

and then the leading-order streamwise momentum equation assumes the form

$$i\bar{\alpha}\big[\lambda_-(\bar{z})Y_- - \bar{c}\big]u_W^{(-)} + \lambda_- v_W^{(-)} + \frac{d\lambda_-}{d\bar{z}}Y_- w_W^{(-)} + i\bar{\alpha}p_W^{(-)} = \frac{\partial^2 u_W^{(-)}}{\partial Y_-^2}.$$

We note that at this stage the amplitude ε of the waves is left undetermined. This is because the balances for the wave in the sublayer are essentially linear, with the wave being driven by the unknown near-wall shear distribution $\lambda_-(\bar{z})$ arising from the streak flow. If we now substitute the roll/streak/wave expansions (5.4.16) into the normal component of the Navier-Stokes equations (5.1.1c) and use (5.4.18), then we find that at leading order we simply have

$$\frac{\partial p_W^{(-)}}{\partial Y_-} = 0,$$

which means that the wave pressure is independent of depth within the viscous sublayer. Finally we turn to the spanwise momentum equation (5.1.1d). Substitution of (5.4.16) into (5.1.1d) yields

$$i\bar{\alpha}\big[\lambda_-(\bar{z})Y_- - \bar{c}\big]w_W^{(-)} + \frac{\partial p_W^{(-)}}{\partial \bar{z}} = \frac{\partial^2 w_W^{(-)}}{\partial Y_-^2}.$$

Exactly the same processes occur in the viscous layer near the upper wall ($y = 1$) where we write

$$y = 1 - \delta Y_+,$$

so that $Y_+ > 0$ in this region, and define

$$\lambda_+(\bar{z}) = -\left.\frac{\partial U_S}{\partial y}\right|_{y=1}. \tag{5.4.19}$$

We can then pose an identical roll/streak/wave expansion to (5.4.16) in the upper layer, except that the flow quantities are now labelled with $+$ rather than $-$. Together, the leading-order wave equations in both layers may be written as

$$i\bar{\alpha}u_W^{(+)} + \frac{\partial v_W^{(\pm)}}{\partial Y_\pm} + \frac{\partial w_W^{(\pm)}}{\partial \bar{z}} = 0, \tag{5.4.20a}$$

$$i\bar{\alpha}\big[\lambda_\pm(\bar{z})Y_\pm - \bar{c}\big]u_W^{(\pm)} + \lambda_\pm v_W^{(\pm)} + \frac{d\lambda_\pm}{d\bar{z}}Y_\pm w_W^{(\pm)} + i\bar{\alpha}p_W^{(\pm)} = \frac{\partial^2 u_W^{(\pm)}}{\partial Y_\pm^2}, \tag{5.4.20b}$$

$$i\bar{\alpha}\big[\lambda_\pm(\bar{z})Y_\pm - \bar{c}\big]w_W^{(\pm)} + \frac{d p_W^{(\pm)}}{d\bar{z}} = \frac{\partial^2 w_W^{(\pm)}}{\partial Y_\pm^2}. \tag{5.4.20c}$$

Here, for convenience, we define the normal velocity component $v_W^{(+)}$ to be directed into the flow from the upper wall.

Equations (5.4.20) are to be solved subject to the usual no-slip condition of viscous flow, namely

$$u_W^{(\pm)} = v_W^{(\pm)} = w_W^{(\pm)} = 0 \quad \text{on} \quad Y_\pm = 0.$$

In order to evaluate how the near-wall flow affects the roll and streak in the core we need to find the asymptotic behaviour of the solution of (5.4.20) at large $Y_\pm$. For two-dimensional flow, this task was performed in Section 3.1.3, leading to formulae (3.1.33). Here we shall write these formulae as

$$\left.\begin{aligned} u_W^{(\pm)} &= B^{(\pm)}(\bar{z}) + \cdots, \\ v_W^{(\pm)} &= -i\bar{\alpha}B^{(\pm)}(\bar{z})Y_\pm + \cdots \end{aligned}\right\} \quad \text{as} \quad Y_\pm \to \infty, \tag{5.4.21}$$

where $B^{(\pm)}(\bar{z})$ are unknown displacement functions. In order to determine the behaviour of the third velocity component one needs to consider the spanwise momentum equation (5.4.20c). At large values of $Y_\pm$ the viscous term on the right-hand side may be disregarded, and we find that

$$w_W^{(\pm)} = -\left[i\bar{\alpha}\lambda_\pm(\bar{z})\right]^{-1} \frac{dp_W^{(\pm)}}{d\bar{z}} Y_\pm^{-1} + \cdots \quad \text{as} \quad Y_\pm \to \infty, \tag{5.4.22}$$

so that the spanwise component of the wave decays algebraically at the edges of the sublayers. This decay ensures that the wave behaves in a quasi-two-dimensional manner in the core, as we shall see shortly.

Before we return to the core flow and arrange for a strong wave feedback upon the rolls, we note that it is possible to simplify the set of equations (5.4.20) and reduce it to a single equation for the wave pressure. The manipulations behind this simplification are discussed in Problems 1 and 2 in Exercises 17 where it is shown that the wave pressures in the two viscous layers are related to the displacement functions $B^{(\pm)}$ via the equations

$$\frac{d^2 p_W^{(\pm)}}{d\bar{z}^2} - \frac{1}{\lambda_\pm(\bar{z})} \frac{d\lambda_\pm}{d\bar{z}} \mathcal{F}^{(\pm)}(\xi_0^{(\pm)}) \frac{dp^{(\pm)}}{d\bar{z}} - \bar{\alpha}^2 p^{(\pm)}$$
$$= \bar{\alpha}^{5/3} \left(\lambda_\pm(\bar{z})\right)^{2/3} \mathcal{G}^{(\pm)}(\xi_0^{(\pm)}) B^{(\pm)}(\bar{z}). \tag{5.4.23}$$

The functions $\mathcal{F}^{(\pm)}, \mathcal{G}^{(\pm)}$ are given in terms of the Airy function Ai as

$$\mathcal{F}^{(\pm)}(\xi_0^{(\pm)}) = \frac{3}{2} + \frac{\xi_0^{(\pm)}}{2Ai(\xi_0^{(\pm)})} \left[\xi_0^{(\pm)} \varkappa(\xi_0^{(\pm)}) + Ai'(\xi_0^{(\pm)})\right], \tag{5.4.24a}$$

$$\mathcal{G}^{(\pm)}(\xi_0^{(\pm)}) = \frac{i^{5/3} Ai'(\xi_0^{(\pm)})}{\varkappa(\xi_0^{(\pm)})}, \tag{5.4.24b}$$

with

$$\xi_0^{(\pm)} = -\frac{i^{1/3}\bar{\alpha}^{1/3}\bar{c}}{\left[\lambda_\pm(\bar{z})\right]^{2/3}}, \qquad \varkappa(\xi) = \int_\xi^\infty Ai(s)\,ds. \qquad (5.4.25)$$

The pressure equation (5.4.23) is to be solved assuming that the sought functions are periodic in $\bar{z}$. This equation is analogous to the Rayleigh pressure equation (5.3.6) derived earlier and, once linked to the outer core flow, will constitute an eigen-value problem from which the scaled wave number and phase speed of the wave can be calculated for a given disturbance size. Now that the dynamics of the viscous wall layers have been elucidated, we examine how this behaviour feeds back into the core flow and enables and sustains a roll flow.

5.4.3   Wave feedback on the roll/streak core flow

The asymptotic behaviour (5.4.21), (5.4.22) together with the wall layer scalings (5.4.16) allow us to deduce the sizes of the wave components in the core region which will supplement the streak and roll expansions given in (5.4.3). Because the streamwise velocity component $u_W^{(\pm)}$ in the viscous sublayer becomes independent of the normal coordinate $Y_\pm$ as we approach the core, this component will remain of the same magnitude $O(\varepsilon)$ in the core. In contrast, the normal wave component $v_W^{(\pm)}$ grows proportional to distance from the wall and therefore will be amplified by a factor δ^{-1}: it therefore increases to $O(\varepsilon\beta)$, in view of the scaling (5.4.16b). Since the wave pressure is independent of normal coordinate throughout both wall layers, it will remain at $O(\varepsilon\delta)$. Finally, the spanwise component $w_W^{(\pm)}$ is inversely proportional to distance on exit from the viscous regions and is therefore reduced by a factor δ to size $O(\varepsilon\delta)$ in the core. On the basis of these arguments we can pose a core flow expansion, in which the roll/streak flow already established is supplemented by a viscous wave perturbation, of the form

$$u = U_S(y,\bar{z}) + \cdots + \varepsilon\left[u_W(y,\bar{z})E + c.c.\right] + \cdots, \qquad (5.4.26a)$$

$$v = Re^{-1}V_R(y,\bar{z}) + \cdots + \varepsilon\beta\left[v_W(y,\bar{z})E + c.c.\right] + \cdots, \qquad (5.4.26b)$$

$$w = Re^{-1}\beta^{-1}\overline{W}_R(y,\bar{z}) + \cdots + \varepsilon\delta\left[w_W(y,\bar{z})E + c.c.\right] + \cdots, \qquad (5.4.26c)$$

$$p = -\frac{2}{Re}x + Re^{-2}\beta^{-2}\overline{P}_R(y,\bar{z}) + \cdots + \varepsilon\delta\left[p_W(y,\bar{z})E + c.c.\right] + \cdots, \qquad (5.4.26d)$$

with E given again by (5.4.17).

Thus far we have one relation (5.4.18) connecting the viscous sublayer thickness δ, the Reynolds number Re and the spanwise wave number β. From the balancing of the wave terms within the core region we will be able to find a second relation. Substituting the expansion (5.4.26) into the continuity equation (5.1.1a) and working with the $O(\varepsilon\beta)$ terms, we find

$$i\bar{\alpha}u_W + \frac{\partial v_W}{\partial y} + \delta\frac{\partial w_W}{\partial \bar{z}} = 0.$$

Since $\delta \ll 1$, the third term may be disregarded which reduces the continuity equation to the following quasi-two-dimensional form:

$$i\bar{\alpha}u_W + \frac{\partial v_W}{\partial y} = 0. \tag{5.4.27}$$

Turning to the streamwise momentum equation now, substitution of (5.4.26) into (5.1.1b) yields

$$\varepsilon\beta\underbrace{\left[i\bar{\alpha}(U_S - \delta\bar{c})u_W + v_W\frac{\partial U_S}{\partial y}\right]}+Re^{-1}\varepsilon\left[V_R\frac{\partial u_W}{\partial y} + \overline{W}_R\frac{\partial u_W}{\partial \overline{z}}\right]$$

$$+ \varepsilon\beta\delta\left[w_W\frac{\partial U_S}{\partial \overline{z}} + i\bar{\alpha}p_W\right] = Re^{-1}\varepsilon\left[\frac{\partial^2 u_W}{\partial y^2} + \beta^2\left(\frac{\partial^2 u_W}{\partial \overline{z}^2} - \bar{\alpha}^2 u_W\right)\right].$$

Clearly, the under-braced term dominates in this equation provided $\beta \gg O(Re^{-1})$, which is what we assumed earlier in (5.4.2). Taking further into account that δ is small, we see that to leading order, the x-momentum equation is simply

$$i\bar{\alpha}U_S u_W + v_W\frac{\partial U_S}{\partial y} = 0. \tag{5.4.28}$$

The equation (5.4.28) should be paired with (5.4.27) and solved for u_W and v_W. One can use for this purpose the procedure described when solving the equations (3.1.40). It is found that

$$u_W = A(\overline{z})\frac{\partial U_S}{\partial y}, \qquad v_W = -i\bar{\alpha}A(\overline{z})U_S, \tag{5.4.29}$$

where the function $A(\overline{z})$ is unknown at this stage except it may be linked to the displacement function $B^{(\pm)}(\overline{z})$ by matching with the solution (5.4.21) in the wall layers. We have

$$B^{(\pm)}(\overline{z}) = \mp\lambda_{\pm}A(\overline{z}).$$

Finally, we consider the wall normal momentum equation (5.1.1c). Substituting (5.4.26) into (5.1.1c) and working with the wave terms we have

$$\varepsilon\beta^2\underbrace{\left[i\bar{\alpha}(U_S - \delta\bar{c})v_W\right]}_{\text{term 1}} + Re^{-1}\varepsilon\beta\underbrace{\left[V_R\frac{\partial v_W}{\partial y} + v_W\frac{\partial V_R}{\partial y} + \overline{W}_R\frac{\partial v_W}{\partial \overline{z}} + \delta w_W\frac{\partial V_R}{\partial \overline{z}}\right]}_{\text{term 2}}$$

$$= -\varepsilon\delta\underbrace{\frac{\partial p_W}{\partial y}}_{\text{term 3}} + Re^{-1}\varepsilon\beta\underbrace{\left[\frac{\partial^2 v_W}{\partial y^2} + \beta^2\left(\frac{\partial^2 v_W}{\partial \overline{z}^2} - \bar{\alpha}^2 v_W\right)\right]}_{\text{term 4}}.$$

Under the assumption $\beta \gg Re^{-1}$, terms 2 and 4 are much smaller than term 1 and can be safely neglected. Also, we can take advantage of the fact that $\delta \ll 1$ and disregard $\delta\bar{c}$ in term 1, which simplifies the above equation to

$$\varepsilon\beta^2 i\bar{\alpha}U_S v_W = -\varepsilon\delta\frac{\partial p_W}{\partial y}.$$

To avoid further degeneration of this equation, we have to set

$$\delta = \beta^2. \qquad (5.4.30)$$

Solving equations (5.4.18), (5.4.30) fixes the Reynolds number dependence of δ and β as

$$\delta = Re^{-2/7}, \qquad \beta = Re^{-1/7}, \qquad (5.4.31)$$

and results in the leading-order equation

$$i\bar{\alpha} U_S v_W = -\frac{\partial p_W}{\partial y}. \qquad (5.4.32)$$

In passing we note that the scaling for β in (5.4.31) is consistent with the assumption (5.4.2) and that these scalings are the familiar large Reynolds number scalings for the lower branch of the linear neutral curve for plane Poiseuille flow; see Section 2.5.1.

Substituting for v_W from (5.4.29) and integrating (5.4.32) between the two walls we find that the difference in the wall pressures can be expressed in terms of $A(\bar{z})$ and the properties of the streak flow by the relation

$$p_W(1, \bar{z}) - p_W(-1, \bar{z}) = -\bar{\alpha}^2 A(\bar{z}) \int_{-1}^{1} \left[U_S(y, \bar{z}) \right]^2 dy. \qquad (5.4.33)$$

It is now evident that equation (5.4.23), together with (5.4.33), constitutes a nonlinear eigen-value problem to determine the streamwise wave number $\bar{\alpha}$ and phase speed $\bar{c}$ of the travelling wave. The nonlinearity in (5.4.23) is not immediately apparent but is due to the occurrence of the streak shear $\lambda_{\pm}(\bar{z})$, both explicitly and through the functions $\xi_0^{(\pm)}(\bar{z})$, and expresses the feedback of the roll/streak flow on the wave.

It remains to consider the spanwise momentum equation (5.1.1d). Substitution of (5.4.26) into (5.1.1d) yields

$$i\bar{\alpha} U_S w_W = -\frac{\partial p_W}{\partial \bar{z}}. \qquad (5.4.34)$$

Now we shall determine the wave amplitude ε such that the nonlinear self-interaction of the wave in the core provides a strong feedback on the roll flow, in the same way as in the finite-Reynolds-number SSP of Section 5.2. As in that section, we introduce the interaction into the normal momentum equation (5.4.4b) for the roll flow. This equation is derived at order $Re^{-2}\beta^{-2}$, and therefore we require the contribution from the wave arising from an inertial term such as $v\partial v/\partial y$ to be also of this order. Since the magnitude of the normal wave component in the core is $O(\varepsilon\beta)$ from (5.4.26b), this requires the balance

$$\varepsilon^2 \beta^2 \sim Re^{-2}\beta^{-2},$$

which in view of (5.4.31), fixes the Reynolds number dependence of the wave amplitude as

$$\varepsilon = Re^{-5/7}.$$

With this value of ε, if we substitute the roll/streak/wave expansions (5.4.26) into the normal momentum equation (5.1.1c) and equate the terms proportional to E^0, the leading-order balance for the roll is now adjusted from the unforced form (5.4.4b) to[9]

$$i\bar{\alpha}u_W^* v_W + v_W \frac{\partial v_W^*}{\partial y} + c.c. = -\frac{\partial \overline{P}_R}{\partial y}. \tag{5.4.35}$$

In addition, from consideration of the spanwise momentum equation (5.1.1d), the spanwise roll equation is modified from (5.4.4c) to

$$V_R \frac{\partial \overline{W}_R}{\partial y} + \overline{W}_R \frac{\partial \overline{W}_R}{\partial \bar{z}} + \left(i\bar{\alpha}u_W^* w_W + v_W \frac{\partial w_W^*}{\partial y} + c.c. \right) = -\frac{\partial \overline{P}_R}{\partial \bar{z}} + \frac{\partial^2 \overline{W}_R}{\partial y^2}, \tag{5.4.36}$$

while the streak equation (5.4.4d) and the roll continuity equation (5.4.4a) are unaltered to leading order. Cross-differentiating (5.4.35) and (5.4.36) to eliminate the roll pressure and using the continuity equation (5.4.4a), we obtain

$$V_R \frac{\partial^2 \overline{W}_R}{\partial y^2} - \overline{W}_R \frac{\partial^2 V_R}{\partial y^2} + \frac{\partial}{\partial y} \left(i\bar{\alpha}u_W^* w_W + v_W \frac{\partial w_W^*}{\partial y} + c.c. \right)$$
$$- \frac{\partial}{\partial \bar{z}} \left(i\bar{\alpha}u_W^* v_W + v_W \frac{\partial v_W^*}{\partial y} + c.c. \right) = \frac{\partial^3 \overline{W}_R}{\partial y^3}. \tag{5.4.37}$$

Equation (5.4.37) should be compared with its finite-Reynolds-number counterpart (5.2.34). One of the differences is that in our current theory we have nonlinear contributions on the left-hand-side from both the roll and the wave, while in the SSP approximation considered earlier the roll was assumed numerically small and nonlinear roll effects neglected as a result. Equation (5.4.37) can be further simplified by observing that the inertial contributions from the wave can be written in terms of the streak flow as

$$\left(i\bar{\alpha}u_W^* w_W + v_W \frac{\partial w_W^*}{\partial y} + c.c. \right) = \bar{\alpha}^2 U_S^2 \frac{d}{d\bar{z}}(AA^*) + 2\bar{\alpha}^2 AA^* \frac{\partial}{\partial \bar{z}}(U_S^2), \tag{5.4.38}$$

$$\left(i\bar{\alpha}u_W^* v_W + v_W \frac{\partial v_W^*}{\partial y} + c.c. \right) = 4\bar{\alpha}^2 AA^* U_S \frac{\partial U_S}{\partial y}. \tag{5.4.39}$$

When deducing (5.4.38) and (5.4.39) we used the explicit expressions (5.4.29) for u_W and v_W as well as equation (5.4.34). The latter was solved for w_W and after differentiating w_W with respect to y, equation (5.4.32) was used to substitute for $\partial p_W / \partial y$.

Substituting (5.4.39) and (5.4.38) into (5.4.37) we obtain our final equation describing the nonlinear feedback of the waves on the roll:

$$V_R \frac{\partial^2 \overline{W}_R}{\partial y^2} - \overline{W}_R \frac{\partial^2 V_R}{\partial y^2} = 2\bar{\alpha}^2 \frac{d}{d\bar{z}}(AA^*) U_S \frac{\partial U_S}{\partial y} + \frac{\partial^3 \overline{W}_R}{\partial y^3}. \tag{5.4.40}$$

This is to be solved in conjunction with the roll continuity equation

---

[9]Remember that $*$ denotes complex conjugate.

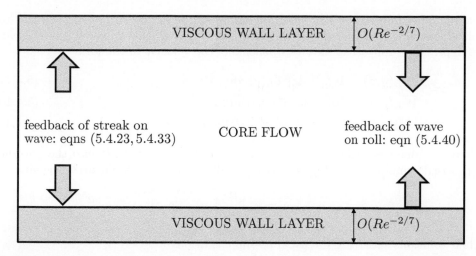

Fig. 5.17: A sketch of the structure of the vortex-viscous wave interaction for plane Poiseuille flow.

$$\frac{\partial V_R}{\partial y} + \frac{\partial \overline{W}_R}{\partial \overline{z}} = 0,$$ (5.4.41)

the streak equation

$$V_R \frac{\partial U_S}{\partial y} + \overline{W}_R \frac{\partial U_S}{\partial \overline{z}} = 2 + \frac{\partial^2 U_S}{\partial y^2},$$ (5.4.42)

and the no-slip boundary conditions

$$U_S = V_R = \overline{W}_R = 0 \quad \text{on} \quad y = \pm 1.$$ (5.4.43)

Equation (5.4.40) describes how the wave affects the roll flow, while equations (5.4.23), (5.4.33) describe the back effect of the roll/streak flow on the wave via the occurrence of the near-wall streak shear contribution $\lambda_\pm(\overline{z})$ and the integrated streak profile in these equations. A sketch of the vortex-viscous wave interaction for plane Poiseuille flow is given in Figure 5.17.

In general the solution of this system is a challenging numerical one, but before describing solutions of the full problem, some insight can be gained by first performing a small-amplitude analysis of the system.

5.4.4   Solution for small amplitude: weakly nonlinear theory

In view of the fact that this particular roll/streak/wave interaction is founded upon the linear neutral stability of a viscous travelling wave, it is possible to study the small amplitude development of the interaction in an analytic fashion. We start by assuming that the wave has a linear neutral wave number denoted by $\bar{\alpha}_0$, writing

$$A = A_0 \cos\left(\bar{\beta}\overline{z}\right),$$

and supposing that $|A_0| \ll 1$. Here the order one quantity $\bar{\beta}$ is the prescribed spanwise wave number corresponding to a long spanwise wavelength $Re^{1/7}2\pi/\bar{\beta}$ in view of the scaling found in (5.4.31). The quadratic nature of the forcing in (5.4.40) will lead to a roll/streak flow with a spanwise wavelength exactly half that of the wave. Therefore, we express the roll/streak flow components of the velocity in the form

$$V_R(y, \bar{z}) = 2\bar{\alpha}_0^2 \bar{\beta}^2 |A_0|^2 \bar{v}_R(y) \cos(2\bar{\beta}\,\bar{z}) + \cdots, \tag{5.4.44a}$$

$$\overline{W}_R(y, \bar{z}) = 2\bar{\alpha}_0^2 \bar{\beta} |A_0|^2 \overline{w}_R(y) \sin(2\bar{\beta}\,\bar{z}) + \cdots, \tag{5.4.44b}$$

$$U_S(y, \bar{z}) = 1 - y^2 + 2\bar{\alpha}_0^2 \bar{\beta}^2 |A_0|^2 \bar{u}_S(y) \cos(2\bar{\beta}\,\bar{z}) + \cdots, \tag{5.4.44c}$$

where $\cdots$ denotes higher powers of $|A_0|^2$. Substituting (5.4.44) into the governing roll/streak equations (5.4.40), (5.4.41), (5.4.42), we obtain the linearized equations

$$\frac{d\bar{v}_R}{dy} + 2\overline{w}_R = 0, \quad -2y\bar{v}_R = \frac{d^2\bar{u}_S}{dy^2}, \quad -2y(1-y^2) = \frac{d^3\overline{w}_R}{dy^3}.$$

Their solutions satisfying the usual no-slip conditions

$$\bar{u}_S = \bar{v}_R = \overline{w}_R = 0 \quad \text{on} \quad y = \pm 1$$

are easily found to be

$$\bar{u}_S = (1 - y^2)(59 + 59y^2 - 91y^4 + 41y^6 - 4y^8)/37800, \tag{5.4.45a}$$

$$\bar{v}_R = y(1 - y^2)^2(5 - y^2)/210, \tag{5.4.45b}$$

$$\overline{w}_R = -(1 - y^2)(5 - 28y^2 + 7y^4)/420. \tag{5.4.45c}$$

Since the expression for $\bar{u}_S$ is even about $y = 0$, it is evident that the quantities $\lambda_\pm = \mp \partial U_S / \partial y$ on $y = \pm 1$ are equal to $O(|A_0|^2)$, and given by

$$\lambda(\bar{z}) = \lambda_\pm(\bar{z}) = 2\big[1 + |A_0|^2 \overline{\lambda}_1 \cos(2\bar{\beta}\,\bar{z}) + \cdots\big], \tag{5.4.46}$$

with

$$\overline{\lambda}_1 = \frac{16}{4725} \bar{\alpha}_0^2 \bar{\beta}^2.$$

Streamlines of the roll flow and a contour plot of the perturbation to the streak profile, calculated from (5.4.44), (5.4.45), are shown in Figure 5.18.

It is also possible to calculate the effect of weak nonlinearity on the wave number and phase speed of the wave by examining the pressure equation (5.4.23). In view of the fact that the perturbation to the laminar parabolic velocity profile in Poiseuille flow is of $O(|A_0|^2)$, from (5.4.44c), it follows that the appropriate expansions for these wave properties are

$$\bar{\alpha} = \bar{\alpha}_0\big(1 + |A_0|^2 \bar{\alpha}_1 + \cdots\big), \qquad \bar{c} = \bar{c}_0\big(1 + |A_0|^2 \bar{c}_1 + \cdots\big). \tag{5.4.47}$$

In the algebra that follows it is easier to work in terms of the wave number and scaled frequency $\overline{\Omega} = \bar{\alpha}\bar{c}$ of the wave with the latter expanding in the form

$$\overline{\Omega} = \overline{\Omega}_0\big(1 + |A_0|^2 \overline{\Omega}_1 + \cdots\big), \tag{5.4.48}$$

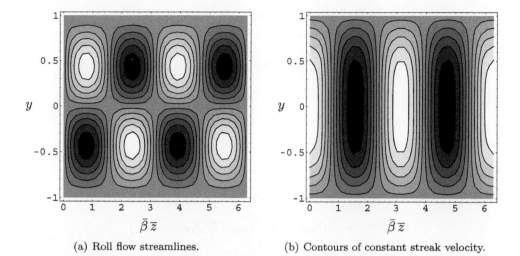

(a) Roll flow streamlines.   (b) Contours of constant streak velocity.

Fig. 5.18: Roll flow and streak perturbations according to the predictions of the weakly nonlinear theory (5.4.44), (5.4.45).

with $\overline{\Omega}_0 = \bar{\alpha}_0 \bar{c}_0$ and $\overline{\Omega}_1 = \bar{\alpha}_1 \bar{c}_0 + \bar{\alpha}_0 \bar{c}_1$. Given that, to the order we are working, we have $\lambda_+ = \lambda_-$, it follows from (5.4.23) that

$$p_W^{(+)} = p_W^{(-)},$$

and this allows us to combine the upper and lower wall pressure equations into one equation, namely

$$\frac{d^2 Q}{d\bar{z}^2} - \frac{1}{\lambda}\frac{d\lambda}{d\bar{z}}\mathcal{F}\big(\xi_0(\bar{z})\big)\frac{dQ}{d\bar{z}} - \bar{\alpha}^2 Q = \frac{2\lambda^{5/3}}{\bar{\alpha}^{1/3} I}\mathcal{G}\big(\xi_0(\bar{z})\big) Q, \qquad (5.4.49)$$

where we have defined

$$Q(\bar{z}) = p_W^{(+)} - p_W^{(-)}, \qquad I = \int_{-1}^{1}\big[U_S(y, \bar{z})\big]^2 dy. \qquad (5.4.50)$$

In view of the expansion for $\lambda(\bar{z})$ in (5.4.46) it is clear that we will generate higher harmonics in $\bar{z}$ from the second term on the left-hand side of (5.4.49). We therefore expand Q in the form

$$Q = Q_0 \cos\big(\bar{\beta}\,\bar{z}\big) + \big|A_0\big|^2 Q_1 \cos\big(\bar{\beta}\,\bar{z}\big) + \big|A_0\big|^2 Q_3 \cos\big(3\bar{\beta}\,\bar{z}\big) + \cdots. \qquad (5.4.51)$$

It is also necessary to expand the other quantities in (5.4.49) in a similar way. For instance it can be seen that

$$I = I_0 + \big|A_0\big|^2 I_1 \cos\big(2\bar{\beta}\,\bar{z}\big) + \cdots,$$

where, from consideration of (5.4.44) and (5.4.50):

$$I_0 = \int_{-1}^{1} (1 - y^2)^2 dy = \frac{16}{15},$$

$$I_1 = \int_{-1}^{1} 4(1 - y^2)\bar{\alpha}_0^2\bar{\beta}^2\overline{u}_S(y) \, dy = \frac{4096}{567567}\bar{\alpha}_0^2\bar{\beta}^2, \qquad (5.4.52)$$

while

$$\xi_0^{(\pm)} = \xi_0 = \xi_{00}\big(1 + |A_0|^2\, \xi_{01} + \cdots\big), \qquad (5.4.53)$$

with

$$\xi_{00} = -\frac{i^{1/3}\overline{\Omega}_0}{(2\bar{\alpha}_0)^{2/3}}, \qquad \xi_{01} = \overline{\Omega}_1 - \frac{2}{3}\bar{\alpha}_1 - \frac{2}{3}\overline{\lambda}_1 \cos\big(2\bar{\beta}\,\bar{z}\big). \qquad (5.4.54)$$

The functions $\mathcal{F}$ and $\mathcal{G}$ can then be expanded in Taylor series as

$$\mathcal{F}(\xi_0) = \mathcal{F}(\xi_{00}) + |A_0|^2\xi_{00}\xi_{01}\mathcal{F}'(\xi_{00}) + \cdots,$$

$$\mathcal{G}(\xi_0) = \mathcal{G}(\xi_{00}) + |A_0|^2\xi_{00}\xi_{01}\mathcal{G}'(\xi_{00}) + \cdots. \qquad (5.4.55)$$

The expansions (5.4.51)–(5.4.55) are then inserted into the pressure equation (5.4.49) and like powers of $|A_0|^2$ are compared. At leading order we obtain the lower branch eigenrelation

$$-\gamma_0^2 = \frac{2^{8/3}}{\bar{\alpha}_0^{1/3} I_0}\mathcal{G}(\xi_{00}), \qquad \gamma_0^2 \equiv \bar{\beta}^2 + \bar{\alpha}_0^2, \qquad (5.4.56)$$

which is a three-dimensional generalization of that given in (2.5.26). This yields the neutral criteria

$$\mathcal{G}(\xi_{00}) = -d_1, \qquad \xi_{00} = -i^{1/3}d_2,$$

where the constants $d_1 \simeq 1.001$, $d_2 \simeq 2.297$; see (2.5.28). For a given scaled spanwise wave number $\bar{\beta}$, equations (5.4.56) determine the leading-order linear neutral wave number $\bar{\alpha}_0$ and frequency $\overline{\Omega}_0$ of the viscous wave. The behaviour of $\bar{\alpha}_0$ and frequency $\overline{\Omega}_0$ as functions of $\bar{\beta}$ is shown in Figure 5.19.

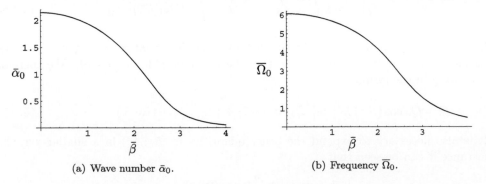

(a) Wave number $\bar{\alpha}_0$.   (b) Frequency $\overline{\Omega}_0$.

Fig. 5.19: Linear scaled neutral values of $\bar{\alpha}_0$ and $\overline{\Omega}_0$ versus scaled spanwise wave number $\bar{\beta}$.

In order to determine how the corrections to the wave number and frequency vary with amplitude we need to proceed to next order in the pressure equation (5.4.49). For this purpose it is useful to consider the perturbations to each of the coefficients in (5.4.49) in turn. First we note that to leading order:

$$-\frac{1}{\lambda}\frac{d\lambda}{d\bar{z}}\mathcal{F}(\xi_0) = \left|A_0\right|^2 2\bar{\beta}\,\overline{\lambda}_1\mathcal{F}(\xi_{00})\sin\left(2\bar{\beta}\,\bar{z}\right),$$ (5.4.57)

while, correct to second order:

$$\frac{2\lambda^{5/3}}{\bar{\alpha}_0^{1/3}I}\mathcal{G}(\xi_0) = -\gamma_0^2\left\{1 + \left|A_0\right|^2\left[\frac{5}{3}\overline{\lambda}_1\cos\left(2\bar{\beta}\,\bar{z}\right) - \frac{\bar{\alpha}_1}{3}\right.\right.$$
$$\left.\left. - \frac{I_1}{I_0}\cos\left(2\bar{\beta}\,\bar{z}\right) + \xi_{00}\xi_{01}\frac{\mathcal{G}'(\xi_{00})}{\mathcal{G}(\xi_{00})}\right]\right\},$$ (5.4.58)

where we have made use of (5.4.56) to simplify this expression. We substitute (5.4.57), (5.4.58) into (5.4.49), along with (5.4.51) for Q and use the forms for $\bar{\alpha}$, ξ_{01} calculated in (5.4.47), (5.4.54). The terms proportional to Q_1 cancel out, in view of (5.4.56). We then find, upon equating coefficients of $\left|A_0\right|^2 Q_0\cos\left(\bar{\beta}\,\bar{z}\right)$, that the wave number and frequency corrections are related by an equation of the form

$$\theta_1\bar{\alpha}_1 + \theta_2\overline{\Omega}_1 + \theta_3 = 0,$$ (5.4.59)

with complex-valued coefficients given by

$$\theta_1(\bar{\beta}) = 2\bar{\alpha}_0^2 + \gamma_0^2\left[\frac{1}{3} + \frac{2}{3}\xi_{00}\frac{\mathcal{G}'(\xi_{00})}{\mathcal{G}(\xi_{00})}\right], \qquad \theta_2(\bar{\beta}) = -\gamma_0^2\xi_{00}\frac{\mathcal{G}'(\xi_{00})}{\mathcal{G}(\xi_{00})},$$

$$\theta_3(\bar{\beta}) = \bar{\beta}^2\overline{\lambda}_1\mathcal{F}(\xi_{00}) - \frac{1}{2}\gamma_0^2\left[\frac{5}{3}\overline{\lambda}_1 - \frac{I_1}{I_0} - \frac{2}{3}\overline{\lambda}_1\xi_{00}\frac{\mathcal{G}'(\xi_{00})}{\mathcal{G}(\xi_{00})}\right].$$

For a prescribed scaled spanwise wave number $\bar{\beta}$, the leading-order streamwise wave number $\bar{\alpha}_0$ can be calculated from (5.4.56) and then the coefficients θ_1, θ_2, θ_3

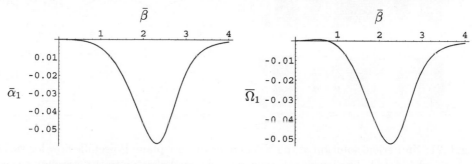

(a) Wave number correction $\bar{\alpha}_1$.　　　　(b) Frequency correction $\overline{\Omega}_1$.

Fig. 5.20: Weakly nonlinear theory: wave number and frequency corrections versus $\bar{\beta}$ calculated from the amplitude equation (5.4.59).

follow. The resulting solutions for $\bar{\alpha}_1$ and $\overline{\Omega}_1$, calculated from a numerical solution of (5.4.59), are shown in Figure 5.20 for a range of values of $\bar{\beta}$. These results can then be used as a starting guess for, and partial check on, computational solutions of the full equations (5.4.40), (5.4.41), (5.4.42), (5.4.23), (5.4.33), (5.4.43) for increasing amplitude.

5.4.5   Full solution of the nonlinear interaction equations

Full numerical solutions have been computed using spectral methods involving Chebyshev/Fourier bases in the wall normal/spanwise directions. More details of the numerical technique can be found in Dempsey *et al.* (2016) where it is also shown that the solution of these VWI equations is in full agreement with corresponding Navier–Stokes simulations provided the Reynolds number is sufficiently large. The degree of nonlinearity present in the flow field is characterized by the quantity

$$\mathcal{A}^2 = \int_0^{2\pi/\bar{\beta}} \left| A(\bar{z}) \right|^2 d\bar{z}. \tag{5.4.60}$$

The solutions computed have several symmetries: as a result the streak shear stresses on the upper and lower walls are equal, i.e. we have

$$\lambda(\bar{z}) = \lambda_-(\bar{z}) = \lambda_+(\bar{z}), \tag{5.4.61}$$

with $\lambda_\pm$ defined in (5.4.12), (5.4.19).

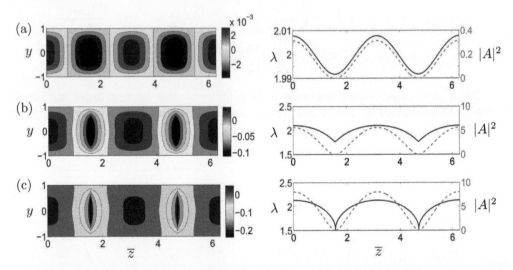

Fig. 5.21: Numerical solution of the VWI equations for plane Poiseuille flow for various wave amplitudes $\mathcal{A}$. The left panels represent the perturbation to the streak in the core defined in (5.4.62), while the right panels show the accompanying wall shear stress as a solid curve and the function $|A(\bar{z})|^2$ as a dotted line: (a) $\mathcal{A}^2 = 1$; (b) $\mathcal{A}^2 = 20$; (c) $\mathcal{A}^2 = 30$.

Representative numerical solutions of the VWI system are shown in Figure 5.21 for a scaled spanwise wave number $\bar{\beta} = 1$. In this figure we plot the perturbation to the streak in the core, given by

$$U_S(y, \bar{z}) - (1 - y^2), \tag{5.4.62}$$

and the wall shear stress $\lambda(\bar{z})$. Figure 5.21(a) shows the solution in the weakly nonlinear regime where we see that a small spanwise modulation gives rise to a uniform roll structure. Increasing the amplitude further results in stronger wave forcing, inducing larger perturbations to the flow and fully nonlinear interactions arise. The solutions shown in Figure 5.21(b) for $\mathcal{A}^2 = 20$ become localized in the spanwise direction, with the wall shear stress developing steep gradients at particular spanwise locations. The flow undergoes further localization for $\mathcal{A}^2 = 30$ as shown in Figure 5.21(c), and an ever larger number of Fourier modes is required to achieve satisfactory resolution.

Exercises 17

1. *Properties of the 'forced' Airy equation*

    The function $\mathcal{R}$ satisfies

$$\mathcal{R}'' - \xi\mathcal{R} = 1, \quad \mathcal{R}'(\xi_0) = 0, \quad \mathcal{R}(\infty) = 0.$$

    (a) Use the method of reduction of order to show that

$$\mathcal{R}(\xi) = Ai(\xi)q(\xi),$$

    where $q(\xi)$ satisfies

$$q'(\xi) = \left[Ai(\xi)\right]^{-2}\left[\int_{\xi_0}^{\xi} Ai(s)\,ds - Ai(\xi_0)Ai'(\xi_0)q(\xi_0)\right].$$

    By applying the boundary condition $\mathcal{R}(\infty) = 0$, determine the value of $q(\xi_0)$ and show that

$$\mathcal{R}(\xi_0) = \frac{1}{Ai'(\xi_0)}\int_{\xi_0}^{\infty} Ai(s)\,ds. \tag{5.4.63}$$

    (b) Verify the following properties of $\mathcal{R}$ and Ai:

    (i) $D^2\mathcal{R}' = \mathcal{R}$, (ii) $D^2 Ai' = Ai$,

    (iii) $D^2\left(\dfrac{\xi^2}{4}\mathcal{R} - \dfrac{1}{2}\mathcal{R}' + \dfrac{\xi}{4}\right) = \xi\mathcal{R}'$, (iv) $D^2\left(\dfrac{\xi^2}{4}Ai - \dfrac{1}{2}Ai'\right) = \xi Ai'$,

    where the operator D^2 is defined by

$$D^2 \equiv \frac{d^2}{d\xi^2} - \xi.$$

(c) Using the properties derived in (b), and integration by parts, establish the following integral results:

(i) $\displaystyle\int_{\xi_0}^{\infty} \left[\mathcal{R}(\xi)\right]^2 d\xi = -\mathcal{R}(\xi_0)\left[2 + \xi_0 \mathcal{R}(\xi_0)\right],$

(ii) $\displaystyle\int_{\xi_0}^{\infty} \mathcal{R}(\xi) Ai(\xi)\, d\xi = -Ai(\xi_0)\left[1 + \xi_0 \mathcal{R}(\xi_0)\right],$

(iii) $\displaystyle\int_{\xi_0}^{\infty} \xi \mathcal{R}(\xi) \mathcal{R}'(\xi)\, d\xi = \mathcal{R}(\xi_0),$

(iv) $\displaystyle\int_{\xi_0}^{\infty} \xi \mathcal{R}(\xi) Ai'(\xi)\, d\xi = \frac{1}{4}\left[3 Ai(\xi_0) - \xi_0 Ai'(\xi_0) - \xi_0^2 Ai'(\xi_0)\mathcal{R}(\xi_0)\right].$

2. *Derivation of the viscous wave pressure equation (5.4.23)*

Consider the viscous wall layer equations

$$i\alpha u + \frac{\partial v}{\partial Y} + \frac{\partial w}{\partial z} = 0, \tag{5.4.64a}$$

$$i\alpha\left[\lambda(z)Y - c\right]u + \lambda v + \frac{d\lambda}{dz}Yw + i\alpha p = \frac{\partial^2 u}{\partial Y^2}, \tag{5.4.64b}$$

$$i\alpha\left[\lambda(z)Y - c\right]w + \frac{dp}{dz} = \frac{\partial^2 w}{\partial Y^2}, \tag{5.4.64c}$$

subject to

$$u = v = w = 0 \qquad \text{at} \quad Y = 0, \tag{5.4.65a}$$

$$u = B(z), \quad w = 0 \quad \text{at} \quad Y = \infty, \tag{5.4.65b}$$

as posed in (5.4.20), (5.4.21). In (5.4.64), (5.4.65) we have removed the annotations used on the variables in section 5.4.2 for clarity.

The aim of this question is to derive a solvability condition for this system which depends only on the wave pressure $p(z)$ and the streak shear $\lambda(z)$.

(a) *The solution for the spanwise velocity*

Show that the solution for w may be expressed in terms of the variable

$$\xi(Y, z) = \left[i\alpha\lambda(z)\right]^{1/3}\left[Y - \frac{c}{\lambda(z)}\right],$$

in the form

$$w = \left[i\alpha\lambda(z)\right]^{-2/3}\frac{dp}{dz}\left[\mathcal{R}(\xi) - \gamma_0(z)Ai(\xi)\right],$$

where

$$\gamma_0(z) = \frac{\mathcal{R}(\xi_0)}{Ai(\xi_0)}, \quad \xi_0 = \xi(0, z),$$

and the function $\mathcal{R}$ satisfies

$$\mathcal{R}'' - \xi\mathcal{R} = 1, \qquad \mathcal{R} \to 0 \quad \text{as} \quad \xi \to \infty.$$

Explain why it is possible to take $\mathcal{R}'(\xi_0) = 0$ without loss of generality.

(b) *The solution for the shear*

By eliminating the normal velocity v between equations (5.4.64a), (5.4.64b), establish that u satisfies

$$\left(\frac{\partial^2}{\partial \xi^2} - \xi \right) \frac{\partial u}{\partial \xi} = -(i\alpha\lambda)^{-1} \left[\lambda \frac{\partial w}{\partial z} - \frac{d\lambda}{dz} \left(w + Y \frac{\partial w}{\partial Y} \right) \right]. \qquad (5.4.66)$$

Using the form for w deduced in (a), show that the right-hand side of (5.4.66), $S(\xi, z)$ say, can be rewritten in the form

$$S(\xi, z) = (i\alpha\lambda)^{-5/3} \left(\nu_1(\xi) \lambda \frac{d^2 p}{dz^2} + \nu_2(\xi) \frac{d\lambda}{dz} \frac{dp}{dz} \right),$$

where

$$\nu_1 = \gamma_0 Ai(\xi) - \mathcal{R}(\xi),$$

$$\nu_2 = \frac{5}{3} \left[\mathcal{R}(\xi) - \gamma_0 Ai(\xi) \right] + \frac{2}{3} \left[\frac{\xi_0 \mathcal{R}(\xi_0) Ai'(\xi_0)}{\left[Ai(\xi_0) \right]^2} Ai(\xi) \right.$$

$$\left. + \xi \mathcal{R}'(\xi) - \gamma_0 \xi Ai'(\xi) \right].$$

(c) *The solvability condition*

Show by integrating by parts and using (5.4.64b), (5.4.65) that

(i) $\displaystyle \int_{\xi_0}^{\infty} \mathcal{R}(\xi) \left(\frac{\partial^2}{\partial \xi^2} - \xi \right) \frac{\partial u}{\partial \xi} \, d\xi = -\mathcal{R}(\xi_0)(i\alpha\lambda)^{-2/3} i\alpha p + B(z),$

(ii) $\displaystyle \int_{\xi_0}^{\infty} \mathcal{R}(\xi) S(\xi, z) \, d\xi = (i\alpha\lambda)^{-5/3} \left(\lambda \frac{d^2 p}{dz^2} \mu_1 + \frac{d\lambda}{dz} \frac{dp}{dz} \mu_2 \right),$

where

$$\mu_n(\xi_0) = \int_{\xi_0}^{\infty} \mathcal{R}(\xi) \nu_n(\xi) \, d\xi, \quad n = 1, 2.$$

Hence deduce that the pressure p satisfies the ordinary differential equation

$$\frac{d^2 p}{dz^2} + \frac{1}{\lambda} \frac{d\lambda}{dz} \frac{\mu_2}{\mu_1} \frac{dp}{dz} - \frac{\mathcal{R}(\xi_0)}{\mu_1} \alpha^2 p = \frac{(i\alpha\lambda)^{5/3}}{\lambda} \frac{B(z)}{\mu_1}. \qquad (5.4.67)$$

(d) *Evaluation of the coefficients μ_1, μ_2*

Use the results established in Question 1(c) to show that

$$\mu_1(\xi_0) = \mathcal{R}(\xi_0),$$

$$\mu_2(\xi_0) = -\mathcal{R}(\xi_0) \left[\frac{3}{2} + \frac{1}{2} \frac{\xi_0 Ai'(\xi_0)}{Ai(\xi_0)} + \frac{1}{2} \frac{\xi_0^2 \mathcal{R}(\xi_0) Ai'(\xi_0)}{Ai(\xi_0)} \right],$$

and using (5.4.63) deduce that the pressure equation (5.4.67) can be put in the form given in (5.4.23), (5.4.24).

3. *A vortex-TS wave interaction in a boundary-layer flow*

In a boundary layer subject to a strong adverse pressure gradient, a vortex-wave interaction involving a three-dimensional TS wave can be formulated (Walton *et al.*, 1994). The wall shear $\lambda(x, z)$ satisfies

$$\lambda \frac{\partial \lambda}{\partial x} = -\frac{1}{2} + Q^2 \frac{\partial W}{\partial z},$$

where the function W represents a spanwise forcing exerted by the wave on the roll flow and is given by

$$W(x, z) = -\frac{1}{\alpha^2 \lambda^2} \frac{\partial}{\partial z} \left(\alpha^2 |p|^2 + \left| \frac{\partial p}{\partial z} \right|^2 \right),$$

while Q is the (real) wave amplitude. The TS wave pressure $p(x, z)$ satisfies a similar equation to that given in (5.4.23), and derived in Question 2(c), for the case of plane Poiseuille flow, namely

$$\frac{\partial^2 p}{\partial z^2} - \frac{1}{\lambda} \frac{\partial \lambda}{\partial z} \mathcal{F}(\xi_0) \frac{\partial p}{\partial z} - \alpha^2 p - (\alpha \lambda)^{5/3} \mathcal{G}(\xi_0) A(x, z) = 0. \tag{5.4.68}$$

The functions $\mathcal{F}, \mathcal{G}$ and ξ_0 are as defined in (5.4.24), (5.4.25) (ignoring the $\pm$ superscripts) and the feedback of the roll on the wave is again manifested through the shear λ.

In contrast to the channel flow problem however, the displacement function A is related to the wave pressure p through the solution of the following equation for the pressure $\bar{p}(x, y, z)$ in the upper deck:

$$\frac{\partial^2 \bar{p}}{\partial y^2} + \frac{\partial^2 \bar{p}}{\partial z^2} - \alpha^2 \bar{p} = 0.$$

This equation should be solved with the following condition of matching with the solution in the lower deck

$$\left. \frac{\partial \bar{p}}{\partial y} \right|_{y=0} = -\alpha^2 A(x, z),$$

and the perturbations attenuation condition at the edge of the upper deck:

$$\bar{p} \to 0 \quad \text{as} \quad y \to \infty.$$

Once $\bar{p}$ is found, the pressure $p(x, z)$ in the lower deck can be calculated by simply setting

$$\bar{p}(x, 0, z) = p(x, z).$$

Here you can restrict your attention to solutions with spanwise periodicity $2\pi/\beta$ and suppose that the interaction is initiated by inputting a wave of the form

$$p = \cos \beta z$$

at $x = -1$ with uniform shear $\lambda = 1$ at this location.

(a) Show that in the absence of a wave, i.e. with $Q \equiv 0$, the shear λ encounters Goldstein's singularity at $x = 0$; see Section 5.3.2 in Part 3 of this book series.

(b) Seek a solution for the upper deck pressure $\bar{p}$ of the form

$$\bar{p}(x, y, z) = \sum_{n=1}^{\infty} \bar{p}_n(x, y) \cos(n\beta z),$$

and show that the wave pressure p and displacement A are related via

$$A(x, z) = \frac{2\beta}{\alpha^2 \pi} \sum_{n=1}^{\infty} (n^2\beta^2 + \alpha^2)^{1/2} \cos(n\beta z) \int_0^{\pi/\beta} p(x, s) \cos(n\beta s)\, ds. \quad (5.4.69)$$

Investigate how the interaction develops near the position of wave input at $x = -1$. For this purpose define

$$\varepsilon = x + 1 \ll 1,$$

and suppose that the wave acquires the form

$$p = \cos\beta z + \varepsilon p_1(z) + \cdots, \qquad Q = Q_0 + \varepsilon Q_1 + \cdots,$$

with the TS wave number expanding as

$$\alpha = \alpha_0 \left(1 + \varepsilon\alpha_1 + \cdots\right).$$

You may assume that the TS wave propagates at a fixed frequency $\Omega = \alpha c$.

(c) Show that the spanwise forcing and wall shear expand as

$$W = \beta \left(1 - \frac{\beta^2}{\alpha_0^2}\right) \sin(2\beta z) + \cdots,$$

$$\lambda = 1 + \varepsilon \left[-\frac{1}{2} + \frac{2\beta^2 Q_0^2}{\alpha_0^2}(\alpha_0^2 - \beta^2)\cos(2\beta z)\right] + \cdots.$$

(d) Substitute the expression for λ into the pressure equation (5.4.68) and use the pressure-displacement relation (5.4.69). Show that at leading order:

$$-\beta^2 - \alpha_0^2 - \alpha_0^{-1/3}\mathcal{G}(\xi_{00})(\beta^2 + \alpha_0^2)^{1/2} = 0, \qquad \xi_{00} = -i^{1/3}\Omega/\alpha_0^{2/3}.$$

As explained in Section 5.4.4 this is only possible if $\mathcal{G}(\xi_{00}) = -d_1$ and $\xi_{00} = -i^{1/3}d_2$ where $d_1 \simeq 1.001$, $d_2 \simeq 2.297$, see equation (2.5.28).

(e) Now proceed to the next order in the pressure equation and show that the coefficient of $\varepsilon \cos\beta z$ takes the form

$$\theta_1\alpha_1 + \theta_2 Q_0^2 + \theta_3,$$

where the complex-valued coefficients are given by

$$\theta_1 = -\alpha_0^{-2/3} d_1 \chi_0 - 3\alpha_0^2 - 2\beta^2,$$

$$\theta_2 = \frac{2\beta^2}{\alpha_0^2}(\beta^2 - \alpha_0^2)\left(\beta^2 \mathcal{F}(\xi_{00}) + \frac{1}{2}\alpha_0^{-2/3} d_1 \chi_0\right),$$

$$\theta_3 = \frac{1}{2}\alpha_0^{-2/3} d_1 \chi_0,$$

and χ_0 is a known complex constant. Deduce that in order for $p_1(z)$ to be periodic in z we must have

$$\theta_1 \alpha_1 + \theta_2 Q_0^2 + \theta_3 = 0.$$

Hence find an expression for the starting amplitude Q_0 in terms of the real and imaginary parts of $\theta_1, \theta_2, \theta_3$.

5.5 More Recent Developments

In this chapter we have seen how the distinct flow structures that emerge in turbulent simulations of shear flows can be shown to be modelled by a self-sustaining process in which a roll flow, which would otherwise decay under the action of viscosity, is maintained by the nonlinear interaction of three-dimensional travelling waves. The rolls then induce a streaky distortion to the basic flow with this modified profile re-invigorating the very waves that underpin the structure. If the Reynolds number is moderate in size, the wave forcing occurs throughout the flow, while at large Reynolds number the forcing is confined to either the critical layer or layers adjacent to solid boundaries.

Although we have concentrated on plane Couette flow and plane Poiseuille flow for relative simplicity, exact coherent states have been calculated for many other flows, including pipes possessing cross-sections that are rectangular (Wedin *et al.*, 2009) or elliptical (Ozcakir and Hall, 2021).

Coherent structures in pipes of circular cross-section have been extensively studied, starting with the almost simultaneous studies of Faisst and Eckhardt (2003) and Wedin and Kerswell (2004). Numerous exact travelling wave solutions possessing a range of symmetry properties have been calculated, with some existing down to Reynolds numbers of several hundred. At high Re, close agreement between some of these states and asymptotic solutions computed using the vortex-wave interaction theory of Section 5.3 was established by Ozcakir *et al.* (2016), with the latter authors also discovering a different type of exact coherent structure in the form of a centre mode where at large Reynolds number, unlike the structures considered in this chapter, there is no scale separation between the roll/streak and wave components of the flow.

References

Abramowitz, M. and Stegun, I. A. (1965). *Handbook of Mathematical Functions* (third edn). Dover Publications.

Balmforth, N. J. and Morrison, P. J. (1999). A necessary and sufficient instability condition for inviscid shear flow. *Stud. Appl. Math.*, **102**, 309–344.

Benney, D. J. (1984). The evolution of disturbances in shear flows at high Reynolds numbers. *Stud. Appl. Math.*, **70**, 1–19.

Benney, D. J. and Bergeron, R. F. (1969). A new class of nonlinear waves in parallel flows. *Stud. Appl. Math.*, **48**, 181–204.

Blasius, H. (1908). Grenzschichten in flüssigkeiten mit kleiner reibung. *Z. Math. Phys.*, **56**(1), 1–37. (Engl. transl. NACA TM 1256).

Bowles, R. I. and Smith, F. T. (1993). On boundary-layer transition in transonic flow. *J. Eng. Maths*, **27**, 309–342.

Brennan, G. S., Gajjar, J. S. B., and Hewitt, R. E. (2021). Tollmien–Schlichting wave cancellation via localised heating elements in boundary layers. *J. Fluid Mech.*, **909**, A16.

Brown, P. G., Brown, S. N., Smith, F. T., and Timoshin, S. N. (1993). On the starting process of strongly nonlinear vortex/Rayleigh-wave interactions. *Mathematika*, **40**, 7–29.

Burgers, J. M. (1924). The motion of a fluid in the boundary layer along a plane smooth surface. In *Proc. of the First Internat. Congress for Appl. Mech.*, p. 113. Delft.

Burkhalter, J. E. and Koschmieder, E. L. (1974). Steady supercritical Taylor vortices after sudden starts. *Phys. Fluids*, **17**, 1928–1935.

Choudhari, M. (1994). Roughness-induced generation of crossflow vortices in three-dimensional boundary layers. *Theor. Comp. Fluid Dyn.*, **6**, 1–30.

Davis, R. E. (1969). On the high Reynolds number flow over a wavy boundary. *J. Fluid Mech.*, **36**, 337–346.

De Tullio, N. and Ruban, A. I. (2015). A numerical evaluation of the asymptotic theory of receptivity for subsonic compressible boundary layers. *J. Fluid Mech.*, **771**, 520–546.

Deguchi, K. and Hall, P. (2014). The high-Reynolds-number asymptotic development of nonlinear equilibrium states in plane Couette flow. *J. Fluid Mech.*, **750**, 99–112.

Deguchi, K. and Walton, A. G. (2013). A swirling spiral wave solution in pipe flow. *J. Fluid Mech.*, **737**, R2.

Dempsey, L. J., Deguchi, K., Hall, P., and Walton, A. G. (2016). Localized vortex/Tollmien–Schlichting wave interaction states in plane Poiseuille flow. *J. Fluid Mech.*, **791**, 97–121.

Denier, J. P., Hall, P., and Seddougui, S. O. (1991). On the receptivity problem for Görtler vortices: vortex motion induced by wall roughness. *Phil. Trans. Roy. Soc. Lond*, **A 335**, 51–85.

Dong, M., Liu, Y., and Wu, X. (2020). Receptivity of inviscid modes in supersonic boundary layers due to scattering of free-stream sound by localised wall roughness. *J. Fluid Mech.*, **896**, A23.

Dryden, H. L. (1947). Some recent contributions to the study of transition and turbulent boundary layers. *NACA TM* 1168.

Duck, P. W., Ruban, A. I., and Zhikharev, C. N. (1996). The generation of Tollmien–Schlichting waves by free-stream turbulence. *J. Fluid Mech.*, **312**, 341–371.

Faisst, H. and Eckhardt, B. (2003). Traveling waves in pipe flow. *Phys. Rev. Lett.*, **91**, 224502.

Fjortoft, R. (1950). Application of the integral theorems in deriving criteria of instability for laminar flows and for the baroclinic circular vortex. *Geofys. Publ., Oslo*, **17**(6), 1–52.

Foote, J. R. and Lin, C. C. (1950). Some recent investigations in the theory of hydrodynamic stability. *Q. Appl. Math.*, **8**, 265–280.

Gajjar, J. S. B. (1996). Nonlinear stability of nonstationary cross-flow vortices in compressible boundary layers. *Stud. Appl. Math.*, **96**, 53–84.

Gajjar, J. S. B. and Cole, J. W. (1989). The upper-branch stability of compressible boundary-layer flows. *Theor. Comp. Fluid Dyn.*, **1**, 105–123.

Gibson, J. F. and Brand, E. (2014). Spanwise-localized solutions of planar shear flows. *J. Fluid Mech.*, **745**, 25–61.

Gibson, J. F., Halcrow, J., and Cvitanovic, P. (2008). Visualizing the geometry of state space in plane Couette flow. *J. Fluid Mech.*, **611**, 107–130.

Goldstein, M. E. (1985). Scattering of acoustic waves into Tollmien–Schlichting waves by small streamwise variations in surface geometry. *J. Fluid Mech.*, **154**, 509–529.

Gregory, N., Stuart, J. T., and Walker, W. S. (1955). On the stability of three-dimensional boundary layers with application to the flow due to a rotating disk. *Phil. Trans. Roy. Soc. Lond.*, **A 248**(943), 155–199.

Haberman, R. (1972). Critical layers in parallel shear flows. *Stud. Appl. Math.*, **51**, 139–161.

Hall, P. and Sherwin, S. (2010). Streamwise vortices in shear flows: harbingers of transition and the skeleton of coherent structures. *J. Fluid Mech.*, **661**, 178–205.

Hall, P. and Smith, F. T. (1988). The nonlinear interaction of Tollmien–Schlichting waves and Taylor–Görtler vortices in curved channel flows. *Proc. Roy. Soc. Lond.*, **A417**, 255–282.

Hall, P. and Smith, F. T. (1991). On strongly nonlinear vortex/wave interactions in boundary-layer transition. *J. Fluid Mech.*, **227**, 641–666.

Hamilton, J. M., Kim, J., and Waleffe, F. (1995). Regeneration mechanisms of near-wall turbulence structures. *J. Fluid Mech.*, **287**, 317–348.

Herbert, T. (1977). Finite amplitude stability of plane parallel flows. In *Laminar-Turbulent Transition: Proc. AGARD Conf. No. 224, Lyngby.*

Hickernell, F. J. (1984). Time-dependent critical layers in shear flows on the beta-plane. *J. Fluid Mech.*, **142**, 431–449.

Hof, B., van Doorne, C., Westerweel, J., Nieuwstadt, F., Faisst, H., Eckhardt, B., Wedin, H., Kerswell, R, and Waleffe, F. (2004). Experimental observations of non-linear travelling waves in the turbulent pipe flow. *Science*, **305**, 1594–1598.

Howard, L. N. (1961). Note on a paper of John W. Miles. *J. Fluid Mech.*, **10**, 509–512.

Itano, T. and Toh, S. (2001). The dynamics of bursting process in wall turbulence. *J. Phys. Soc. Jpn.*, **70**, 703–716.

Kachanov, Yu. S., Kozlov, V. V., and Levchenko V. Ya. (1982). *The Appearance of Turbulence in the Boundary Layer*. Nauka, Novosibirsk.

Kelvin, W. (1871). Hydrokinetic solutions and observations. *Phil. Mag.*, **42**(4), 362–377.

Klebanoff, P. S. and Tidstrom, K. D. (1959). Evolution of amplified waves leading to transition in a boundary layer with zero pressure gradient. *NASA TN* D-195.

Klebanoff, P. S., Tidstrom, K. D., and Sargent, L. M. (1962). The three-dimensional nature of boundary-layer instability. *J. Fluid Mech.*, **12**, 1–34.

Kline, S. J., Reynolds, W. C., Schraub, F. A., and Runstadler, P. W. (1967). The structure of turbulent boundary layers. *J. Fluid Mech.*, **30**, 741–773.

Kobayashi, R., Kohama, Y., and Takamadate, C. (1980). Spiral vortices in boundary layer transition regime on a rotating disk. *Acta Mech.*, **35**, 71–82.

Kreilos, T. and Eckhardt, B. (2012). Periodic orbits near onset of chaos in plane Couette flow. *Chaos: An Interdisciplinary Journal of Nonlinear Science*, **22**(4), 047505.

Landau, L. D. (1944). On the problem of turbulence. *Dokl. Akad. Nauk SSSR*, **44**(8), 339–349.

Lee, S.-J. and Lee, S.-H. (2001). Flow field analysis of a turbulent boundary layer over a riblet surface. *Experiments in Fluids*, **30**, 153–166.

Lin, C. C. (1946). On the stability of two-dimensional parallel flows. Part 3. Stabilty in a viscous fluid. *Q. Appl. Math.*, **3**, 277–301.

Lin, C. C. (1955). *The Theory of Hydrodynamic Stability*. Cambridge University Press.

Maslowe, S. A. (1986). Critical layers in shear flows. *Ann. Rev. Fluid Mech.*, **18**(1), 405–432.

Messiter, A. F. (1970). Boundary-layer flow near the trailing edge of a flat plate. *SIAM J. Appl. Math.*, **18**(1), 241–257.

Nakayama, Y. (1988). *Visualized Flow*. Pergamon, Oxford.

Neiland, V. Ya. (1969). Theory of laminar boundary layer separation in supersonic flow. *Izv. Akad. Nauk SSSR, Mech. Zhidk. Gaza* (4), 53–57.

Neiland, V. Ya., Bogolepov, V. V., Dudin, G. N., and Lipatov, I. I. (2008). *Asymptotic Theory of Supersonic Viscous Gas Flows*. Elsevier.

Ng, L. L. and Crouch, J. D. (1999). Roughness-induced receptivity to crossflow vortices on a swept wing. *Phys. Fluids*, **11**(2), 432–438.

Ozcakir, O. and Hall, P. (2021). Travelling waves in elliptic pipe flow. *J. Fluid Mech.*, **923**, R3.

Ozcakir, O., Tanveer, S., Hall, P., and Overman, E. A. (2016). Travelling wave states in pipe flow. *J. Fluid Mech.*, **791**, 284–328.

Pierce, D. (1961). Photographic evidence of the formation and growth of vorticity behind plates accelerated from rest in still air. *J. Fluid Mech.*, **11**, 460–464.

Prandtl, L. (1904). Über flüssigkeitsbewegung bei sehr kleiner Reibung. In *Verh. III. Intern. Math. Kongr., Heidelberg*, pp. 484–491. Teubner, Leipzig, 1905.

Rayleigh, J. W. S. (1880). On the stability, or instability, of certain fluid motions. *Proc. Lond. Math. Soc.*, **11**, 57–70.

Rayleigh, J. W. S. (1916). On the dynamics of revolving fluids. *Proc. Lond. Math. Soc. A*, **93**, 148–154.

Reynolds, O. (1883). An experimental investigation of the circumstances which determine whether the motion of water shall be direct or sinuous, and of the law of resistance in parallel channels. *Phil. Trans. Roy. Soc. Lond. A*, **186**, 123–164.

Rosenbluth, M. N. and Simon, A. (1964). Necessary and sufficient condition for the stability of plane parallel inviscid flow. *Phys. Fluids*, **7**, 557–558.

Ruban, A. I. (1981). A singular solution of the boundary-layer equations which can be extended continuously through the point of zero skin friction. *Izv. Akad. Nauk SSSR, Mech. Zhid. Gaza* (6), 42–52.

Ruban, A. I. (1982). Asymptotic theory of short separation bubbles at the leading edge of a thin airfoil. *Izv. Akad. Nauk SSSR, Mech. Zhid. Gaza* (1), 42–52.

Ruban, A. I. (1984). On Tollmien–Schlichting wave generation by sound. *Izv. Akad. Nauk SSSR Mekh. Zhidk. Gaza* (4), 44–52.

Ruban, A. I., Bernots, T., and Kravtsova, M. A. (2016). Linear and nonlinear receptivity of the boundary layer in transonic flows. *J. Fluid Mech.*, **786**, 154–189.

Ruban, A. I., Bernots, T., and Pryce, D. (2013). Receptivity of the boundary layer to vibrations of the wing surface. *J. Fluid Mech.*, **723**, 480–528.

Ruban, A. I., Keshari, S. K., and Kravtsova, M. A. (2021). On boundary-layer receptivity to entropy waves. *J. Fluid Mech.*, **929**, A17.

Saeed, T. I., Mughal, M. S., and Morrison, J. F. (2016). The interaction of a swept-wing boundary layer with surface excrescences. In *54th AIAA Aerospace Science Meeting*. AIAA2016-2065. Publisher: American Institute of Aeronautics and Astronautics.

Schlichting, H. (1933). Zur entstehung der turbulenz bei der plattenströmung. *Nachr. Ges. Wiss. Göttingen, Math. Phys. Klasse*, 181–208.

Schneider, S. P. (2001). Effects of high-speed tunnel noise on laminar-turbulent transition. *Journal of Spacecraft and Rockets*, **38**(3), 323–333.

Schneider, W. (1974). Upstream propagation of unsteady disturbances in supersonic boundary layers. *J. Fluid Mech.*, **63**, 465–485.

Schrader, L.-U., Brandt, L., and Henningson, D. S. (2009). Receptivity mechanisms in three-dimensional boundary-layer flows. *J. Fluid Mech.*, **618**, 209–241.

Schubauer, G. B. and Skramstad, H. K. (1948). Laminar boundary layer oscillations and transition on a flat plate. *NACA Rep.* 909.

Skufca, J. D., Yorke, J. A., and Eckhardt, B. (2006). Edge of chaos in a parallel shear flow. *Phys. Rev. Lett.*, **96**(17), 174101.

Smith, F. T. (1979a). Nonlinear stability of boundary layers for disturbances of various sizes. *Proc. Roy. Soc. Lond.*, **A 368**, 573–589.

Smith, F. T. (1979b). On the nonparallel flow stability of the Blasius boundary layer. *Proc. Roy. Soc. Lond.*, **A 366**, 91–109.

Smith, F. T. and Bodonyi, R. J. (1982a). Amplitude-dependent neutral modes in the Hagen–Poiseuille flow through a circular pipe. *Proc. Roy. Soc. Lond.*, **A 384**, 463–489.

Smith, F. T. and Bodonyi, R. J. (1982b). Nonlinear critical layers and their development in streaming-flow stability. *J. Fluid Mech.*, **118**, 165–185.

Squire, H. B. (1933). On the stability of three-dimensional disturbances of viscous flow between parallel walls. *Proc. Roy. Soc. Lond. A*, **142**, 621–628.

Stewartson, K. (1969). On the flow near the trailing edge of a flat plate. *Mathematika*, **16**(1), 106–121.

Stewartson, K. (1981). Marginally stable inviscid flows with critical layers. *IMA J. Appl. Math.*, **27**(2), 133–176.

Stewartson, K., Smith, F. T., and Kaups, K. (1982). Marginal separation. *Stud. Appl. Math.*, **67**(1), 45–61.

Stewartson, K. and Williams, P. G. (1969). Self-induced separation. *Proc. Roy. Soc. Lond.*, **A 312**, 181–206.

Stuart, J. T. (1960). On the non-linear mechanics of wave disturbances in stable and unstable parallel flows. Part 1. The basic behaviour in plane Poiseuille flow. *J. Fluid Mech.*, **9**(3), 353–370.

Sychev, V. V. (1972). Laminar separation. *Izv. Akad. Nauk SSSR, Mech. Zhidk. Gaza* (3), 47–59.

Sychev, V. V., Ruban, A. I., Sychev, Vic. V., and Korolev, G. L. (1998). *Asymptotic Theory of Separated Flows*. Cambridge University Press.

Taylor, G. I. (1923). Stability of a viscous liquid contained between two rotating cylinders. *Phil. Trans. Roy. Soc. Lond. A*, **223**, 289–343.

Terent'ev, E. D. (1981). Linear problem for a vibrator in subsonic boundary layer. *Prikl. Mat. Mekh.*, **45**, 1049–1055.

Terent'ev, E. D. (1984). The linear problem for a vibrator performing harmonic oscillations at supercritical frequencies in a subsonic boundary layer. *Prikl. Mat. Mekh.*, **48**, 184–191.

Terent'ev, E. D. (1987). On the formation of a wave packet in a boundary layer on a flat plate. *Prikl. Mat. Mech.*, **51**, 814–819.

Timoshin, S. N. (1990). Asymptotic form of the lower branch of the neutral curve in transonic flow. *Uch. Zap. TsAGI*, **21**(6), 50–57.

Tollmien, W. (1929). Über der entstehung der turbulenz. *Nachr. Ges. Wiss. Göttingen, Math. Phys. Klasse*, 21–44. (Engl. transl. NACA TM 609).

Waleffe, F. (1997). On a self-sustaining process in shear flows. *Phys. Fluids*, **9**, 883–900.

Waleffe, F. (2002). Exact coherent structures and their instabilities: toward a dynamical-system theory of shear turbulence. In *Proceedings of the International*

Symposium on 'Dynamics and Statistics of coherent structures in turbulence: roles of elementary vortices', Tokyo, Japan, pp. 115–128.

Walton, A. G., Bowles, R. I., and Smith, F. T. (1994). Vortex-wave interaction in separating flows. *Eur. J. Mech.*, **B 13**, 629–655.

Watson, J. (1960). On the non-linear mechanics of wave disturbances in stable and unstable parallel flows. Part 2. The development of a solution for plane Poiseuille flow and for plane Couette flow. *J. Fluid Mech.*, **9**(3), 371–389.

Wedin, H., Bottaro, A., and Nagata, M. (2009). Three-dimensional traveling waves in a square duct. *Phys. Rev. E*, **79**, 065305.

Wedin, H. and Kerswell, R. R. (2004). Exact coherent structures in pipe flow: travelling wave solutions. *J. Fluid Mech.*, **508**, 333–371.

Werlé, H. (1980). Transition et décollement:visualisations au tunnel hydrodynamique de l'ONERA. *Rech. Aerosp.* (1980-5), 331–345.

Wu, X. (1999). Generation of Tollmien–Schlichting waves by convecting gusts interacting with sound. *J. Fluid Mech.*, **397**, 285–316.

Wu, X. (2001). Receptivity of boundary layers with distributed roughness to vortical and acoustic disturbances; a second order asymptotic theory and comparison with experiments. *J. Fluid Mech.*, **431**, 91–133.

Wu, X. (2004). Non-equilibrium, nonlinear critical layers in laminar-turbulent transition. *Acta Mech. Sinica*, **20**, 327–339.

Zhigulev, V. N. and Tumin, A. M. (1987). *The Appearance of Turbulence*. Nauka, Novosibirsk.

Zhuk, V. I. and Ryzhov, O. S. (1980). Free interaction and stability of the boundary layer in an incompressible fluid. *Dokl. Akad. Nauk SSSR*, **253**(6), 1326–1329.

Index

Figure Acknowledgements

The authors hold the copyright for all the figures in this volume with the following exceptions:

Fig. I. 1: Osborne Reynolds, Public domain, via Wikimedia Commons.

Fig. I. 2: Credit: N. H. Johannesen & C. Lowe.

Fig. I.4: Reprinted from Frontiers of Fluid Mechanics, H. WERLÉ, pp. 180–185, ©1988, with permission from Elsevier.

Fig. 1.23: D. Pierce, 'Photographic evidence of the formation and growth of vorticity behind plates accelerated from rest in still air', The Journal of Fluid Mechanics, **11**, 3, 460–464, reproduced with permission.

Fig. 1.25: Figure courtesy of Prof. J. F. Morrison.

Fig. 1.27: Reprinted by permission from Springer Nature GmbH. Springer, Acta Mechanica, 'Spiral vortices in boundary layer transition regime on a rotating disk', Prof. Dr. R. Kobayashi et al, Copyright ©1969, Springer-Verlag.

Fig. 1.33: Reprinted from J. E. Burkhalter, E. L. Koschmieder "Steady supercritical Taylor vortices after sudden starts", The Physics of Fluids **17**, 1929–1935 (1974) https://doi.org/10.1063/1.1694646, with the permission of AIP Publishing.

Fig. 1.38: G. I. Barenblatt, The Journal of Fluid Mechanics, Visualized Flow. Compiled by the Japan Society of Mechanical Engineers. Edited by Y. NAKAYAMA, W. A. WOODS and D. G. CLARK. Pergamon Press, 1988, pp. 658–659, reproduced with permission.

Fig. 4.6: Figure courtesy of Dr. R. Kumar.

Fig. 5.1a: S. J. Kline, W. C. Reynolds, F. A. Schraub, P. W. Runstadler, The structure of turbulent boundary layers, The Journal of Fluid Mechanics, **30**, 4, 741–773, reproduced with permission.

Fig. 5.1b: Reprinted by permission from Springer Nature GmbH: Springer, Experiments in Fluids, 'Flow field analysis of a turbulent boundary layer over a riblet surface', S.-J. Lee et al, Copyright ©2001, Springer-Verlag Berlin Heidelberg.

Fig. 5.2: James M. Hamilton, John Kim, Fabian Waleffe, 'Regeneration mechanisms of near-wall turbulence structures', The Journal of Fluid Mechanics, **287**, 317–348, reproduced with permission.

Fig. 5.11: Figure courtesy of Prof. F. Waleffe.

Fig. 5.12: Kengo Deguchi, Philip Hall, 'Free-stream coherent structures in parallel boundary-layer flows', The Journal of Fluid Mechanics, **752**, 602–625, reproduced with permission.

Fig. 5.16: Figure courtesy of Dr. J. Maestri.